Hydrology and Water Supply
for
Pond Aquaculture

Hydrology and Water Supply for Pond Aquaculture

Kyung H. Yoo

Department of Agricultural Engineering
Auburn University, Alabama

Claude E. Boyd

Department of Fisheries and Allied Aquacultures
Auburn University, Alabama

An **avi** Book

CHAPMAN & HALL
New York • London

An AVI Book

This edition published by
Chapman & Hall
One Penn Plaza
New York, NY 10119

Published in Great Britain by
Chapman & Hall
2-6 Boundary Row
London SE1 8HN

Printed in the United States of America

Library of Congress Cataloging in Publication Data

Yoo, Kyung H., 1945-

    Hydrology and water supply for pond aquaculture/Kyung
H. Yoo and Cluade E. Boyd.
      p. cm.
    "An AVI Book."
    Includes bibliographical references and index.
    ISBN 0-442-00268-8
    1. Fish ponds—Design and construction.   2. Fish ponds—Water-
supply.   3. Water-supply engineering. I. Boyd, Claude E.
II. Title
SH157.85.F52Y66   1993
639.3'11—dc20   92-26643
                                    CIP

British Library Cataloguing in Publication Data available

Please send your order for this or any **Chapman & Hall book to Chapman &
Hall, 29 West 35th Street, New York, NY 10001, Attn: Customer Service
Department.** You may also call our Order Department at 1-212-244-3336 or fax
your purchase order to 1-800-248-4724.

For a complete listing of Chapman & Hall's titles, send your requests to
Chapman & Hall, Dept. BC, One Penn Plaza, New York, NY 10119.

Contents

Preface

In 1979, several graduate students in the Department of Fisheries and Allied Aquacultures at Auburn University met with one of the authors (CEB) and asked him to teach a new course on water supply for aquaculture. They felt that information on climatology, hydrology, water distribution systems, pumps, and wells would be valuable to them. Most of these students were planning to work in commercial aquaculture in the United States or abroad, and they thought that such a course would better prepare them to plan aquaculture projects and to communicate with engineers, contractors, and other specialists who often become involved in the planning and construction phases of aquaculture endeavors. The course was developed, and after a few years it was decided that more effective presentation of some of the material could be made by an engineer. The other author (KHY) accepted the challenge, and three courses on the water supply aspects of aquaculture are now offered at Auburn University. A course providing background in hydrology is followed by courses on selected topics from water supply engineering. Most graduate programs in aquaculture at other universities will eventually include similar coursework, because students need a formal introduction to this important, yet somewhat neglected, part of aquaculture.

We have written this book to serve as a text for a course in water supply for aquaculture or for individual study. The book is divided into two parts. Part I, "Hydrology, Morphometry, and Soils," is concerned with hydrological phenomena that affect pond design, construction, and management. Part II, "Design of Water Supply and Pond Systems," deals primarily with engineering techniques used in design and construction of pond aquaculture facilities.

The information in this book is not intended for specialists in the fields of water supply engineering or hydrology. It was written to provide individuals involved in pond aquaculture projects a background on the major points that must be addressed in site selection, design of ponds and water supply systems, and operation of ponds within restraints imposed by local hydrology and other environmental factors. Nevertheless, water supply specialists who have never been involved in aquaculture projects may find this book useful as an orientation to the kinds of problems involved in water supply engineering for aquaculture.

In the United States, and in many other nations, nearly all practical work in hydrology and engineering uses the U.S. Customary System of units, and since most of the information available to us for preparing this text was in these units, we used them throughout the text. However, for a few topics, such as properties of water and consumption of electricity, we used metric units because they are almost always given that way. We see no benefit in one system of units over another in practical work, because there is no international standard for measurements in daily endeavors. For example, in shrimp farming, production is reported in pounds per hectare in Ecuador and in kilograms per rai (a unit equal to $1600 \, m^2$) in Thailand. Nearly anywhere in the world, you are likely to hear pipe sizes reported in inches and motor power in horsepower. However, discharge may be given in cubic meters per second. We find this slightly bewildering mixture of units rather refreshing. After all, variety is a very useful thing; it keeps us alert and prevents monotony. Readers accustomed to other systems of measurement can readily convert from U.S. Customary units to their system. Appendix A lists factors for converting U.S. Customary units to metric units, and vice versa.

The authors are indebted to many individuals and organizations in preparing this book. We are grateful to Nathan Stone, Barnaby Watten, Delbert Fitzsimmons, George Bloomsburg, Bryan Duncan, Eugene Rochester, and others who provided information or reviewed chapters of the manuscript. We also appreciate the assistance of Teresa Rodriguez and Steve Etheridge with the illustrations. The typing and editorial help provided by June Burns and Margaret Tanner was extremely valuable.

The authors are also indebted to the following U.S. government agencies for use of their published materials; U.S. Soil Conservation Service, U.S. Bureau of Reclamation, U.S. Environmental Protection Agency, and U.S. National Weather Service.

Most of all the authors wish to thank their families for their patience and encouragement during this difficult task.

Symbols

Chapter 1

A = area, cross-sectional area of flow (length^2)
D = depth of fluid
F = force
g = gravitational acceleration (32.2 ft/sec^2 or 9.8 m/sec^2)
H_L = head loss between two points
H_v = velocity head (ft or m)
i = hydraulic gradient (ft/ft)
K = hydraulic conductivity (length/time)
n = porosity
P = pressure
Q = discharge (length^3/time)
v = velocity of topmost layer (ft/sec or m/sec)
Z = elevation of a point above the reference plane
v = kinematic viscosity
ρ = density
γ = unit weight of water (62.4 lb/ft^3 or 1000 kg/m^3)
η = absolute viscosity

Chapter 2

e_a = actual vapor pressure
e_s = saturation vapor pressure
RH = relative humidity in percent

Chapter 3

b = empirical coefficient(s)
C_p = pan coefficient
D = deep percolation of water that drains from soil
E = evaporation (in./hr)
e = vapor pressure (in. Hg)
e_a = actual vapor pressure (in. Hg)
e_{a5} = actual vapor pressure at 5 ft above ground level (in. Hg)
E_{day} = daily evaporation (in./day)
E_L = evaporation from lake
E_p = evaporation from pan
ET = evapotranspiration
ET_D = daily evapotranspiration (mm)
ET_M = monthly evapotranspiration (mm)

e_s = saturation vapor pressure at average air temperature (in. Hg)
e_{s5} = saturation vapor pressure at 5 ft above ground level (in. Hg)
H_C = conduction and convection of energy from water surface to atmosphere
H_E = energy used for evaporation
H_L = net change of energy in water body from long-wave radiation
H_R = net change of energy storage within the body of water
H_S = gain of energy in body of water from shortwave solar radiation
H_W = net gain or loss of energy through inflow and outflow
I = heat index for a 12-month period, inflows, water added by irrigation
L = latent heat of vaporization, wind run 6 in. above rim of evaporation pan (miles/day)
O = outflows
P = precipitation, annual precipitation (mm)
P_a = average atmospheric pressure (in. Hg)
PET = potential evapotranspiration (in.)
RO = overland flow into or out of the lysimeter
SM = soil moisture (mm)
T = temperature (°F), mean annual air temperature (°C)
T_a = average air temperature (°C)
Tm = mean monthly temperature (°C)
T_w = average temperature of water surface (°C)
V = wind velocity (mph)
v = wind velocity at unspecified height (mph)
v_{30} = wind velocity at 30 ft height (mph)
X = monthly Class A pan evaporation (in.)
Y = monthly pond evaporation (in.)
Z = height of wind measurement (ft)
ΔS = change in storage
ΔW = change in water content of soil in lysimeter
κ = von Kármán's constant

Chapter 4

A = watershed area (acres)
C = runoff coefficient (dimensionless)
I = storm rainfall (in.)
i = intensity of rainfall for duration equal to concentration time of watershed for desired return period (in./hr)
KE = kinetic energy
L = maximum length of flow (ft)
M = mass
P = rainfall for storm (cm)
Q = peak discharge (ft³/sec), direct runoff (in.), overland flow for a particular storm (in.)
R = hydrologic response (dimensionless)
S = watershed gradient (ft vertical/ft horizontal), maximum potential difference between rainfall and runoff (in.)
Tc = time of concentration (min)

Chapter 5

A_u = area (acres) within upper contour
A_l = area (acres) within lower contour
E = pond evaporation
h = vertical distance (ft) between contours
I = inflow from well, stream, or reservoir
OF = spillway overflow
P = precipitation
PX = $P - E$ or the precipitation excess
RO = runoff
S = seepage
SD_{max} = maximum storage depth (in.)
V = volume (acre-ft) between the upper and lower contour lines
WL_B = water-level elevation at beginning of period of measurement, water level at beginning of month
WL_E = water-level elevation at end of period of measurement, water level at end of month
ΔV = change in storage volume

Chapter 7

a = area of lake (ft^2)
h_{ave} = average depth
h_{max} = maximum depth
s = length of shore line (ft)

Chapter 8

A = cross-sectional area of flow (ft^2)
C = flow resistance coefficient, seepage coefficient of channel material
EL_u = elevation of channel bottom at upstream point (ft)
EL_d = elevation of channel bottom at downstream point (ft)
$F.B.$ = freeboard (ft)
G = erosion coefficient
h = drop height (ft), effective flow depth above weir crest (ft)
h_L = head loss (ft)
L = distance between two points (ft), required distance between two drop structures (ft)
L_w = length of weir crest (ft)
n = Manning's surface roughness coefficient of channel (dimensionless)
P = wetted perimeter (ft)
Q = flow rate (ft^3/sec)
R = hydraulic radius (ft)
S = slope of channel (ft/ft)
S_o = original slope of channel (ft/ft)
S_p = permissible maximum slope (ft/ft)
SEEP = seepage rate (cfs/mile)

v = mean flow velocity (ft/sec)
v_{max} = maximum permissible velocity (ft/sec)
y = depth of flow (ft), normal depth of flow (ft)
Z = elevation head (ft)
θ = angle of channel side slope

Chapter 9

A = cross-sectional area of culvert flow (ft^2)
BP = brake power (hp)
C = Hazen–Williams coefficient, orifice coefficient
c = speed of the pressure wave in a water pipe
D = inside diameter of pipe (ft)
d = inside diameter of pipe (in.)
e = absolute roughness (ft)
E_p = pump efficiency in fraction
f = pipe friction factor (dimensionless)
H = water depth, pressure head, total head (ft), effective head causing flow (ft)
H_e = elevation head (ft)
H_f = head loss due to friction or friction loss (ft)
H_L = hydraulic head loss (ft)
H_m = minor loss (ft)
H_p = pressure head (ft)
H_v = velocity head (ft)
k_c = friction loss coefficient for culvert flowing full
k_m = minor loss coefficient
k_x = exit coefficient
L = pipe length between two points considered (ft), culvert length (ft)
n = Manning's roughness coefficient of culvert
p = static pressure
P_h = water-hammer pressure at $t_c > t_m$ (psi)
Q = flow rate (gpm or cfs)
R = hydraulic radius of square culvert (ft)
R_e = Reynolds number (dimensionless)
r = inside radius of pipe (ft)
S = slope of energy grade line (ft/ft)
S_n = neutral slope (ft/ft)
t_c = actual closure time of valve (sec)
t_m = minimum time of closure to reduce water-hammer pressure (sec)
v = initial velocity of pipe flow (ft/sec)
WP = water power (hp)
ρ = density of water (1.94 slug/ft^3 at 70°F)
ν = kinematic viscosity of water (ft^2/sec)
ω = unit weight of water
ΔP = increased pressure by water-hammer effect (psi)
θ = slope angle of culvert
ε = relative roughness

Chapter 10

A = cross-sectional area of flow (ft^2)
C = coefficient of pipe orifice, flow coefficient, triangular weir coefficient, velocity correction factor
c_1 = concentration of tracer in water injected at an upstream point (ppm)
c_2 = concentration of tracer in water sampled at a downstream point (ppm)
c_0 = concentration of tracer in water before injection (ppm)
d = vertical distance from reference point
h = effective hydraulic head (ft)
h_d = head at the downstream (ft)
h_i = head reading of inner tube (ft)
h_o = head reading of outer tube (ft)
h_u = head reading along the sloping side of the flume upstream (in.), head at the upstream (ft)
h_v = velocity head (ft)
i = segment number
j = point of measurement
K = coefficient for free-flow cutthroat flume
n = exponent for free-flow cutthroat flume
L_w = length of weir notch at the crest (ft)
n = total number of segments
Q = flow rate (cfs or gpm)
q = flow rate of a segment, injection rate of water mixed with tracer (gpm)
t = time of flow (sec)
V = volume of flow (ft^3)
v = mean velocity of flow (ft/sec)
W = width of throat (ft)
y = flow depth
ε = effective head correction factor for V-notch angle
Δh = head difference

Chapter 11

A_w = cross-sectional area of a wire (Cm)
BP = brake power (hp)
$BP(Q)_i$ = brake power for pump i at Q
$BP(Q)_s$ = combined brake power at Q
$BP(\text{TDH})_i$ = brake power for pump i at TDH
$BP(\text{TDH})_p$ = combined brake power at TDH
C = variable cost per hour of operation ($/hr)
D = impeller diameter, permissible voltage drop rate in wire, diameter of flat-face pulley or pitch diameter of V-belt pulley
E_L = line voltage
E_P = pump efficiency in decimal
E_D = drive train efficiency in decimal
E_M = motor efficiency in decimal
$E_p(Q)_s$ = combined pump efficiency in decimal at Q
$E_p(\text{TDH})_p$ = combined pump efficiency in decimal at TDH

Fs	= safety factor
H_b	= expected absolute barometric pressure
H_s	= total suction head
H_{vp}	= vapor pressure of water
I_L	= line current
K	= average power output per unit volume of fuel, resistance factor (ohm-Cm/ft)
L	= length of wire (ft)
n	= total number of pumps
Q	= discharge rate of pump (gpm)
$Q(TDH)_i$	= Q for pump i at TDH
$Q(TDH)_p$	= combined Q at TDH
R	= cost of electricity ($/kWh), cost of fuel ($/unit volume)
RPM	= rotational speed of pump shaft in revolutions per minute (rpm)
T	= torque (ft · lb)
TDH	= total dynamic head (ft)
$TDH(Q)_i$	= TDH for pump i at Q
$TDH(Q)_s$	= combined TDH at Q
V	= voltage (V)
WP	= water power (hp)
$WP(Q)_i$	= water power for pump i at Q
$WP(TDH)_i$	= water power for pump i at TDH

Chapter 12

A	= flow direction cross-sectional area of aquifer (ft^2)
BP	= brake power (hp)
C	= variable cost per hour of operation ($/hr)
D_i	= drawdown at the point by well i
D_t	= total drawdown at a given point
d	= thickness of the confined layer (ft)
E_D	= drive train efficiency in decimal
E_M	= motor efficiency in decimal
E_P	= pump efficiency in decimal
H	= initial water level above the top of impervious layer (ft), initial piezometric level above the top of the aquifer (ft)
h	= final piezometric level above the top of the aquifer (ft), final water level at the well above the impervious layer (ft)
H_d	= discharge head (ft)
H_e	= elevation head (ft)
h_e	= elevation head measured at the top of aquifer above a reference plane (ft)
H_f	= friction head (ft)
H_p	= pressure head (ft)
h_p	= pressure head in the aquifer (ft)
H_s	= suction head (ft)
h_t	= total head (ft)
H_v	= velocity head (ft)

K = hydraulic conductivity of aquifer material (ft/day)
n = total number of wells
Q = flow rate in aquifer (cfs or gpm), well yield (gpm)
R = cost of electricity ($/kWh), radius of influence (ft)
r = radius of well (ft)
TDH = total dynamic head (ft)
V_b = bulk volume
V_r = volume of retained water
V_y = volume of water drained
V_v = void volume
Δh = hydraulic head change between two test wells (ft), change in height
ΔL = distance between two test wells (ft)
ΔV = change in volume

Chapter 13

A = area of excavation at ground surface (ft^2), cross-sectional area of
        sample (ft^2), submerged area (ft^2)
A_1 = cross-sectional area at one end of segment (ft^2)
A_2 = cross-sectional area at the other end of segment (ft^2)
B = area of excavation at mid-depth (ft^2), width of triangle section (ft)
C = area of excavation at pond bottom (ft^2)
D = distance between length of segment along the embankment (ft),
        pond depth (ft)
D_f = fetch length (ft)
F = total force (lb)
H = height of embankment (ft)
h = water depth (ft), wave height (ft), hydraulic head (ft)
i = segment number
L = length of sample (ft)
n = total number of segments
P = pressure due to water depth (lb/ft^2)
t = time of flow (min)
V = volume of earth fill in a segment (yd^3), volume of water passed for
        time t (ft^3), volume of excavation (yd^3)
W = width of middle section (ft)
ω = unit weight of water (62.4 lb/ft^3 or 1000 kg/m^3)

Chapter 14

A = pounds nitrogen per 1000 lb of feed
D = depth
L = length, the length is 10 times the width
P = decimal fraction of protein in the diet
T = temperature (°C)
V = volume
W = width
Y = nitrification rate (lb ammonia-N/ft^2 of media surface per day)

Abbreviations

AC, alternating current
AWG, American wire gage
BOD, biochemical oxygen demand
BP, barometric pressure; brake power
Btu, British thermal unit
°C, degrees Celsius
cfs, cubic feet per second
Cm, circular mil
DC, direct current
EPA, Environmental Protection
    Agency
°F, degrees Fahrenheit
F.B., free board
fps, feet per second
ft, foot
gal, gallon
gph, gallons per hour
gpm, gallons per minute
hp, horsepower
hr, hour
in., inch
kWh, kilowatt-hour
lb, pound

LE, linear coefficient of expansion
LPG, liquified petroleum gas
min, minute
m.s.l., mean sea level
NEMA, National Electrical
    Manufacturers Association
pf, power factor
ppm, parts per million
ppt, parts per thousand
psi, pounds per square inch
PTO, power take-off
RPM, revolution per minute
SCS, Soil Conservation Service
sec, second
TDH, total dynamic head
USBR, U.S. Bureau of Reclamation
USCS, Unified Soil Classification
    Systems
USDA, United States Department of
    Agriculture
W, watt
WP, water power
yd, yard

Customary Metric Conversion Factors

Length

	in.	ft	yd	mile	mm	m	km
1 in.	1	0.083	0.0278	—	25.4	0.0254	—
1 ft	12	1	0.33	—	305	0.305	—
1 yd	36	3	1	—	910	0.91	—
1 mile	—	5,280	1,760	1	—	1,610	1.61
1 mm	0.0394	—	—	—	1	0.001	—
1 m	39.37	3.28	1.1	—	1,000	—	0.001
1 km	—	3,280	1,093	0.620	—	1,000	—

Area

	in.^{2}	ft^2	yd^2	acre	cm^2	m^2	ha
1 in.^{2}	1	0.007	—	—	6.45	0.000645	—
1 ft^2	144	1	0.11	—	—	0.093	—
1 yd^2	1,296	9	1	—	—	0.836	—
1 acre	—	43,560	4,840	1	—	4,050	0.405
1 cm^2	0.155	—	—	—	1	0.0001	—
1 m^2	1,555	10.8	1.2	—	10,000	1	.0001
1 ha	—	107,600	11,955	2.47	—	10,000	1

Volume

	in.^{3}	ft^3	gal	liter	m^3	ac-ft	ha-m
1 in.^{3}	1	0.00058	0.00433	0.0164	—	—	—
1 ft^3	1,728	1	7.48	28.3	0.0283	—	—
1 gal	231	0.134	1	3.8	0.0038	—	—
1 liter	60.5	0.035	0.264	1	0.001	—	—
1 m^3	—	35.3	264.0	1,000	1	—	—
1 ac-ft	—	43,560	325,850	—	1,233	1	0.1233
1 ha-m	—	353,300	—	—	10,000	8.11	1

Hydrology and Water Supply
for
Pond Aquaculture

Introduction

Pond aquaculture is a rapidly growing enterprise in many nations. For example, in the United States there are about 120,000 acres of channel catfish ponds, and the annual expansion of area devoted to channel catfish farming is usually about 5–10%. An even larger aquacultural effort is the culture of marine shrimp in many tropical nations. There are more than 2,500,000 acres worldwide devoted to pond culture of marine shrimp, and shrimp farming is rapidly expanding in many places (Rosenberry 1990). For example, between 1980 and 1989 the area devoted to shrimp ponds increased in Thailand from 64,000 to 198,000 acres, in China from 23,000 to 358,000 acres, and in Ecuador from 46,000 to 250,000 acres.

Channel catfish and marine shrimp are two of the better-known success stories in pond aquaculture. However, many other species of fish and crustaceans are cultured in ponds. The area devoted to pond aquaculture is large and expanding, because consumption of fish, shrimp, and other aquatic animals is rising and there has been little increase in the capture of fish from natural environments in the past decade. The greater demand for fish, shrimp, and other aquatic animals must be supplied through aquaculture. There is significant production of aquatic animals in raceways, pen, and cages, but the greatest amount comes from pond culture.

The construction of ponds is expensive, ranging from $1,000 to more than $5,000 per acre in 1991, depending on the site and the design. Therefore, every effort should be made to ensure that ponds are designed and constructed in such a manner that they maintain sufficient water of adequate quality to provide favorable culture conditions for the target species. One cannot expect to develop a profitable aquacultural operation

1

Harvesting fish from a channel-catfish pond. *(Courtesy of the International Center for Aquaculture and Aquatic Environment.)*

if a lot of money is spent constructing production ponds that cannot be maintained and managed properly.

AMOUNT OF WATER USE

Aquaculture is a water-intensive endeavor, as illustrated by the data in Table I-1. The least water use is for static ponds that only discharge water after heavy storms. If undrained, water use in a levee pond in a humid climate such as Alabama's is only about 2.5 ft/year. Annual draining will increase the use to about 7.5 ft/year. A watershed pond has more over-flow than a levee pond, and average water use is about 15 ft in a humid climate. Of course, water for watershed ponds comes from runoff and does not have to be pumped from an aquifer, as is often the case with

TABLE I-1   **Amounts of Water Used in Selected Aquaculture Systems**

Aquaculture System	Water Use	
	ft/yr	gal/1b production
Channel catfish ponds (Alabama)		
Levee ponds		
Undrained	2.5	150–200
Drained annually	7.5	450–600
Watershed ponds	15	800–1,200
Brackish-water shrimp ponds (tropics)		
Semi-intensive (5% pond volume daily water exchange)	70	6,000–12,000
Intensive (20% pond volume daily water exchange)	250	5,000–10,000
Trout raceway		
Unaerated	—	10,000–14,000
Mechanically aerated	—	2,000–5,000

levee ponds. Water use increases drastically when daily water exchange is employed as in brackish-water shrimp ponds. Semi-intensive shrimp ponds probably use around 70 ft/year, whereas intensive ponds use about four times as much water. Depth of water cannot be meaningfully calculated for raceways, but typical production in unaerated raceways is 20,000 lb of trout per year for each cubic foot per second (450 gpm) of flow. Each pound of fish or shrimp produced in aquaculture requires from 150 to 14,000 gal of water.

Average water application rates to major crops in the United States by irrigation range from 0.8 ft for soybeans to 2.9 ft for rice (van der Leeden et al. 1990). Only channel catfish production in undrained, levee ponds uses less water than the most heavily irrigated crop. Average water consumed by beef, pork, and poultry does not exceed 5–10 gal/lb of production. Of course, the amount of water used in producing feed and maintaining sanitary conditions increases total water requirements. Nevertheless, the amount of water required per pound of production of aquacultural product is often much greater than that necessary for the production of a pound of other types of agricultural products. This factor is significant because the demand placed upon natural water supplies by an acre of aquaculture crop is generally greater than that of an acre of some other crop.

SOURCES OF WATER

The primary sources of water for inland aquaculture ponds are storm runoff, springs, streams, lakes or storage reservoirs, and ground water from wells. Of course, the ultimate source of all water for inland ponds is precipitation. The precipitation at any location exhibits a typical annual pattern with some months being much drier or wetter than other months. In addition, there will be year-to-year variation in patterns and annual totals of precipitation. Locations may differ greatly in precipitation patterns, and although the reasons for these differences usually can be readily explained, nothing can be done to change them. The water delivered to the land surface by precipitation may be divided into several fractions: soil moisture, ground water, surface runoff to streams and finally to the sea, and inland surface-water storage. Water is constantly returned to the atmosphere by evaporation and transpiration. The water available for aquaculture at a site depends on local terrain, soil properties, geologic features, vegetation, amount of precipitation, patterns of precipitation, solar radiation and temperature, and various other factors, including competition from other activities. For example, in areas where aquaculture ponds are supplied by ground water, the same aquifers used to supply water to the ponds may also be used for domestic, municipal, industrial, and agricultural purposes.

Traditionally, aquaculture has been conducted primarily in well-watered locations, but in recent years there has been significant expansion of aquacultural operations into semiarid and arid regions. Even areas with high annual precipitation totals may have long periods each year without significant precipitation, in which shortages of surface water occur. Well water is a highly valued source of water for ponds, but if too many wells are developed in an area, ground-water supplies may be depleted. Water for coastal aquaculture ponds is taken from estuaries or from the sea. Water levels fluctuate with tidal action, but these sources are usually of more constant quantity than inland water supplies.

Even where water quantity is sufficient, water quality may be substandard. The water supply may be contaminated with toxic substances, excessive amounts of suspended soil particles, or enriched with nutrients. In some coastal areas or in arid regions, water supplies may not have acceptable levels of salinity for intended purposes.

In a new aquaculture project, one of the most important considerations is the examination of the water supply. The amount and quality of the water available must be determined. The availability and quality of water at many locations will vary seasonally and from year to year. The worse-case scenario (lowest availability and quality) is the key condition for planning purposes. In a water supply assessment, the suitability of the water supply for future expansion of the proposed facility should be included, and predictions about how future development of competing water uses in the area may impact the aquaculture facility are desirable.

PONDS

Water Requirements

The three basic pond types for aquaculture are watershed, excavated, and levee. Watershed ponds are formed by building a dam across a natural watercourse where topography permits water storage behind the dam. The dam is usually constructed between two hills that constrict the watershed. Watershed ponds may store only overland flow, or they may receive some combination of overland flow, stream flow, and ground-water inflow. Excavated ponds are formed by digging a hole in the ground. They may be filled by ground-water inflow where the water table is near the land surface by overland flow if constructed in a low-lying area, or by well water. The water in levee ponds is impounded in an area surrounded by levees. Little runoff enters levee ponds, so they must be filled by water from wells, storage reservoirs, streams, or estuaries.

There are few places where a pond can be filled and its water level maintained by rain falling directly into it. Therefore, runoff or some other source of water must normally be available. Furthermore, rainfall is not always dependable, and there may be long periods between rains that are large enough to generate runoff. Watershed ponds filled entirely by runoff may have low water levels during drier months even in humid climates without a pronounced dry season. For example, water levels in watershed ponds in Alabama may decline by 2–3 ft during the summer and fall even though there are 2 or more inches of rain each month during this period. We have seen ponds in Israel and India that were 10 ft or more deep at the end of the wet season but only 3 or 4 ft deep at the end of the dry season.

Ponds for aquaculture need a fairly stable water level, and a water supply sufficient to rapidly refill them is desirable. Excessive decreases in water levels may adversely affect fish through destruction of spawning areas and benthic food organisms and by causing crowding. During long dry periods, dissolved substances are concentrated, whereas in wet periods dissolved substances are diluted. If ponds are drained for fish harvest but cannot be refilled rapidly, part or all of a growing season may be lost. Weeds may form dense stands on pond bottoms where refilling is slow. Of course, rapid flushing of water through ponds is undesirable in most situations, because flushing removes plankton and nutrients. Large watershed ponds in the southeastern United States may receive enough runoff during winter and spring to exchange pond water several times.

The water budget for a pond may be expressed as follows:

$$\text{Inflows} = \text{outflows} \pm \Delta \text{ storage}$$

Inflows include

1. Precipitation falling directly into the pond
2. Runoff from the watershed
3. Seepage into the pond where the bottom extends below the water table
4. Regulated inflow of water from wells, streams, estuaries, and so on, through water-control devices

Outflows include

1. Evaporation and transpiration
2. Seepage through bottom or under levees
3. Overflow following storms
4. Intentional discharge through water-control devices
5. Consumptive use for irrigation, livestock watering, domestic use, and so on

Most ponds do not have all possible inflows and outflows. Major considerations in aquaculture ponds are (1) availability of water to fill ponds and maintain water levels, (2) avoiding excessive flushing of ponds by runoff, (3) preventing excessive seepage from ponds, and (4) ability to exchange water in certain types of culture (e.g., marine shrimp).

Sites

Site selection is of paramount importance. The topography must be suitable for pond construction, soils must have suitable properties for forming stable, watertight levees and bottoms, and the water supply must be of sufficient quantity to maintain desired water levels and water-exchange rates throughout the culture period. If in situ properties of soil and water at the site under consideration are not suitable, then special construction techniques may often be used. For example, clay may be trucked in to provide a clay blanket over a sandy area, rough terrain may be leveled to provide a better surface for pond construction, a well may be developed to supplement an undependable supply of surface water, or canals may be constructed to bring in water from a distant source. Normal construction is expensive, and of course special construction techniques increase costs. However, it is better to spend more money on construction than to build ponds that cannot be managed properly. In addition, acceptable construction techniques should be used, for money saved by construction shortcuts often proves to be false economy. At some sites it may be too expensive or even impossible to overcome soil or water limitations. Such sites should be avoided, for the project is destined to fail.

It is not sufficient to consider only the effects of soil and water properties on engineering aspects of an aquaculture project. Soils may have properties that make them undesirable substrates for pond bottoms. For example, high levels of acidity may result from oxidation of iron pyrite in levees and bottoms when ponds are constructed in an area with potential acid sulfate soils. Soils with large concentrations of organic matter often are not good for pond bottoms because the organic matter will decompose and exert a high oxygen demand.

ALTERNATIVES TO PONDS

Fish and other aquatic animals can be cultured in flowing-water systems, recirculating systems, or in cages or pens placed in water bodies (Soderberg 1986). These systems may be used in places where pond culture is not possible. For example, if water supply is adequate, flowing-water systems allow the production of large amounts of fish in small surface areas of water at places where the terrain does not permit conventional pond culture. Recirculating systems permit water reuse and may be used where there is inadequate water for pond culture. Cages and pens permit confinement of animals for controlled culture in natural bodies of water (Schmittou 1991).

WATER SUPPLY DEVELOPMENT

Water used in watershed ponds may enter largely by runoff. The major concern is to provide sufficient watershed area to generate adequate runoff to fill and maintain ponds without excessive flushing. Also, dams must be constructed with overflow structures to prevent their failure after unusually heavy storms. Levee ponds and most excavated ponds must be filled and maintained by water from an external source. This water reaches the pond through pipelines or canals and is released through various types of water-control structures (e.g., valves, gates, or monks). The water may flow by gravity from a water supply at higher elevation, but usually it must be pumped into the distribution system from a well, stream, or other source.

For an adequate water supply at reasonable cost, distribution systems, wells, and pump facilities must be designed and constructed by accepted techniques. For example, if it is desired to construct ponds and supply them with well water, the amount of water that can be obtained from the local aquifer at reasonable cost must be determined. Once this is known, wells, pumps, and pipelines can be designed. The amount of pond area that can be sustained by the water supply can then be determined, and ponds may be designed.

Ponds must have reliable overflow structures to prevent water from topping and possibly causing levees and dams to fail. It is usually necessary to drain ponds to allow fish or other aquatic animals to be harvested. A controlled-discharge drain structure is necessary. This device must allow rapid and predictable lowering of the water level so that harvest of aquatic animals may be performed according to schedule. The pond bottom should be sloped and the drain structure intake positioned so that the pond can be completely drained. This facilitates elimination of all aquatic animals from the pond and allows for pond bottom treatments to improve the pond soil.

Water measurement also is essential in a good aquacultural pond management scheme. A knowledge of the amount of water used during each crop will permit calculations of water cost, allow for planning to conserve water or to expand operations, and provide data to assess the relative impact of an aquacultural facility on local water supplies. Pond effluents are viewed as pollutants in many places, and the regulatory agencies may require measurements of pond discharge. Often it is desired to measure the discharge of individual pumps, pipes, canals, or other devices. Simple techniques generally may be used to obtain measurements of volume or flow rate that are sufficiently accurate for planning and design purposes.

Aquaculturists usually hire engineering and construction specialists to design and build ponds. Nevertheless, aquaculturists need to have a general understanding of the factors considered in site selection and in pond design and construction to communicate effectively with engineers, hydrologists, soil scientists, contractors, and other technicians who may be involved in the initial phases of an aquaculture project. Also, after ponds have been constructed, characteristics of local hydrologic conditions and features of the water supply will have implications in pond management. Thus, an aquaculturist who has a knowledge of hydrology and the basic characteristics of water supply systems is better equipped to manage ponds in harmony with the local hydrologic cycle.

ENVIRONMENTAL IMPACTS

Poor environmental conditions negatively influence production of aquatic animals. On the other hand, environmentalists have often suggested that aquaculture can be damaging to surrounding ecosystems. Statements about the negative aspects of aquaculture have largely been ignored or denied by the aquaculture industry. However, this approach can no longer be taken, because it is clear that aquaculture can be detrimental to the environment.

The two most apparent influences of pond aquaculture are destruction of wetlands and eutrophication. Wetland destruction results simply because natural wetlands often are good sites for pond construction due to their close proximity to water. Applications of fertilizers and feeds to promote fish and shrimp production cause high concentrations of nutrients and organic matter in pond water. Pond effluents may contribute to eutrophication of the receiving water bodies. Other possible negative impacts of aquaculture on ecosystems include redistribution of local water regimes, contamination of ground water, introduction of non-native species, disposal of sediment in natural waterways, and alteration of habitat. In most situations, it is possible to mitigate negative influences to a socially acceptable degree. In the future, aquaculture projects will have to be managed to minimize possibilities for environmental perturbations.

I

Hydrology, Morphometry, and Soils

Repelon Aquaculture Station in Colombia, SA. The large freshwater lake in the background serves as the water supply. (*Courtesy of the International Center for Aquaculture and Aquatic Environments.*)

1

Physical Properties of Water and Water Cycle

1.1 INTRODUCTION

The purpose of this chapter is to provide some background information on the physical properties of water and factors affecting the flow of water. This material will be helpful in understanding the development of many of the principles and techniques presented in the following chapters. A review of the hydrologic or water cycle and some comments on the distribution of water are given. Water is the medium for aquaculture, and we feel that those involved in aquaculture should have a good conception of the hydrosphere. Some readers may be familiar with this material and prefer to go directly to Chapter 2.

1.2 PHYSICAL PROPERTIES

1.2.1 Molecular Structure

Water is a unique substance with unexpected physical properties for a compound of simple structure and low molecular weight. A water molecule consists of two hydrogen atoms covalently bonded to one oxygen atom (Fig. 1-1). The angle formed by lines through the centers of the hydrogen nuclei and the oxygen nucleus is 105°. The distance between hydrogen and oxygen nuclei is 0.96×10^{-8} cm. An oxygen nucleus is heavier than a hydrogen nucleus, so electrons are pulled relatively closer to the oxygen nucleus. This gives the oxygen molecule a small negative charge and each of the hydrogen molecules a slight positive charge. Water is said to be a polar substance because one end has a positive charge and the other end has a negative charge (Fig. 1-1).

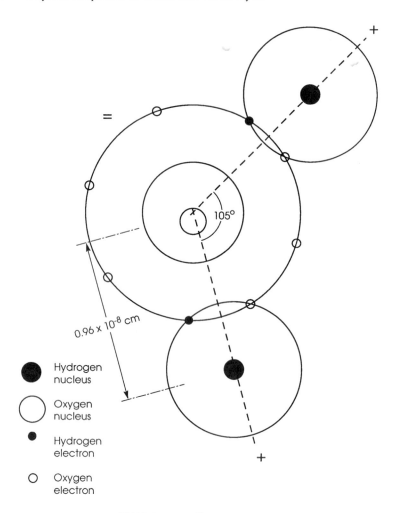

FIGURE 1-1.   The water molecule.

All molecules exert attractions for each other through van der Waals forces. The negatively charged electrons of one molecule and the positively charged nuclei of another molecule attract each other. This attraction is almost, but not completely, neutralized by repulsion of electrons by electrons and nuclei by nuclei; hence, van der Waals forces are weak. Electronic van der Waals attraction between molecules increases in intensity with increasing molecular weight, because the number of electrons and nuclei tend to increase as molecular weight increases.

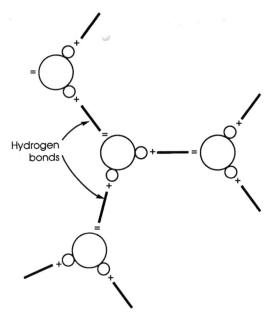

FIGURE 1-2.   Hydrogen bonds between molecules.

The positively charged end of one water molecule attracts the negatively charged end of another to form a hydrogen bond (Fig. 1-2). For
simplicity, we only show hydrogen bonding in two dimensions; however,
hydrogen bonding in water is three-dimensional. A water molecule can be
bonded to another water molecule with the axis of attraction extending in
any direction. Hydrogen bonds are not as strong as covalent bonds or
ionic bonds, but they are much stronger than van der Waals attractions.
Hence, the degree of molecular attraction in water is greater than in
nonpolar substances. Furthermore, because of hydrogen bonding, water
actually has the structure $(H_2O)_n$ rather than H_2O. The number of
associated molecules (n) is greatest in ice and decreases as temperature
increases. In water vapor all hydrogen bonds are broken, and each water
molecule exists as a separate entity—H_2O.

1.2.2 Solvent Action

In an electric field, such as the charged plates of a condenser, water
molecules orient themselves, pointing their positive ends toward the
negative plate and their negative ends toward the positive plate (Fig. 1-3).

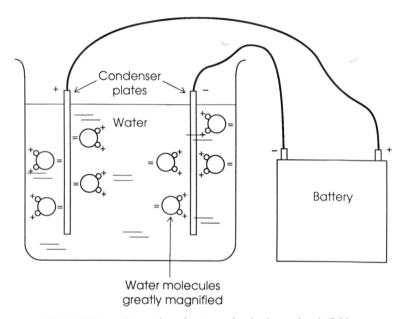

FIGURE 1-3.   Orientation of water molecules in an electric field.

The orientation of molecules neutralizes part of the charge applied to the condenser plates. The voltage required to produce a given voltage on condenser plates is a measure of the dielectric constant of the substance surrounding the condenser plates. In a vacuum, 1 V applied to the plates produces a voltage of 1 V. In contrast, the dielectric constants of air and water are 1.0006 V and 81 V, respectively.

A crystal of salt maintains its structure in air because of electrical attraction between anions and cations, such as $Na^+ + Cl^- = NaCl$. Salt dissolves readily in water because the attractive forces between anions and cations are 81 times less in water than in air. Water insulates ions from each other because water molecules are attracted to dissolved ions. Each anion attracts the positive ends of several water molecules, and each cation attracts the negative ends of other water molecules (Fig. 1-4). Ions can each attract several water molecules because ionic charges are much stronger than the charges on opposite sides of a water molecule. Water is said to hydrate ions, and hydration neutralizes charges on ions just as the orientation of water molecules neutralizes the charge on a condenser plate. Because of its high dielectric constant and polar properties, water is an excellent solvent for most inorganic and many organic substances (Kemmer 1979).

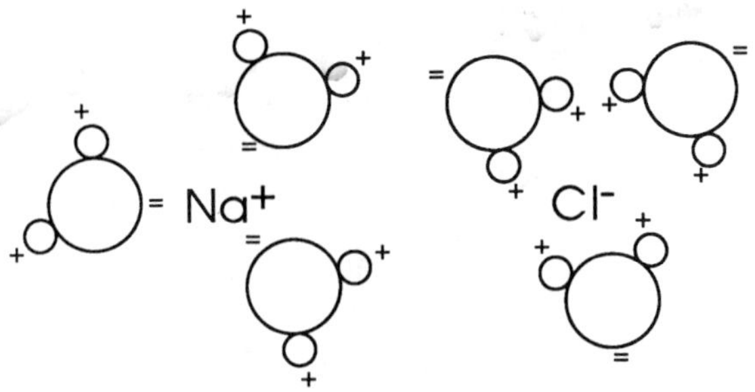

FIGURE 1-4.   Hydration of dissolved ions by water molecules.

1.2.3 Thermal Characteristics and Phase Change

Water is a liquid between 0 and 100°C (32 and 212°F) at standard atmospheric pressure (760 mm Hg or 1 atm). The freezing and boiling points of water, 0 and 100°C, respectively, are much higher than the freezing and boiling points of other hydrogen compounds of low molecular weight; for example, methane (CH_4), ammonia (NH_3), phosgene (PH_3), and hydrogen sulfide (H_2S) are gases at ordinary temperatures on the earth's surface. The aberrant behavior of water results because its molecules form hydrogen bonds with each other. The molecules of other common hydrogen compounds do not form hydrogen bonds; they are joined only by van der Waals attractions. Considerable thermal energy is required to break hydrogen bonds and convert ice to liquid water or to change liquid water to water vapor.

Depending on its internal energy content, water exists in a solid, liquid, or gaseous phase. In ice all hydrogen atoms are bonded; in liquid water a portion of the hydrogen atoms is bonded; in water vapor there are no hydrogen bonds. An increase in the internal energy content of water agitates molecules, causes hydrogen bonds to stretch and break, and increases temperature; the opposite effect results when the energy content of water declines.

The amount of energy (or heat) required to raise the temperature of a substance by 1°C is termed the *specific heat* of the particular substance. The specific heat of ice is 0.47 (cal/g)/°C, and the specific heat of water is 1.0 (cal/g)/°C. Water has a high specific heat when compared with many other substances. For example, specific heat values in calories per gram for some common substances found on watersheds (Satterlund 1972) are as follows: dry mineral soil, 0.18–0.22; moist sandy soil, 0.21–0.36; moist loam soil, 0.25–0.5; wet organic soil, 0.70–0.80; air, 0.24; peat and humus, 0.45; ice and snow, 0.49; liquid water, 1.00; fresh leaves, 0.80–0.95; twigs, bark, and wood, 0.61–0.77.

Water freezes when its energy content declines and molecular motion slows to the point that enough hydrogen bonds can form to produce the solid structure of ice. Likewise, ice melts when its energy content increases and molecular motion speeds up to the point that too few hydrogen bonds are present to maintain the solid, crystalline structure of ice. The freezing point of water is 0°C (32°F). If the temperature of water is decreased to 0°C, an additional 80 cal of heat must be removed from each gram of water to cause it to freeze with no change in temperature. Likewise, to melt 1 g of ice at 0°C with no change in temperature requires an input of 80 cal. The energy necessary to cause the phase change between liquid water and ice is termed the *latent heat of fusion*.

Water changes from liquid to vapor when it attains enough internal energy and molecular motion to break all hydrogen bonds. Water vapor condenses to form liquid water when it loses energy and molecular motion decreases to permit formation of hydrogen bonds. The amount of energy necessary to cause the liquid–vapor phase change is 540 cal/g. This quantity of energy is known as the *latent heat of vaporization*.

The heat required to convert 10 g of ice at −15°C to water vapor at 100°C is calculated as follows:

(i) To raise temperature to 0°C:

$$10\,g \times 15°C \times 0.47\,(cal/g)/°C = 70.5\,cal$$

(ii) To melt ice at 0°C:

$$10\,g \times 80\,cal/g = 800\,cal$$

(iii) To raise temperature to 100°C:

$$10\,g \times 100°C \times 1.0\,(cal/g)/°C = 1,000\,cal$$

(iv) To convert water to vapor at 100°C:

$$10\,g \times 540\,cal/g = 5,400\,cal$$

(v) The required heat is thus

$$70.5\,cal + 800\,cal + 1,000\,cal + 5,400\,cal = 7,270.5\,cal$$

Water can change directly from ice to vapor by sublimation. For example, wet clothes hung outdoors on a sunny day with below-freezing temperature will freeze and then dry. Energy for sublimation is supplied by the sun. The heat of sublimation of ice is 677 cal/g.

1.2.4 Temperature and Heat Content

The temperature of water is a measure of its internal, thermal energy content. It is a property that can be sensed and measured directly with a thermometer. The heat content is a capacity property that must be calculated. The heat content is usually considered as the amount of energy above that held by liquid water at 0°C. Therefore, the heat content is a function of temperature and volume. A 1-L (1.06-q or 0.25-gal) bottle of water at 20°C has a heat content of 20,000 cal [1,000 mL × 20°C × 1 (cal/g)/°C]. A 1-ha (2.471-acre) pond of 1 m (3.25 ft) average depth and 20°C (68°F) has a heat content of 2×10^{11} cal [10,000 m^3 × 10^6 ml/m^3 × 20°C × 1 (cal/g)/°C]. The two water bodies have the same temperature but much different heat contents.

1.2.5 Density

Water molecules in ice form a tetrahedral lattice through hydrogen bonding. Orderly spacing of molecules in the lattice creates a lot of open space, so a given volume of ice weighs less than the same volume of liquid water. Thus, ice is less dense than liquid water and thus floats. The density of water continues to increase as the temperature rises until maximum density is attained at 4°C. Further warming decreases its density. Two processes influence density as water warms above 0°C. Remnants of the crystalline lattice of ice break up and increase density, and bonds stretch and decrease density. From 0 to 4°C, destruction of the lattice remnants has the greatest influence on density, and density increases to a maximum at 4°C. Further warming causes density to decrease through bond stretching. Densities of water at different temperatures are given in Table 1-1.

1.2.6 Vapor Pressure

Vapor pressure is the pressure exerted when a substance is in equilibrium with its own vapor. If a beaker of water is placed inside a chamber filled with dry air, water molecules enter the air until equilibrium is reached. At equilibrium the same number of water molecules enter the air from the water as enter the water from the air; in other words, there is no net movement of water molecules. The pressure of the water molecules in the air (water vapor) acting down on the surface of water in the beaker is the vapor pressure of water. This pressure increases as temperature rises. For

TABLE 1-1   Density of Water at Different Temperatures

Temperature (°C)[a]	Density (g/mL)[b]	Temperature (°C)	Density (g/mL)
0	0.99987	50	0.98807
3.98	1.00000	55	0.98573
5	0.99999	60	0.98324
10	0.99973	65	0.98059
15	0.99913	70	0.97781
20	0.99823	75	0.97489
25	0.99707	80	0.97183
30	0.99567	85	0.96865
35	0.99406	90	0.96534
40	0.99224	95	0.96192
45	0.99025	100	0.95838

[a] °F = 1.8 × °C + 32.
[b] Multiply by 62.4 to convert g/mL to lb/ft³.

pure water at standard atmospheric pressure, the vapor pressure reaches atmospheric pressure (760 mm Hg) at 100°C. Bubbles then form and push back the atmosphere to break the water surface; this is the boiling point. Atmospheric pressure varies with altitude and weather conditions, so the boiling point of water is not necessarily 100°C. At low pressure, water may boil at room temperature. The vapor pressure of water is considered further in Chapters 2 and 3.

1.2.7 Surface Phenomena

The rise of water in small-bore tubes or soil pores is called *capillary action*. To explain capillary action, one must consider cohesion, adhesion, and surface tension. *Cohesive* forces result from attraction between like molecules. Water molecules exhibit cohesion because they form hydrogen bonds with each other. *Adhesive* forces result from attraction between unlike molecules. Water adheres to a solid surface if the surface molecules form hydrogen bonds with water. Such a surface is called *hydrophilic* because it wets easily. Adhesive forces between water and the solid surface are greater than cohesive forces between water molecules. Water will bead on a hydrophobic surface and run off, for cohesion is stronger than adhesion. For example, raw wood wets readily, but a coat of paint causes wood to shed water. The net cohesive force on molecules in a body of water is zero, because the cohesive forces cannot act above the surface, and the molecules of the surface layer are subjected to an inward cohesive force from molecules below the surface. The surface molecules act as a skin over the surface to provide *surface tension*, which permits insects and spiders to walk over the surface of water, and needles and razor blades to float when gently placed on a water surface. The strength of the surface film decreases with increasing temperature, and increases when electrolytes are added to water. Soap and most other organic substances decrease surface tension when dissolved in water.

Water rises to considerable heights in a thin glass tube or in clay or fine-silt soils. Capillary action is the combined effect of surface tension, adhesion, and cohesion. In a thin tube, water adheres to its walls and spreads upward over as much surface as possible. Water moving up the wall is attached to the surface film, and molecules in the surface film are joined by cohesion to molecules below. As adhesion drags the surface film upward, it pulls a column of water up the tube against the force of gravity. Of course, the column of water below the surface film is under tension because the water pressure is less than the atmospheric pressure. Capillary rise is inversely proportional to tube diameter or soil pore size. Space exists among soil particles because they do not fit together perfectly. This space can function in much the same manner as the thin glass tube, and permits water to rise in a dry soil. Capillarity in soil increases in the following order: sand < silt < silty clay < clay. Of course, if capillary tube walls or soil pores are hydrophobic, capillary action can be downward instead of upward.

1.2.8 Viscosity

Fluids have an internal resistance to flow. This resistance to shear or angular deformation is termed *viscosity*, and it is the capacity of a fluid to convert kinetic energy to heat energy. Viscosity results from cohesion between fluid particles and interchange of particles between layers of different velocities. In viscous fluids such as water, low velocity or laminar flow exhibits a layered pattern with little exchange of molecules between layers. At higher velocities, flow becomes turbulent, the layers break up, and individual molecules move randomly.

Viscosity may be visualized by considering a deck of playing cards in which each card represents a layer of fluid. If the deck is gently shoved on a table top and released, the upper cards slide faster than the lower cards. In fact, the bottommost card may not move at all. For laminar flow, viscosity is illustrated in Figure 1-5, and defined as

$$\eta = \frac{F}{A}\left(\frac{D}{v}\right) \tag{1-1}$$

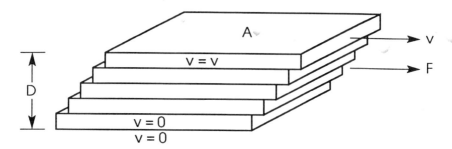

FIGURE 1-5.   Illustration of viscosity: v = velocity, A = area, D = depth, and F = force.

where η = absolute viscosity
       F = force
       A = area
       D = depth of fluid
       v = velocity of topmost layer

The example of viscosity in laminar flow (Fig. 1-5) is very similar to the analogy of the cards. The water can be thought of as moving in layers. The bottom layer is in direct contact with the container and may adhere to it; in addition, there is friction between this layer and the bottom. The layers above are less affected by the bottom, but friction exists between layers. The upper layers also press on the lower layers. Therefore, if a uniform force is supplied to a mass of liquid as illustrated in Figure 1-5, the bottommost layer remains stationary ($v = 0$) and the velocity of each layer above increases with respect to the layer below it until maximum velocity ($v = v$) is attained by the topmost layer. When turbulent flow occurs, the situation is more complex, because water no longer moves in layers.

Viscosity often is reported in centipoise (cp). In terms of force, 100 cp is equivalent to 1 g/cm · sec or 0.1 N · sec/m². Viscosity decreases as temperature rises. To illustrate, the absolute viscosity of water at 0, 20, and 50°C is 1.78, 1.00, and 0.547 cp, respectively. Water flow in pipes and channels and seepage through soil are favored slightly by warmth, because viscous shear losses decrease with decreasing viscosity.

Viscosity as just described above is called *absolute* or *dynamic*. Scientists and engineers often use kinematic viscosity (**v**), which is the ratio of dynamic viscosity and fluid density:

$$\mathbf{v} = \frac{\eta}{\rho} \tag{1-2}$$

where $\mathbf{v}$ = kinematic viscosity and ρ = density. The dimensions commonly used with kinematic viscosity are meters squared per second or feet squared per second. Kinematic viscosities for selected temperatures are provided in Table 1-2.

1.3 PRESSURE

1.3.1 Pressure at a Point

Matter is often classified as solid or fluid. A *fluid* is a substance with little or no elasticity of shape and readily conforms to the shape of its container. This means that fluids are liquids, and the ideal liquid offers no resistance to shearing forces. Unless completely confined, a liquid has a free surface, which is always horizontal (level) except at the edges. Liquids are considered to be essentially incompressible. Obviously, water can be treated as a fluid.

To evaluate the pressure on a small area under water, consider Figure 1-6. The water column of height h pressing down on a small area ΔA has volume $h \Delta A$, and the weight of water or force (F) is

$$F = \gamma h \Delta A \qquad (1\text{-}3)$$

where γ = weight of water per unit volume.

Pressure (P) is a force acting over a unit area:

$$P = \frac{F}{\Delta A} \qquad (1\text{-}4)$$

Thus, the pressure on the point may be expressed as

TABLE 1-2   Kinematic Viscosity of Water at Selected Temperatures

Temperature (°F)	Kinematic Viscosity (ft²/sec)	Temperature (°C)	Kinematic Viscosity (m²/sec)
32	1.93×10^{-5}	0	1.80×10^{-6}
40	1.67×10^{-5}	5	1.52×10^{-6}
50	1.41×10^{-5}	10	1.31×10^{-6}
60	1.21×10^{-5}	15	1.14×10^{-6}
70	1.05×10^{-5}	20	1.00×10^{-6}
80	0.93×10^{-5}	25	0.89×10^{-6}
90	0.83×10^{-5}	30	0.80×10^{-6}
100	0.74×10^{-5}	35	0.73×10^{-6}
120	0.61×10^{-5}	40	0.66×10^{-6}

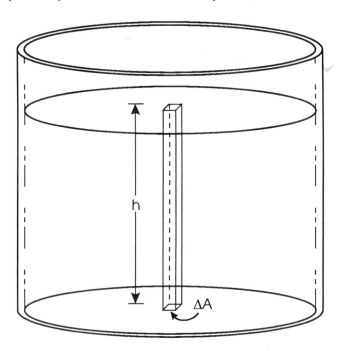

FIGURE 1-6.   Pressure of water on a surface (ΔA) beneath some depth of water (h).

$$P = \frac{\gamma h \Delta A}{\Delta A} = \gamma h \qquad (1\text{-}5)$$

This pressure is for the fluid only, and it is the product of the height of the water column and the unit weight of water. To obtain the absolute pressure, we must include the atmospheric pressure. Pressure can be converted to force by multiplying by the area over which the pressure is acting. This force is always normal to the surface.

Atmospheric pressure is the weight of the atmosphere over a point at a specific elevation. The concept is no different than that illustrated in Figure 1-6 for a liquid. In Figure 1-6, the pressure of the liquid is the weight of the liquid above the point. With atmospheric pressure, the actual pressure acting on a point is the weight of the atmospheric gases above the point. At sea level under standard conditions, atmospheric pressure will support a column of mercury 760 mm (29.92 in.) tall or a water column 10.4 m (34 ft) tall (Fig. 1-7). Atmospheric pressure also may be reported in atmospheres; 1 atmosphere (atm) = 760 mm Hg, or in millibars (mb), 1 atm = 1013.3 mb. As elevation increases, the weight

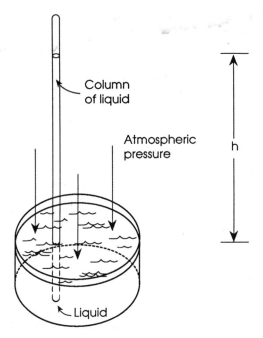

FIGURE 1-7.   Illustration of atmospheric pressure acting on a fluid. The height (*h*) of the column of fluid in the tube is equal to the weight of the atmosphere acting on the liquid in the open vessel.

of the atmosphere above a point decreases, so atmospheric pressure decreases. Standard atmospheric pressures for different altitudes are given in Table 1-3.

1.3.2   Head

Pressure also can result from elevation of water above a reference plane, velocity, or pressure applied by a pump. In hydrology and engineering applications, the term *head* is used to express the energy of water at one point relative to another point or reference plane. Head is expressed as depth of water in feet.

Pressure provided by a pump is expressed in pounds per square inch (psi). Conversions from pressure to head are

$$H \text{ (ft)} = P \text{ (psi)} \times 2.31 \tag{1-6}$$

and

TABLE 1-3   Standard Atmospheric Pressures for Different Elevations above Mean Sea Level

Elevation (ft)	Pressure (in. Hg)[a]	Elevation (ft)	Pressure (in. Hg)
0 (sea level)	29.92	9,000	21.38
500	29.39	10,000	20.58
1,000	28.86	11,000	19.79
2,000	27.82	12,000	19.03
3,000	26.81	13,000	18.29
4,000	25.84	14,000	17.57
5,000	24.89	15,000	16.88
6,000	23.98	16,000	16.21
7,000	23.09	17,000	15.56
8,000	22.22	18,000	14.94

[a] Multiply by 1.13 to convert in. Hg to ft of water.

$$P = H \times 0.43 \tag{1-7}$$

Pressure provided by elevation is the height of the water surface above the point of interest. The elevation may be converted directly to head.

Water velocity can be converted to head in depth of water as follows:

$$H_v = \frac{v^2}{2g} \tag{1-8}$$

where H_v = velocity head (ft or m)
  v = velocity (ft/sec or m/sec)
  g = gravitational acceleration (32.2 ft/sec^2 or 9.8 m/sec^2)

Elevation above a reference plane is normally measured in feet or meters, so no conversion is necessary to obtain head.

Consider a situation where water from a reservoir with a depth of 10 ft is flowing at 5 ft/sec through a pipe with the same elevation as the bottom of the reservoir and with an elevation of 20 ft above a reference plane. The head in the pipe is

$$\text{Velocity head} = \frac{5^2}{2(32.2)} = 0.39 \, \text{ft}$$

Pressure head = depth of water in reservoir = 10 ft
Elevation head = height of pipe above reference = 20 ft
Total head = 30.39 ft

If desired, atmospheric pressure may be added to obtain the absolute head. Standard atmospheric pressure is 34 ft or 14.7 psi.

1.4 WATER FLOW

The flow of water in a pipe, channel, or porous medium can be estimated from the velocity and cross-sectional area of the field of flow:

$$Q = Av \tag{1-9}$$

where Q = discharge (length^3/time)
      A = cross-sectional area of flow (length^2)
      v = velocity (length/time)

Water flows from a point of greater head to a point of lesser head. As water flows, its head relative to a reference plane decreases. This head loss results because water must overcome frictional forces opposing flow. Head may be increased by lifting the water source to a higher elevation or by increasing its pressure with a pump.

1.4.1 Conservation of Energy

The energy conservation equation describes the energy loss (head loss) that occurs when water flows from one point to the next. It is a modification of the well-known Bernoulli equation into which the head loss term has been incorporated. The difference in the total head between two places in a pipe, in an open channel, or in a porous medium such as soil, is the head loss. The equation is

$$\frac{v_1^2}{2g} + \frac{P_1}{\gamma} + Z_1 = \frac{v_2^2}{2g} + \frac{P_2}{\gamma} + Z_2 + H_L \tag{1-10}$$

where v_1, v_2 = velocity at points 1 and 2
            g = gravitational acceleration
      P_1, P_2 = pressure at points 1 and 2
      Z_1, Z_2 = elevation of points 1 and 2 above the reference plane
            γ = unit weight of water (62.4 lb/ft^3 or 1,000 kg/m^3)
          H_L = head loss between points 1 and 2

Equation (1-10) has all terms necessary to describe pipe flow, as illustrated in Figure 1-8. An application of the energy conservation equation is provided in the following section.

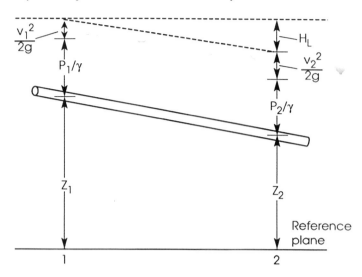

FIGURE 1-8.   Illustration of the energy conservation equation for water flow in a pipe from point 1 to point 2: v = velocity, g = gravitational acceleration, P = pressure, γ = unit weight of water, Z = height above reference plane, and H_L = head loss.

1.4.2 Darcy's Equation

Soils and other geologic materials have interconnected voids (pore space) between their constituent particles. Water may flow through these pores in response to pressure and gravity. Minute channels through which water moves in porous materials are not straight but exhibit a random pattern. Hence, when a molecule of water passes through a porous material, the length of the path followed by the molecule is unknown, but the rate of discharge of water through a cross section of the material may be determined. However, water molecules take various paths through the material, and their velocities differ. The entire cross section of material does not permit flow; flow occurs only through voids. The porosity of soils or other porous media is defined as

$$n = \frac{\text{volume of voids}}{\text{volume of bulk}} \tag{1-11}$$

where n = porosity.

Imagine that the pipe in Figure 1-9 has been filled with soil and that water is passing through the pipe. When the soil is saturated with water, Q_{in} must equal Q_{out}. Equation (1-10) may be used to equate energy at points 1 and 2 along the path of seepage as already illustrated. However,

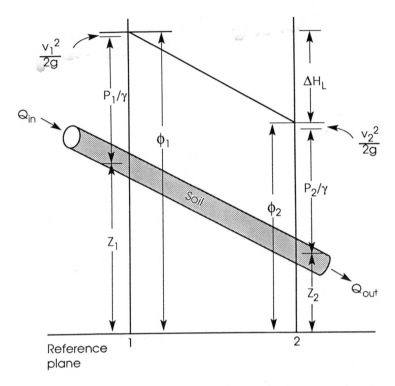

FIGURE 1-9.   Illustration of the energy conservation equation for seepage through a soil-filled tube from point 1 to point 2: Q = discharge, v = velocity, g = gravitational acceleration, P = pressure, γ = unit weight of water, Z = height above reference plane, H_L = head loss, and ϕ = potential.

the seepage velocity is small, and the square of a small number is even smaller. Thus, the velocity head term may be dropped from both sides of the equation. Pressure and elevation energy terms normally are combined, termed *potential*, and denoted by ϕ. This permits equation (1-10) for the system illustrated in Figure 1-9 to be written as

$$\phi_1 - \phi_2 = H_L \qquad (1\text{-}12)$$

Head loss results because energy is lost by viscous shear between water and soil particles.

If we consider a small distance between points 1 and 2 along the soil tube (ΔL), the difference in potential between the two ends of the segment is $\Delta \phi$. But potential difference is head loss, so we may use ΔH_L

rather than $\Delta\phi$. Darcy's equation states that seepage velocity is proportional to a coefficient multiplied by the energy gradient. That is,

$$v = K\frac{\Delta H_L}{\Delta L} = Ki \tag{1-13}$$

where K = hydraulic conductivity (length/time) and i = hydraulic gradient (dimensionless). The *hydraulic gradient* is the head loss per unit length of travel through the porous medium. Consider a pond with a bottom 10 ft above the water table and a water depth of 4 ft. Water seeps through the bottom downward to the water table. The length of flow is 10 ft, and the head loss measured from the pond surface to the water table is 14 ft. The hydraulic gradient is $14 \div 10 = 1.4$. Increasing the hydraulic gradient increases the flow rate through a given material. If the pond bottom were 5 ft above the water table, the hydraulic gradient is $9 \div 5 = 1.8$. Obviously, $K \times 1.8$ is greater than $K \times 1.4$. Also, in comparing two porous materials with the same hydraulic gradient, the material with the larger hydraulic conductivity will have the greater flow rate.

Sometimes Darcy's equation is given in a slightly different form. Because $Q = Av$, we may write

$$\frac{Q}{A} = Ki \tag{1-14}$$

and

$$Q = KiA \tag{1-15}$$

Area is in length squared and K is in length/time; thus, Q is in length cubed/time or volume/time.

The hydraulic conductivity of a porous material depends on the size and arrangement of particles in unconsolidated formations and on the sizes and arrangement of crevices, fractures, and solution caverns in consolidated formations. Any change in material characteristics influence the conductivity. For example, compaction of loose soil reduces K. Hydraulic conductivity is usually determined in the laboratory and expressed as the velocity of seepage through the permeable material under a hydraulic gradient of 1.00. The hydraulic conductivities of various materials are given in Table 1-4.

TABLE 1-4   Hydraulic Conductivity of Soil (K)

Soil Type	K (cm/sec)	K (ft/sec)
Clean gravel	2.5–4.0	0.8–0.13
Fine gravel	1.0–3.5	0.03–0.12
Coarse, clean sand	0.01–1.0	0.0003–0.03
Mixed sand	0.005–0.01	0.0002–0.03
Fine sand	0.001–0.05	0.00003–0.002
Silty sand	0.0001–0.002	0.000003–0.00007
Silt	0.00001–0.0005	0.0000003–0.00002
Clay	10^{-9}–10^{-6}	10^{-10}–10^{-7}

1.5 THE HYDROLOGIC EQUATION

In hydrology, all inputs of water into a unit (watershed, stream, lake, pond, etc.) are called *inflow*, and all outputs are called *outflow*. In calculations necessary to design and manage water supplies, inflows often must be balanced against outflows to prepare water budgets. This effort is simplified by use of the hydrologic equation:

$$\text{Inflow} = \text{outflow} \pm \text{change in storage} \qquad (1\text{-}16)$$

There may be several inflows and outflows, but if the magnitudes of all but one of the variables can be determined, the magnitude of the unknown variable may be estimated. For example, consider a small research pond that must be filled by rain falling directly into it and by inflow from a well. It is desired to maintain a fairly stable water level in the pond. The amount of well water necessary to maintain the water level for a year may be estimated if annual values for seepage, evaporation, and rainfall are available. Rainfall and well water are inflows; seepage and evaporation are outflows. Because well water will be introduced to maintain the water level during dry weather, the change in storage will be assigned a value of zero. The hydrologic equation may be written in the following form so that the amount of well water required per year to keep the pond full may be estimated:

$$\text{Rainfall} + \text{well water} = \text{seepage} + \text{evaporation} \pm 0$$

$$\text{Well water} = \text{seepage} + \text{evaporation} - \text{rainfall}$$

The water budget approach is frequently used in this book. In some instances, solutions require more variables and assumptions than necessary in the preceding example, but the approach is the same. The greatest difficulty in using the hydrologic equation is the acquisition of reliable estimates of certain variables. For instance, it is virtually impossible to obtain highly accurate estimates of seepage to use in designing ponds and water supplies in an area where no other ponds have been constructed.

1.6 WATERSHEDS

By definition, a watershed is an area of land from which all overland flow drains to a particular watercourse. The boundary or water divide of a watershed is established by topography. This boundary is often visible for small watersheds; for larger watersheds it is discernible on topographic maps.

Soil features, type and extent of vegetative cover, and slope influence the proportion of rain converted to runoff by a watershed. Hence, information on size and features of a watershed frequently are required. Topographic maps of specific areas are available from governmental agencies. These maps usually represent large areas, and contour lines are at wide intervals; they have limited value in work with small watersheds for aquaculture ponds. However, topographic maps such as the one in Figure 1-10 may be extremely valuable in aquaculture applications. Preparation of outline maps and topographic maps is not difficult if one is familiar with basic surveying and mapping procedures.

1.7 THE WATER CYCLE

Water in nature is continually in motion, and these movements of water comprise the familiar water cycle or hydrologic cycle (Fig. 1-11). Water evaporates from free-water surfaces of oceans, lakes, ponds, and streams and from moist soil surfaces; water also is transpired by plants. When water evaporates, it changes from a liquid to a gas (water vapor). Energy necessary to change liquid water to water vapor originates from the sun, so solar radiation is the engine that drives the water cycle. Water vapor is unconfined and transported by general atmospheric circulation. Certain events—differential heating of land surfaces, uplifting over topographic barriers, and convergence of cold- and warm-air masses—cause air to rise. When air rises, it expands and cools. Cold air cannot hold as much water vapor as warm air, so when air rises and cools, some of the water vapor may condense into tiny droplets of liquid water. These droplets form clouds. When water droplets in clouds grow too large to be buoyed up by air currents, they fall as rain. If temperatures are below freezing, water in clouds freezes to become sleet, snow, or hail.

FIGURE 1-10.   A topographic map of a pond and its watershed. Contour lines are in feet.

When rain falls on the earth's surface, some of it strikes vegetation, buildings, and other objects. This rain is said to be *intercepted*. If it rains enough to fill the interception capacity of objects above the ground, water runs down or drips from objects to the ground. Intercepted water returns to the atmosphere through evaporation. Direct rainfall is not intercepted; it directly strikes land or water surfaces.

Some of the water reaching the ground wets the soil surface and is detained (detention storage). After it rains for a while, the ground surface becomes saturated as water begins to form puddles in depressions on the surface. At the same time, water begins to pass through the soil surface and move downward through soil pores by infiltration. If the rate

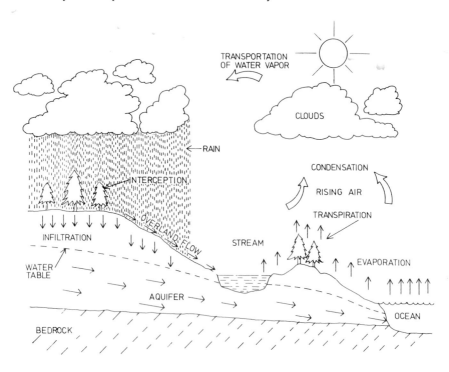

FIGURE 1-11.   The water cycle.

at which rain reaches the ground exceeds the rate of infiltration, the storage capacity of surface depressions is exceeded and water begins to flow downslope. Movement of water over the land surface during and soon after a rain is known by several terms: storm runoff, direct runoff, and overland flow. The term *overland flow* is used in this text. When rainfall is heavy, some of the water that infiltrates the soil flows downslope through the soil surface layer. Water moving through this pathway is called *subsurface storm drainage*, but it is difficult to separate it from overland flow.

Water from a storm enters one of four compartments: (1) some remains on or above watershed surfaces because of interception, detention storage, or depression storage; (2) some passes through the watershed to be stored in soil (soil moisture); (3) some infiltrates deeper to the water table to enter the ground-water zone; (4) some becomes overland flow. For a given storm, the distribution of water among the different compartments depends on watershed conditions, temperature, and storm characteristics.

Overland flow progresses downslope and enters streams. Because overland flow only occurs during and immediately after rains, it does not sustain streams during dry weather. Stream beds cut below the water table, and seepage of ground water into streams (base flow) causes streams to flow during dry weather. Streams in most regions eventually reach the ocean, but in arid areas some streams flow into closed basin lakes or simply dry up.

A description of the water cycle is different for regions with appreciable winter snow cover. Snow accumulates, and infiltration and runoff do not occur until snow melts. Pond culture of fish and crustaceans is practiced most often in regions where snow is not an important factor in the water cycle. However, ground water or spring discharge used to supply raceways for trout culture may originate from snow melt.

Ground water is beneath most places on the land surface. This water is contained in the voids within the underlying geologic material, and the water-bearing formations are called *aquifers*. Ground water moves slowly in response to gravity, and it seeps into streams, issues forth as springs, and flows into oceans. Much ground water also is removed by wells.

In our discussion we have followed water through its cycle: water vapor, condensation, rain, intercepted water, detention and depression storage, soil moisture, ground water, base flow, runoff, and ocean water. At any place where water is exposed to the atmosphere, it may evaporate and go full circle to become water vapor again. Obviously, a single molecule of water does not pass through all stages in the hydrologic cycle, but if a large area of the earth is considered, water is constantly moving through all parts of the cycle.

1.8 DISTRIBUTION OF WATER ON THE EARTH'S SURFACE

The various compartments of the water cycle contain different amounts of water, and the average renewal time for water in each compartment differs. The distribution of the earth's water according to Baumgartner and Reichel (1975) is as follows: oceans, 97.39%; polar ice caps, icebergs, and glaciers, 2.01%; ground water and soil moisture, 0.58%; lakes and streams, 0.02%; atmosphere 0.001%. The total volume of the earth's water is approximately 1.1×10^{15} acre-ft. Of this, 2.6%, or about 2.9×10^{13} acre-ft, is freshwater. The distribution of freshwater is shown in Table 1-5. Average renewal times for water in selected compartments of the cycle (Wetzel 1975) are 37,000 years for oceans; 16,000 years for ice caps, icebergs, and glaciers; 300 years for ground water; 1 to 100 years for lakes; 280 days for soil moisture; 12 to 20 days for streams; and 9 days for

TABLE 1-5   The Distribution of Freshwater

Compartment	%
Polar ice caps, icebergs, and glaciers	77.23
Ground water to a depth of 2,500 ft	9.86
Ground water from depths of 2,500 to 13,000 ft	12.35
Soil moisture	0.17
Lakes	0.35
Streams	0.003
Hydrated earth minerals	0.001
Plants, animals, and humans	0.003
Atmosphere	0.04

atmospheric moisture. Because water is constantly moving through its cycle, the importance of evaporation and precipitation in maintaining the volume of each compartment is apparent.

From a practical view, total volumes and average renewal times mean little, because the amount of water available from the various compartments of the water cycle differs greatly from place to place on the earth's surface. Aquaculturists should be familiar with the availability of water and both the short- and long-term variability within the water cycle at their locations.

2

Rainfall

2.1 INTRODUCTION

Atmospheric moisture can reach the earth's surface as precipitation from clouds or as condensation on vegetation or other surfaces. Precipitation includes rain, sleet, snow, and hail, and condensation comprises dew or frost. All forms of moisture that reach the earth's surface are of interest in general treatments of hydrology, but this book is intended primarily for warm-water, pond aquaculture that is conducted in relatively warm climates.

The major sources of water for filling and maintaining water levels in ponds are surface runoff and well water, which originate mainly from rainfall. The availability of water for inland aquaculture is closely correlated with rainfall. Sites in arid regions usually have much less water for use in ponds than do sites in humid regions. Even in a humid region, normal rainfall patterns may result in water shortage during certain times of the year, and droughts may also occur. An understanding of local rainfall patterns is of immense value to a pond manager.

2.2 THE PROCESS

Clouds form when air cools to 100% relative humidity and moisture condenses around hydroscopic particles of dust, salts, and acids (called nuclei of condensation) to form tiny droplets or particles of ice. Droplets grow because they bump together and coalesce when the temperature is above freezing. At freezing temperatures, supercooled water and ice crystals exist in clouds. The vapor pressure of supercooled water is

37

greater than that of ice. Ice particles grow at the expense of water droplets because of their lower vapor pressure. Precipitation occurs when water droplets or ice particles grow too large to remain suspended in the air by turbulence caused by air currents. The size of droplets or particles at the initiation of precipitation depends on the degree of air turbulence. In the tops of some thunderheads, turbulence and freezing temperature can lead to the formation of ice particles of several centimeters in diameter, known as hail. Precipitation often begins its descent as sleet, snow, or hail, but frozen precipitation usually melts while falling through warm air. Of course, sometimes ice particles may be so large that they reach the ground as hail even in summer.

Rising air is necessary for precipitation. As air rises, the pressure on it decreases, because atmospheric pressure declines with increasing elevation (Table 1-3). The temperature of an air parcel, like the temperature of any substance, results from the thermal energy of its molecules. Under reduced pressure the air parcel contains the same number of molecules as before expansion, only now its molecules occupy a larger volume. Thus, the temperature of an air parcel must decrease upon expansion. The rate of cooling, which is called the *adiabatic lapse rate*, is 5.5°F/1,000 ft (1°C/ 100 m) for air at less than 100% relative humidity. The word *adiabatic* means that no energy is gained or lost by the rising air mass. If air rises high enough, it will cool until 100% relative humidity is reached, and moisture will condense around the nuclei of condensation to form water droplets and clouds. The elevation at which air reaches the dew point and begins to condense moisture is called the *lifting condensation level*. Once rising air begins to condense moisture, it gains heat, for water vapor must release latent heat of vaporization to condense. The heat from condensation counteracts some of the cooling from expansion, and the adiabatic lapse rate for rising air with 100% relative humidity is 3°F/1,000 ft (0.6°C/ 100 m). It is common to speak of "wet" and "dry" adiabatic lapse rates.

An explanation of relative humidity may benefit some readers. Relative humidity of air is a measure of the percentage saturation of air with water vapor; it is calculated as follows:

$$\text{RH} = \frac{e_a}{e_s} \times 100 \qquad (2\text{-}1)$$

where RH = relative humidity (%)
e_a = actual vapor pressure
e_s = saturation vapor pressure

Saturation vapor pressure (Table 2-1) increases with warmth, so warm air can hold more water vapor than can cool air. Air with vapor pressure

TABLE 2-1   Saturation Vapor Pressure at Different Temperatures

Air Temperature (°F)	Saturation Vapor Pressure (in. Hg)	Air Temperature (°F)	Saturation Vapor Pressure (in. Hg)
0	0.038	50	0.360
5	0.049	55	0.432
10	0.063	60	0.517
15	0.081	65	0.616
20	0.103	70	0.732
25	0.130	75	0.866
30	0.164	80	1.022
32	0.200	85	1.201
35	0.203	90	1.408
40	0.247	95	1.645
45	0.298	100	1.916

equal to the saturation vapor pressure has a relative humidity of 100%. Air at 75°F with a vapor pressure of 0.360 in. Hg reaches 100% relative humidity when cooled to 50°F (see Table 2-1). An air mass at 75°F would cool to 50°F after rising 4,545 ft [(25°F ÷ 5.5°F) × 1,000 ft = 4,545 ft]. If the land had an elevation of 155 ft, the lifting condensation level would be 4,700 ft (4,545 ft + 155 ft).

2.3   TYPES OF RAINFALL

Rising or uplifting of an air mass is necessary for rainfall, and three distinct types of uplifting cause rainfall.

2.3.1   Orographic Rainfall

Air often must rise over topographic barriers such as mountains. As air rises up a mountain, it expands, cools, increases in relative humidity, reaches the condensation level, and precipitation may occur. Therefore, it is not unusual for the windward slopes of mountains to receive much rainfall. Once air passes over a mountain and descends the leeward slope, it warms because of compression and its relative humidity decreases. Thus, conditions are not good for precipitation, and a rain shadow often occurs on the leeward side of a mountain range.

Orographic precipitation is illlustrated in Figure 2-1. Air from the sea that is nearly saturated with water vapor rises over a mountain. As it rises it cools at the dry lapse rate until it reaches saturation at 1,000 ft elevation and rainfall begins. After this, the air cools at the wet lapse rate. The temperature falls from 80°F at sea level to 62.5°F at 5,000 ft. The air has

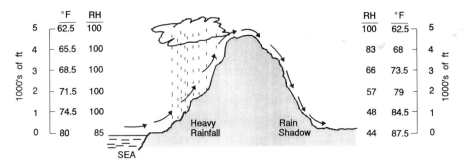

FIGURE 2-1.   Illustration of orographic precipitation showing zone of heavy rainfall and rain shadow. Estimated temperature (°F) and percentage relative humidity (RH) are given for different elevations.

100% relative humidity at 5,000 ft. As it descends the leeward side of the mountain, it warms by compression at the dry rate, because the excess moisture was lost as rainfall on the windward side of the mountain. The temperature increases to 87.2°F and the relative humidity drops to 44% by the time the air reaches ground level, 500 ft, on the leeward side. The falling air does not have enough moisture to generate rainfall, and it has a high capacity for evaporation. Thus, there will be a rain shadow on the leeward side of the mountain.

A good example of this phenomenon is found in the coastal area of Washington and Oregon in the United States. The windward slopes of the coastal mountain ranges receive rainfall up to 100 in./yr, and rain forests exist in places. Progressing eastward across these two states, rainfall drops rapidly and is no more than 10–20 in./yr in central and eastern parts (Fig. 2-2). Orographic precipitation and rain shadows also are found on many islands in the trade-wind zones where winds typically come from one direction. Windward sides of islands often are well watered, but leeward sides are frequently deserts.

2.3.2 Convective Rainfall

Convection of air results when air parcels near the earth's surface are heated by insolation. The warm air is lighter than overlaying air and it rises. Tropical regions have high rates of insolation, and conditions favoring convective uplifting of air are prevalent. Anyone who has spent much time in the tropics is familiar with the almost daily afternoon and early evening rains that occur during the wet season.

Convectional processes also are common in subtropical and temperate regions in the summer. Because of differences in soil color, vegetative

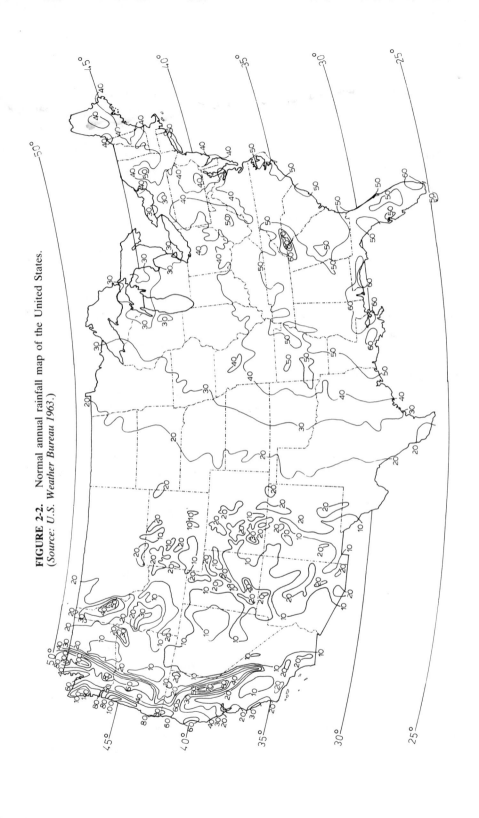

FIGURE 2-2. Normal annual rainfall map of the United States. (*Source: U.S. Weather Bureau 1963.*)

cover, water bodies, expansive areas of pavement, and other factors, some parts of the earth's surface heat faster than others, and thermal updrafts, or convection, result. Of course, when air rises in an updraft, surface air from adjacent areas flows into the zone of the updraft to replace the rising air. This results in the formation of cells of ascending and descending air. Thunderheads may develop from the rising cells of air, and intensive rainfall may result. Thunderstorms often are limited in extent. For example, heavy rainfall may occur at a particular location, but a short distance away, sometimes only a mile or less, there may be no rainfall.

2.3.3 Frontal Rainfall

When the leading edges of two air masses of different temperatures converge, the warmer, lighter air rises over the cooler, heavier air. The resulting uplifting can cause precipitation along the entire zone of convergence of the two air masses. Rainfall can occur over extensive areas if a strong weather front passes through a region. For example, in the southeastern United States, cold fronts frequently travel across the region during winter and early spring. It is not uncommon for heavy rainfall to occur over this entire region as the cold front passes.

Sometimes, a cold-air mass moves into an area and becomes stationary for several days. If warm air continually converges with the stationary cold front, several rainy days may occur in succession. This also is a common event in the southeastern United States when warm air from the Gulf of Mexico converges with a stationary cold front.

Although frontal precipitation is more common in cool or cold seasons, it also can occur in summer when a mass of relatively cool air passes through a region.

2.4 TRANSPORTATION OF WATER VAPOR AND PRECIPITATION

The rainfall process and the distribution and amount of rainfall are closely related to the movement of the atmosphere. The importance of upward movement of air in the rainfall process has been mentioned. Here we consider how other air movements affect rainfall. If all of the moisture in the atmosphere above an area fell as rain, rainfall would be less than 1 in. over this area. Larger rains are possible because moist air flows into the area where rain is falling. For example, air continues to flow up a mountain, surrounding air is swept into thermal updrafts, and warm air continues to converge with and rise over a cold front. On a larger scale,

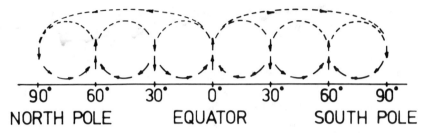

FIGURE 2-3.   Generalized atmospheric circulation patterns of the world.

the general circulation of the atmosphere is important in moving large amounts of water vapor from one place to the other. The two hemispheres of the earth are divided into three more or less independent atmospheric circulation zones (Fig. 2-3). At the equator, strong insolation heats air and causes it to rise. This air moves poleward in both directions from the equator, and a part of it descends at about 30° north and 30° south latitude. Atmospheric pressure tends to be low in equatorial regions, but high-pressure belts are prevalent in subtropical latitudes. Part of the air that descends in subtropical regions flows toward the low-pressure equatorial region to replace the rising air there. Movement of air back toward the equator causes the trade winds. A part of the air from the subtropical high-pressure belts flows poleward toward low-pressure belts that exist at about 60° north and 60° south latitude. Cold, high-pressure air masses exist over the poles, and polar air flows toward the low-pressure belts at 60° north and 60° south latitude. Thus, cold and warm air masses converge at about 60° north and 60° south latitude, and air tends to rise there. The rising air divides, with a part flowing poleward and a part flowing toward the subtropics.

The pattern here described is an oversimplification of atmospheric circulation based on average conditions. At a specific location, prevailing conditions may differ considerably from the global average. Also, as the zones of maximum insolation migrate northward and then southward during the year, the cells of atmospheric circulation also migrate. For example, the Asiatic monsoon results from the transfer of air masses across the equator with the migration of the sun's rays, and there is a complete reversal of wind directions with the season.

The direction of the winds in the cells of atmospheric circulation is affected by the Coriolis force. To understand this force, consider a helicopter hovering above the equator. This aircraft has no north–south velocity, but it is moving toward the East at the same speed as the earth's surface (ca. 1,000 mph). Now, if the helicopter begins to travel northward, it will maintain its eastward velocity caused by the earth's rotation. As the aircraft moves northward, the rotational velocity of the earth's surface decreases because the diameter of the earth decreases as one travels poleward from the equator. This causes the helicopter to deflect to the right relative to the earth's surface. If we consider another helicopter hovering above the earth some distance north of the equator, which begins to travel southward toward the equator, it also will deflect toward the right as it flies southward. This occurs because the rotational speed of points on the earth's surface increase in velocity as the aircraft progresses southward. In the Southern Hemisphere, the same logic prevails, but with the opposite direction of wind deflection. Hence, moving air always tends to deflect to the right in the Northern Hemisphere and to the left in the Southern Hemisphere.

Combining the ideas of cells of atmospheric motion and the Coriolis force, the wind belts of the world can be generalized as shown in Figure 2-4. However, these belts of winds and pressures tend to shift with season, as illustrated in Figure 2-5. These belts shift with season because the land is cooler than the oceans in winter, and cool, heavy air develops over landmass, whereas in summer the land is warmer than the oceans and the cooler, heavier air builds up over the oceans.

Ocean currents (Fig. 2-6) are caused by wind. Where winds cause water to pile up against a landmass, the water must then flow along the landmass, but the direction of this flow is also affected by the Coriolis force. For example, the trade winds cause water to pile up in the Gulf

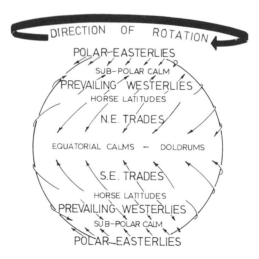

FIGURE 2-4.   Generalized wind belts of the world.

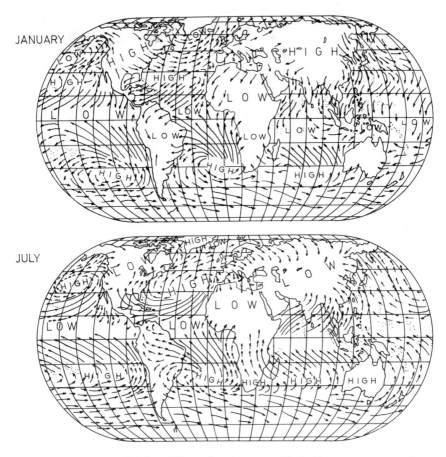

FIGURE 2-5.   Effects of season on world wind belts.

of Mexico along the eastern coast of Central America, and along the northern coast of South America (Fig. 2-6).

This water, the Gulf Stream, then flows northward along the eastern coast of North America and makes a clockwise circulation pattern in the Atlantic Ocean. Ocean currents flowing from tropical and subtropical areas are warm, whereas those coming from polar and subpolar regions are cold. Thus, ocean currents can deliver warm water to cold regions, and vice versa, and greatly modify climates of coastal areas.

The ability of air to hold water vapor or the ability of a water surface to generate water vapor through evaporation increases with increasing temperature (Table 2-1). Thus, the moisture-holding capacity of air generally decreases with increasing latitude. Also, cold ocean currents

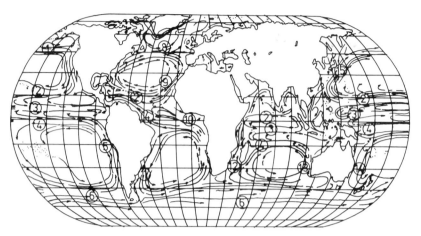

1	JAPAN CURRENT	12	AGULHAS CURRENT
2	NORTH EQUATORIAL CURRENT	13	WEST AUSTRALIAN CURRENT
3	EQUATORIAL COUNTER CURRENT	14	EAST AUSTRALIAN CURRENT
4	SOUTH EQUATORIAL CURRENT	15	KARO SIWA
5	HUMBOLT CURRENT		
6	WEST WIND DRIFT		
7	GULF STREAM		
8	NORTH ATLANTIC DRIFT		
9	CANARIES CURRENT		
10	GUINEA CURRENT		
11	BRAZIL CURRENT		

FIGURE 2-6.   Major ocean currents of the world.

yield less water vapor to overlaying air than do warm ocean currents. Assuming equal temperature, evaporation rates are usually much greater over water than over land, so winds blowing from oceans usually carry more water vapor than winds originating on land.

Several facts may now be combined to provide some basic rules for the amounts of precipitation at a location:

1. Precipitation is normally greater in areas where air tends to rise than in areas where air tends to fall. Compare the coastal regions of Washington and Oregon with the eastern areas of those states (Fig. 2-2).
2. Amounts of precipitation generally decrease with increasing latitude. Compare central Louisiana with central Minnesota in Figure 2-2.
3. The amount of precipitation usually declines from the coast to the interior of a large landmass. Compare the eastern coast of the United States with the interior of the country at the same latitudes (Fig. 2-2).
4. A warm ocean offshore favors high precipitation, and a cold ocean favors low precipitation. Compare southern California with southern Florida (Fig. 2-2).

One can use these principles to identify areas of comparatively low and high precipitation on a physical map of a region or to explain why some areas receive more precipitation than others. Those interested in a detailed analysis of precipitation patterns may consult a climatology text such as Critchfield (1974).

Local air movements that occur within the large, generalized circulation pattern are much more difficult to assess. They are caused by differences in temperature, which in turn affect pressure, and the air moves from high-pressure to low-pressure areas. For example, in coastal areas during the day, the land may heat faster than the sea. Pressure is higher over the sea, and air flows onto the land as a sea breeze. At night, the land cools more quickly than the sea, and when it becomes cooler than the sea, higher pressure over land causes a breeze toward the sea. Mountain and valley breezes are similar. Mountain slopes often cool faster than valleys at night, and air flows down into valleys. During the day, the mountain slopes may warm faster than the valleys, and airflow is reversed. In inland areas, some places warm faster than others, causing convectional circulation. This was mentioned in the discussion of convectional precipitation.

2.5 MEASURING RAINFALL

Of the two types of rain gauges, recording and nonrecording, the nonrecording type is more common. The standard U.S. National Weather Service rain gauge (Fig. 2-7) consists of an 8-in.-diameter by 20-in.-tall brass bucket or overflow can into which a 2-in.-diameter by 20-in.-tall brass inner or collector tube is placed. A removable, brass collector funnel of 8-in.-diameter at the top is mounted on top of the overflow can to direct water into the collector tube. The rain gauge is mounted in a support attached to a wooden or concrete base. Rain is concentrated 10 times in the collector tube for measuring, because the area of the collector funnel is 10 times the area of the collector tube. A calibrated dipstick is inserted in the collector tube, and the wetted distance on the stick indicates rainfall to the nearest 0.01 in. If rainfall exceeds 2 in., the inner tube overflows into the overflow bucket. Water from the overflow can is poured into the inner tube for measurement. Most 24-hr rainfall events do not exceed 2 in.

A tipping-bucket recording rain gauge has a 10-in.-diameter receiving funnel. In the funnel is a bucket with two compartments. One compartment fills and tips as the other moves into position to fill. It takes 0.01 in. of rain to fill a compartment. As the bucket tips, an electrical circuit closes and causes a mark to be made on a graph. A permanent record is

FIGURE 2-8.   Recording rain gauge. (*Source: Belfort Instrument Co., Baltimore, Maryland.*)

FIGURE 2-7.   Standard, nonrecording rain gauge. (*Source: Belfort Instrument Co., Baltimore, Maryland.*)

obtained of the time of occurrence of each 0.01-in. increment of rain; total rainfall also is recorded.

A weighing-type recording rain gauge has a scale that converts the weight of rainfall caught in a bucket beneath an 8-in. circular opening at the top of the gauge into the curvilinear movement of a pen that marks on chart paper (Fig. 2-8). The chart, graduated in hundredths of an inch, is wrapped around a vertical cylinder that is rotated at constant speed by a spring-driven or battery-powered clock. Charts usually are changed at seven-day intervals.

Exposure of a rain gauge is important to measurement accuracy. The following suggestions for rain gauge installation were provided by the U.S. National Weather Service (1972). The gauge should be positioned on fairly level ground, and its top must be level. Trees, buildings, and other high objects should be no closer to the gauge than their height. Nearer objects may intercept rain that otherwise would have been caught by the gauge. Elimination of air turbulence around the gauge is important, for wind tends to blow rain over the top of the gauge. Windbreaks of uniform height and at a distance from the rain gauge of about twice their height serve to reduce air turbulence around the gauge. Windshields consisting of metal slats established in a circle of somewhat larger diameter than the top of the gauge also are effective in reducing air turbulence. Most rain gauges are not protected by windbreaks or shields. In open fields, a single, tall object or a cluster of tall objects near a rain gauge can create serious air turbulence. Vegetation at rain gauge sites must be cut periodically.

2.6 RAINFALL AT A POINT

2.6.1 Daily, Monthly, and Annual Totals

Rainfall data generally are collected on a daily basis and reported as monthly and annual totals. In many nations, precipitation is recorded at numerous sites, and governmental weather services prepare maps with *isohyets*, or lines of equal precipitation, as illustrated for the United States in Figure 2-2. These maps do not contain sufficient detail to depict accurately normal rainfall at a specific locality. Local rainfall records are highly desirable for planning and design purposes, because two places only a few miles apart may receive significantly different amounts of rain.

Normal monthly and annual precipitation for some locations in the world are provided in Table 2-2. Values range from less than 15 in. to more than 80 in. Actually, normal rainfall is an artificial statistic; rarely does a year have normal rainfall. Data for 66 years (1924–1989) at Auburn, Alabama, revealed a normal rainfall of 54.30 in. annually with a standard deviation of 15.75 in. In few years did actual rainfall coincide closely with normal rainfall (Fig. 2-9). Some years were exceptionally drier or wetter than normal, as can be seen by comparing 1963 and 1974 (wet years) with 1927 and 1954 (dry years). Wetter-than-average and drier-than-average years occurred in succession, for example 1949–1951 and 1961–1964, respectively. No cycle of wet and dry years was obvious. Normal rainfall data are useful for purposes where the usual situation is of interest. Sometimes extreme values are more important than average values in planning and designing projects; that is, water supplies should be based on the smallest amount of rainfall expected and hydraulic structures should be designed for the largest expected rainfall.

There are few places where rainfall is distributed regularly throughout the year. Instead, some months are typically wetter or drier than others (Table 2-2). At Auburn, Alabama, winter is the wettest time of year, and fall is the driest; March usually has the most rainfall and October the least

TABLE 2-2   Monthly and Annual Rainfall Totals in Inches for Selected Cities

Location	J	F	M	A	M	J	J	A	S	O	N	D	Total
Bangkok, Thailand	0.35	1.14	1.33	3.50	6.53	6.73	7.00	7.51	12.04	10.03	2.24	0.27	58.74
Bombay, India	0.07	0.03	0.00	0.11	0.62	20.47	27.91	16.49	11.69	3.46	0.82	0.07	81.81
Jakarta, Indonesia	13.18	9.48	7.91	5.55	4.56	3.81	2.40	1.96	3.07	3.58	5.94	7.59	69.09
Manila, Philippines	0.07	0.27	0.23	0.94	4.33	9.29	9.96	18.89	10.66	7.91	5.07	2.20	70.51
New Dehli, India	0.98	0.86	0.66	0.27	0.31	2.55	8.30	6.81	5.90	1.22	0.03	0.19	28.14
Brisbane, Australia	5.62	7.20	5.78	3.07	2.24	2.20	1.92	1.18	1.77	3.03	3.62	5.35	42.99
Perth, Australia	0.27	0.47	0.86	2.04	4.92	7.55	7.20	5.31	2.71	2.12	0.90	0.59	35.00
Edinburgh, Scotland	2.16	1.65	1.73	1.53	2.04	2.08	2.79	3.07	2.36	2.63	2.28	2.24	26.61
Madrid, Spain	1.49	1.10	1.77	1.73	1.73	1.06	0.43	0.55	1.22	2.08	1.85	1.88	17.16
Warsaw, Poland	0.98	1.10	0.78	1.25	1.57	2.36	3.11	1.85	1.61	1.22	1.22	1.45	18.54
Montgomery, Alabama	4.02	4.30	6.02	4.45	3.47	4.03	5.09	3.47	4.41	2.24	3.43	4.93	49.86
Dallas, Texas	2.48	2.40	3.30	4.21	4.48	3.81	2.79	2.99	2.71	2.79	2.71	2.51	37.24
Stockton, California	2.91	2.40	1.96	1.37	0.42	0.07	0.01	0.03	0.17	0.72	1.72	2.68	14.17
Mexico City, Mexico	0.15	0.19	0.43	0.70	1.81	3.93	4.56	4.48	4.01	1.45	0.51	0.19	22.48
Havana, Cuba	2.79	1.81	1.81	2.28	4.68	6.49	4.88	5.31	5.90	6.81	3.11	2.28	48.18
Bogota, Columbia	1.49	1.88	2.59	3.77	4.05	2.36	1.88	1.49	2.08	6.22	5.51	3.66	37.08
Manus, Brazil	10.94	10.94	11.81	11.29	7.59	3.89	2.40	1.61	2.44	4.40	6.49	8.66	82.48
Ankara, Turkey	1.45	1.41	1.41	1.45	1.92	1.18	0.55	0.35	0.66	0.94	1.18	1.69	14.17

Sources: Data from Critchfield (1974), Wallis (1977), Thailand Ministry of Communication (1981).

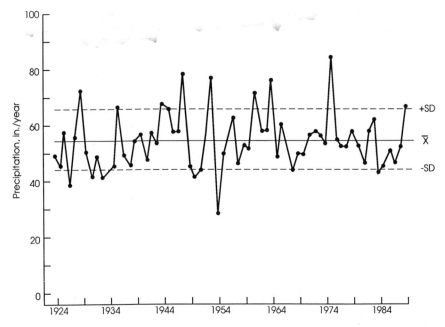

FIGURE 2-9. Annual precipitation at Auburn, Alabama, for the period 1924–1989. The solid line represents normal (mean) precipitation: dashed lines depict ± one standard deviation (SD). (*Source: Data from Southeast Agricultural Weather Service Center, Auburn University, Alabama.*)

(Table 2-3). For a given year, a particular month may be much wetter or drier than average (Table 2-3). The seasonality of precipitation may differ greatly from region to region. In many tropical regions, there are distinct wet and dry seasons, as illustrated in Table 2-2 for several locations, such as Bangkok, Thailand; Bombay, India; and Manila, Philippines.

Daily precipitation is even more variable than monthly or annual precipitation. There are normally more days with measurable rainfall during some seasons than others, but a day with heavy rainfall can come at any time. Storms producing small amounts of rain are more common than those yielding heavy rainfall. This fact is illustrated in Figure 2-10 with a frequency distribution of daily rainfall amounts over a five-year period at Auburn, Alabama, where 24-hr rainfall totals of less than 1 in. are most common.

Rainfall is measured at many locations in the United States. These records are summarized for each state by day, month, and year and published in a pamphlet entitled *Climatological Data* by the National Climatic Data Center in Asheville, North Carolina. Wallis (1977) has

TABLE 2-3   Normal, Minimum, and Maximum Monthly Rainfall Totals in Inches of Depth at Auburn, Alabama

Month	Normal	Minimum	Maximum
January	5.14	0.49	12.00
February	5.34	1.50	17.61
March	6.85	0.30	17.47
April	5.31	0.50	18.07
May	4.23	0.36	10.33
June	3.85	0.57	8.64
July	5.74	1.39	15.73
August	3.59	0.01	11.03
September	4.32	0.36	13.13
October	2.83	0.00	8.41
November	3.42	0.23	17.77
December	5.48	0.82	14.27

Source: Data from Southeast Agricultural Weather Service Center, Auburn University, Alabama.

summarized precipitation data for many locations in the United States. Most nations have weather services that provide records of rainfall at different localities. Nevertheless, rainfall records seldom are available within less than a few miles of where they are needed, and it often is necessary to transpose rainfall data from one place to another.

2.6.2   Transposing Data

Where rainfall records are unavailable, data from the nearest gauging site are used as a substitute. Errors resulting from transposing rainfall tend to increase with distance, and they are greater for individual storms than for monthly or annual rainfall. The size of errors that may result from transposing rainfall data small distances are illustrated by comparisons of daily rainfall amounts between gauges located different distances apart on the Piedmont Plateau near Auburn, Alabama (Fig. 2-11). When monthly rainfall was considered, agreement between catches of gauges improved (Fig. 2-11). On an annual basis, rainfall differed little among gauging stations; the range was 43.70 to 45.14 in.

If there are several rain-gauging stations in the vicinity of an ungauged site, a method presented by Viessman et al. (1972) may be used to estimate rainfall at the ungauged location. This technique is illustrated for the situation in Figure 2-12. Rainfall is known at six places (B–G) in the vicinity of A. The seven points are plotted on a map, and horizontal (X) and vertical (Y) axes are constructed with their origins at A (Fig. 2-12).

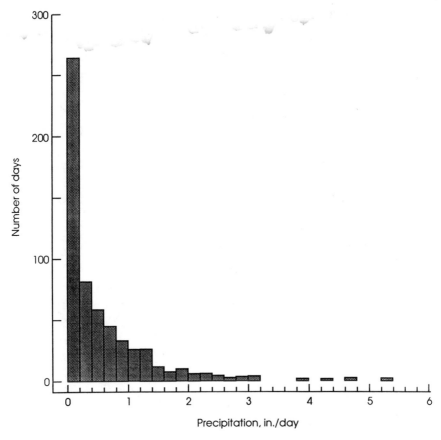

FIGURE 2-10.   Frequency distribution of measurable daily precipitation at Auburn, Alabama (1980–1984).

Vertical and horizontal displacements of each gauging station relative to A are recorded without regard to sign; for example, for F, $X = 8.5$ mile, $Y = 1.5$ mile. For each station, the sums of squares (SS) of X and Y are calculated:

$$SS = X^2 + Y^2 \qquad (2\text{-}2)$$

A factor (W) for weighing rainfall is obtained for each station as follows:

$$W = \frac{1}{SS} \times 10^3 \qquad (2\text{-}3)$$

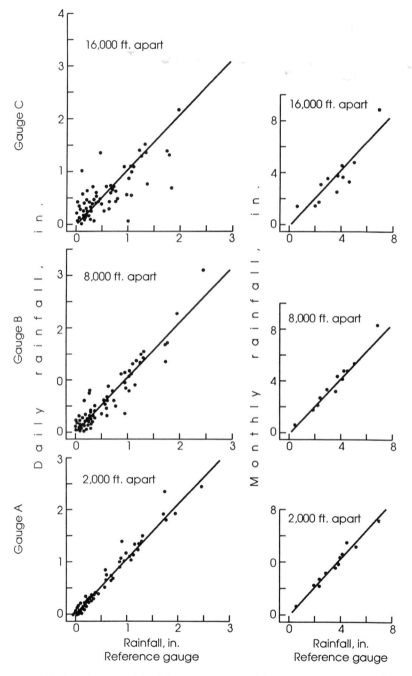

FIGURE 2-11. Agreement in daily and monthly rainfall at gauges different distances apart on the Piedmont Plateau near Auburn, Alabama.

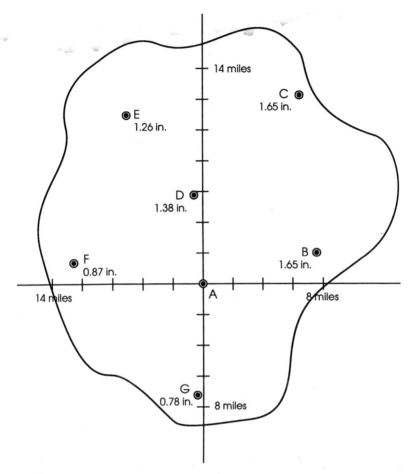

14 miles

C ●
1.65 in.

● E
1.26 in.

D ●
1.38 in.

B ●
1.65 in.

● F
0.87 in.

14 miles   8 miles
A

G ●
0.78 in.   8 miles

FIGURE 2-12.   Data for illustrating a technique for estimating rainfall at an ungauged point from rainfall at points in the vicinity. Point A on map represents ungauged station. See text for calculations.

Rainfall for each station is multiplied by the appropriate W to weigh rainfall. The sum of all weighted rainfall values divided by the sum of W values indicates rainfall at A.

For the example in Figure 2-12, computations necessary for estimating rainfall at A are presented in Table 2-4.

2.6.3   Intensity–Duration–Frequency

Design of water-control structures are based on the largest amount of rainfall expected over a specific length of time, usually a few minutes to a

TABLE 2-4   Estimation of Precipitation at an Ungauged Site, A

Station	Rain (in.)	X	Y	SS	W	Weighted Rainfall
A	—	0	0	—	—	—
B	1.65	7.6	2.0	61.76	16.2	26.73
C	1.50	6.4	12.5	197.21	5.1	7.65
D	1.38	0.5	5.9	35.06	28.5	39.33
E	1.26	5.0	11.0	146.00	6.8	8.57
F	0.87	8.5	1.5	74.50	13.4	11.66
G	0.78	0.4	7.4	54.92	18.2	14.20
Sum					88.2	108.14

$$\text{Estimated rainfall at A} = \frac{108.14}{88.2} = 1.23 \text{ in.}$$

day, for a given return period. The return period refers to the frequency of rain in time, say probable number of years between rainfalls of a given depth. We might be interested in the greatest amount of rain expected in 6 hr every 25 years at a place where rainfall has been recorded hourly for 25 years. Examination of the record shows that the greatest rainfall during 6 hr was 5 in. Thus, the return period for a 5-in. rain in 6 hr is 25 years. The return period is a useful way of expressing the frequency of rains of given sizes. Because rainfall is highly variable, the return period is more reliable if it is based on a record of 50 to 100 years or more. Return periods calculated from a few years of data may be misleading.

Gumbel's equation (Schwab et al. 1981), which is based on the theory of extreme values, is used for computing the probability or return period of rainfall events. If the total number of events is n and the rank of events arranged in descending order of magnitude is m, the frequency or return period (t), in years, is calculated from the equation

$$t = \frac{n+1}{m} \tag{2-4}$$

The probability (P) of occurrence of x in. of rainfall may be computed as

$$P\,(\%) = \frac{m}{n+1} \times 100 \tag{2-5}$$

The return period is reported more often than the frequency.

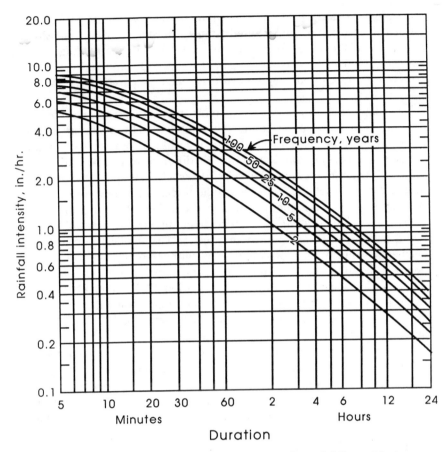

FIGURE 2-13. Intensity–duration–frequency curves for rainfall at Montgomery, Alabama. (*Source: U.S. Weather Bureau 1955.*)

Rainfall intensity–duration–frequency curves are prepared by treating a long record of rainfall events by Gumbel's method. Such a series of rainfall intensity–duration–frequency curves is presented for Montgomery, Alabama (Fig. 2-13). Curves show inches of rainfall per hour expected in 5 min to 24 hr for return periods of 2 to 100 years. To illustrate use of the curves, suppose that you are interested in the maximum rainfall intensity expected in 30 min for a 25-year return period. Follow the 30-min line vertically until it intersects the 25-year return period curve. Project a horizontal line, parallel to the abscissa, until it intersects the ordinate. Read from the ordinate at this intersection the value of 4.3 in./hr. This is the maximum amount of rain expected once or more in 30 min for a 25-

year return period. Intensity–duration–frequency curves for selected stations in the continental United States, Alaska, Hawaii, and Puerto Rico are available (U.S. Weather Bureau 1955). Maps providing 30-min, 1-hr, 6-hr, and 24-hr maximum rainfall for 2-, 10-, and 100-year return periods for the United States have also been prepared (U.S. Weather Bureau 1961).

Only daily rainfall is recorded at most gauging stations, so intensity–duration–frequency curves cannot be computed. However, the Gumbel procedure may be used to prepare a logarithmic plot of daily rainfall versus return period, and 24-hr rains for different return periods may be estimated from the graph. To construct such a graph, examine daily rainfall data for each year and record the largest 24-hr rain for each year. The largest rain is assigned a ranking order (m) of 1, and the second largest rain is taken as $m = 2$. Repeat this procedure until all years of record have been assigned a ranking order. The smallest rain in the list receives a value of $m = n$, where n = number of years in the rainfall record. Use equation (2-4) to compute the return period for the largest rain recorded each year. In solving equation (2-4), remember to substitute the ranking order of a year into the equation for the value of m. Plot return periods on the probability scale and depths of rain on the logarithmic scale (Fig. 2-14).

A note of caution is necessary. Intensity–duration–frequency curves are based on application of probability techniques to a finite rainfall record, usually less than 100 years. If the 12-hr, 25-year rain is 4.92 in. on an intensity–duration–frequency curve, it should not be surprising if three such rains fall within the next 25 years. The reason for this is simply that rainfall is a random phenomenon and our knowledge of the rainfall record is limited. Furthermore, rains larger than the 100-year storm may fall at anytime. Storms with return periods of 200, 500, 1,000, and 10,000 years or longer have occurred without being identified as such, and large storms will occur again.

Fairbridge (1967) presents a graph showing the greatest amounts of rainfall recorded over various intervals of time. A few values were extrapolated from the graph and are given in Table 2-5.

2.7 RAINFALL OVER AN AREA

Variation in amounts of rain delivered by a storm to different places in a watershed depend on watershed area, terrain, and storm characteristics. There is usually little variation in rainfall from an extensive storm over a small watershed with flat terrain. Much greater variation may result from thunderstorms over large, hilly watersheds.

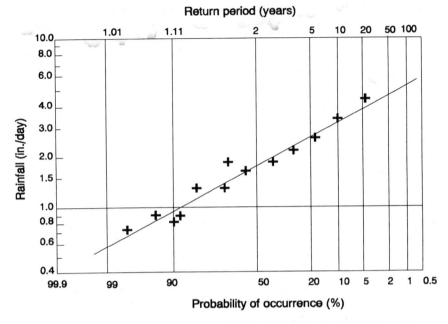

FIGURE 2-14.   Return periods for daily rainfall calculated by the Gumbel procedure.

TABLE 2-5   Record Rainfall Events

Place	Duration	Depth (in.)
Opid's Camp, California	1 min	0.59
Porta Bella, Panama	5 min	2.52
Fussen, Bavaria	8 min	5.00
Curtea DeAnges, Romania	20 min	7.99
Holt, Missouri	40 min	12.0
D'Hanis, Texas	4 hr	22.0
Smethport, Pennsylvania	6 hr	30.0
Baguio, Philippines	1 day	48.0
Silver Hill, Jamaica	8 days	140
Cherrapunji, India	1 month	380
Cherrapunji, India	1 year	950

To carefully assess the amount of rain falling on a watershed, climatologists have established rain gauge networks. Holtan et al. (1962) has recommended densities of rain gauges for research projects on agricultural watersheds (Table 2-4).

For most purposes, a single gauge near the center of a small watershed

TABLE 2-6   Minimum Number of Rain Gauges Based on Watershed Area

Watershed Size (acres)	Minimum Number of Gauges
0–30	1
30–100	3
100–200	1 per 100 acres
500–2,500	1 per 250 acres
2,500–5,000	1 per square mile
Over 5,000	1 per each 3 mile^2

provides a good estimate of average precipitation. A 1,020-acre area near Auburn, Alabama, was fitted with seven rain gauges, and average rainfall over the area was estimated by the Thiessen method (Thiessen 1911). Daily and monthly rainfall for the centermost gauge and the watershed average agreed reasonably well (Fig. 2-15). Annual totals were 44.84 in. and 45.83 in. for watershed average and centermost rain gauge, respectively.

Where several rain gauges are present over an area, their catches may be averaged to estimate the average area rainfall. However, unless rain gauges are evenly spaced and the area is relatively flat, station averaging may lead to appreciable error.

The Thiessen polygon method weighs the catch of each gauge in direct proportion to the area it represents. In this procedure, gauge locations

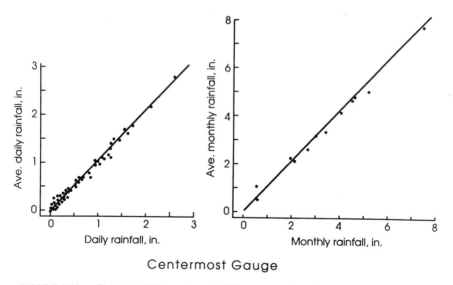

Centermost Gauge

FIGURE 2-15.   Point rainfall at center of 1,020-acre area (*X* axis) near Auburn, Alabama, compared with average precipitation over the area (*Y* axis).

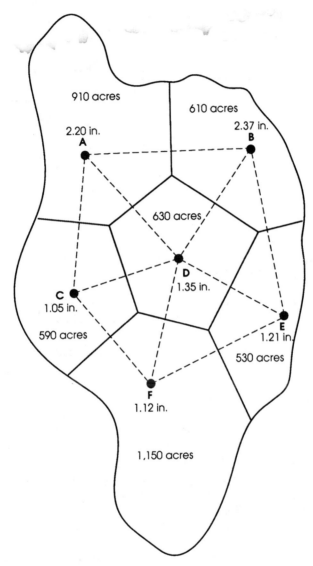

FIGURE 2-16.   Data for illustrating the Thiessen polygon method for estimating average rainfall over an area. See text for calculations.

are plotted on an area map. Dashed lines connect all adjoining stations (Fig. 2-16). Perpendicular bisectors of each dashed line are constructed as solid lines, and at margins these bisectors are extended to the boundary. Each gauge is isolated within an area bounded by perpendicular bisectors

TABLE 2-7   Calculation of Average Rainfall by the Thiessen Polygon Method Using Data from Figure 2-16

Station	Rainfall (in.)	Area (acres)	Rainfall × Area
A	2.20	910	2,002.0
B	4.21	610	1,568.1
C	1.05	590	619.5
D	1.35	630	850.5
E	1.21	530	641.3
F	1.12	1,150	1,288.0
Sum		4,420	7,969.4

$$\text{Average rainfall} = \frac{7,969.4}{4,420} = 1.80 \text{ in.}$$

or perpendicular bisectors and boundary. When the area around a gauge does not reach the boundary, bisectors form a polygon around the gauge (Fig. 2-16). The area around each gauging station is measured, and the catch of each gauge is multiplied by its surrounding area. The sum of products of areas and precipitation values is divided by total area to give average precipitation.

The Thiessen polygon method is illustrated with calculations based on data provided in Figure 2-16 (Table 2-7). The method is not as time-consuming as it may first appear. Once polygons are drawn and their areas determined for an area, information may be used for all subsequent storms.

The isohyetal method also may be used to determine the average precipitation over a land unit (Hjelmfelt and Cassidy 1975). This method requires isohyets to be drawn for each precipitation event, the areas between isohyets to be determined, depths within isohyets to be multiplied by areas within isohyets (weighted precipitation), and the weighted precipitation to be divided by the total area of the land unit of interest to give average precipitation. The isohyetal method is much more time-consuming than the Thiessen method, and it is not more accurate except in areas with rugged terrain. Therefore, we feel that the Thiessen method is more applicable to aquaculture applications.

3

Evaporation

3.1 INTRODUCTION

Water is returned to the atmosphere by evaporation from free-water surfaces, moist soils, and vegetation. Water loss from vegetated areas is termed *evapotranspiration*, because a great deal of the water is lost through transpiration by plants. Evaporation is important in the hydrologic cycle because it recharges atmospheric moisture. Although excessive evaporative losses lead to death in plants and animals, evaporation is important to organisms because it is a cooling process. Evaporation is important in aquaculture because it is a major loss of water from ponds. Evaporation losses from ponds are especially critical in dry periods when this loss is not compensated either by rain falling directly into ponds or by runoff.

Several factors affect evaporation rate, but temperature and water availability are usually the most influential. Ponds always provide a free-water surface, so evaporation is regulated strongly by temperature. Solar radiation, humidity, and wind velocity also influence evaporation.

Evaporation is difficult to measure. It is easy to determine the rate of water loss from an open container that does not leak and is filled with a mass of water (evaporation pan) or with well-watered vegetation (lysimeter). Water budgets for land or water areas are more complex than those of evaporation pans and lysimeters. Thus, estimates of evaporation made with an evaporation pan setting beside a pond or from a lysimeter constructed in the middle of a meadow usually differ considerably from evaporation from the pond or meadow.

3.2 THE PROCESS

Air at a particular temperature and pressure holds a given amount of water vapor. Water molecules are in constant motion and exert a pressure when they strike surfaces. The pressure exerted by water vapor is termed *vapor pressure*, and the vapor pressure in air saturated with water vapor (100% relative humidity) is known as the *saturation vapor pressure*. Increasing temperature causes the saturation vapor pressure to increase (Table 2-1).

If air that is not saturated with water vapor is brought in contact with a water surface, molecules of water bounce from the surface film of the water into the air until the vapor pressure is equal to the pressure of water molecules escaping the surface film. The relative humidity of the air is now at 100%, and evaporation ceases. Water molecules continue to move back and forth across the surface film, but there is no net flow in either direction. The driving force for evaporation is the vapor pressure deficit (VPD):

$$VPD = e_s - e_a \qquad (3\text{-}1)$$

where e_s = saturation vapor pressure and e_a = actual vapor pressure. The greater the vapor pressure deficit, the greater is the potential for evaporation. When $e_s - e_a = 0$, evaporation stops.

To change water from the liquid phase to the gaseous phase at 100°C requires considerable heat energy to provide the latent heat of evaporation. Some may have difficulty reconciling this fact with the observation that water evaporates readily below 100°C. Molecules in a mass of water are in constant motion, and the average degree of molecular motion increases with temperature. All molecules in water of a given temperature are not moving at the same speed. The faster-moving molecules contain more thermal energy than the slower-moving ones, and some of the faster-moving molecules escape the surface to evaporate. Thus, evaporation occurs at temperatures below 100°C. Increasing the temperature favors evaporation, because more molecules gain sufficient energy to escape the water surface.

The energy requirement for evaporating 1 g of water from a pond surface at 30°C is 610 cal, which is the energy needed to raise the temperature of 1 g of water from 30 to 100°C and supply the latent heat of vaporization. The water surface remains essentially at 30°C, but the heat content of the escaping molecules is 610 cal greater than the heat content of 1 g of water molecules remaining in the pond. Evaporation is a cooling process because of the large amount of heat lost as latent heat, which is contained in the escaping molecules.

Wind velocity also influences evaporation. As evaporation continues, the relative humidity of the air above the water surface increases, and the tendency for water molecules to escape the surface declines. Wind blowing over the surface often replaces the humid air with drier air, which favors evaporation. The influence of wind is especially pronounced in an arid region where air generally contains little moisture.

Dissolved salts decrease the vapor pressure of water, and evaporation rates from saltwater are about 5% less than those of freshwater. However, dissolved salt concentrations do not vary enough among freshwaters to influence evaporation rates appreciably. Turbid waters heat faster than clear waters, so an increase in turbidity enhances evaporation. Changes in atmospheric pressure also slightly affect evaporation.

Temperature, however, has the greatest influence on evaporation, and the temperature of a locality is closely related to the amount of incoming solar radiation. The temperature difference between air and water also affects evaporation. A cold water surface cannot generate a high vapor pressure, and cold air does not hold much water vapor. Therefore, cold air over cold water, cold air over warm water, and warm air over cold water are not as favorable for evaporation as warm air over warm water. However, relative humidity must be considered. For example, cold, dry air may take on more moisture than warm, moist air, because the greater moisture-holding capacity of the warm air has already been filled.

A shortage of water never limits evaporation from a free-water surface, but evaporation from soils and transpiration by plants often may be limited by a lack of water. When the surface of a soil dries out, evaporation rates decline markedly. Likewise, transpiration rates in plants fall as the available moisture content of the soil diminishes. Aside from this, evaporation from soils and transpiration by plants are governed by the same factors influencing evaporation from a free-water surface.

3.3 ESTIMATES OF EVAPORATION

Methods for estimating evaporation from a free-water surface include mass-transfer techniques, energy budgets, water budgets, and empirical formulae. None of the methods is perfect, but all can provide reliable data. Only empirical formulae and evaporation pans are considered useful in aquaculture applications, but a brief discussion of other techniques is provided.

3.3.1 Mass Transfer

The mass-transfer approach to assessing evaporation was first mentioned almost two centuries ago by Dalton. He recognized the relationship between evaporation and vapor pressure to be

$$E = b(e_s - e_a) \qquad (3\text{-}2)$$

where E = evaporation and b = empirical coefficient(s).

The value of the empirical coefficient(s) is related to wind velocity, surface roughness, temperature, and air density and pressure. In short, the empirical coefficient(s) must adjust for movement of water vapor by convection. Convection is the transfer of heat by mass movement of air or other heat-containing medium. Water vapor moves with the air, and much of the heat transferred by convection is contained in water vapor as latent heat. Convection also may be called eddy diffusion.

A thin boundary layer exists above a surface where evaporation is occurring. The layer usually extends a few tenths of an inch above the surface, and within this zone airflow is laminar (layered flow). Upward movement of heat within the boundary layer is by molecular movement (diffusion), and the process is called *conduction*. Water vapor, which contains heat, also moves upward in the boundary layer by diffusion. The loss of water vapor and heat from a surface by diffusion of molecules is a slow process. However, above the boundary layer is a thicker layer (up to several feet thick) where air exhibits turbulent flow. In the turbulent layer, movement of heat and water vapor occurs by mass movement of the air or eddy diffusion. The degree of eddy diffusion increases as a function of wind speed, temperature, and vapor pressure gradients between the surface and air. Roughness of the surface also causes eddy diffusion to increase. The rate of heat and water vapor loss from a surface is a function of eddy diffusion. The wind velocity profile above the surface is especially important, for water vapor is assumed to move at the same speed as air.

Several evaporation equations are given in this chapter. Some of these equations are metric or contain a mixture of metric and U.S. Customary units. No attempt is made to achieve consistency in units. It is easier to transform quantities from metric to U.S. Customary units, and vice versa.

A widely used mass-transfer equation is that of Thornthwaite and Holzman (1939):

$$E = \frac{833\kappa^2(e_1 - e_2)(v_2 - v_1)}{(T + 459.4)\log_e(Z_2/Z_1)^2} \qquad (3\text{-}3)$$

where E = evaporation (in./hr)
κ = von Kármán's constant
e = vapor pressure (in. Hg)
v = wind velocity (mph)

T = temperature (°F)

Z = heights of wind measurements (ft)

Subscripts 1 and 2 represent the vertical position of two measurements. Data for solving the Thornthwaite–Holzman equation are difficult to obtain and are seldom available for practical application.

Five simpler equations for estimating daily evaporation by the mass-transfer concept were taken from Hjelmfelt and Cassidy (1975) and are presented here:

Meyer equation

$$E_{day} = 0.5(e_s - e_a)(1 + 0.1v_{30}) \tag{3-4}$$

Horton equation

$$E_{day} = 0.4[(2 - e_a^{-0.2v})(e_s - e_a)] \tag{3-5}$$

Rohwer equation

$$E_{day} = (1.13 - 0.0143P_a)(0.44 + 0.118v)(e_s - e_a) \tag{3-6}$$

Harbeck–Anderson equation

$$E_{day} = 0.06v_{30}(e_s - e_a) \tag{3-7}$$

Kohler equation

$$E_{day} = 0.072v_{30}(e_s - e_a)[1 - 0.03(T_a - T_w)] \tag{3-8}$$

where E_{day} = daily evaporation (in./day)

v = wind velocity at unspecified height (mph)

v_{30} = wind velocity at 30 ft (mph)

P_a = average atmospheric pressure (in. Hg)

e_a = actual vapor pressure (in. Hg)

e_s = saturation vapor pressure at average air temperature (in. Hg)

T_a = average air temperature ('C)

T_w = average temperature of water surface ('C)

These equations do not give the same evaporation rates for identical conditions. To illustrate, suppose that average air and surface-water temperatures are 25'C and 23'C, respectively. Atmospheric pressure is 29.00 in. Hg, relative humidity averages 60%, and wind velocity at 30 ft averages 10 mph. The saturation vapor pressure at 25'C (77'F) is 0.928 in. Hg. We may compute the actual vapor pressure from the relative humidity:

$$RH = (e_a \div e_s)100$$

$$0.60 = (e_a \div 0.928)$$

$$e_a = 0.557 \text{ in. Hg}$$

Solving all five equations gives the following values: Rohwer, 0.43 in./day; Meyer, 0.37 in./day; Kohler et al., 0.25 in./day; Harbeck–Anderson, 0.22 in./day; Horton, 0.12 in./day. The highest estimate is almost four times greater than the lowest. Each equation was developed to fit a set of evaporation data obtained at a specific location. Therefore, it is not surprising that equations fail to provide comparable estimates. When using empirical equations to estimate evaporation, one should use several equations and consider the range of values. Depending on the use of the evaporation estimate, either the highest, lowest, or average estimate might be of most interest.

3.3.2 Energy Budget

In this approach to estimating evaporation from a free-water surface, a thermal budget for the water body must be developed. The equation for energy used in evaporation is

$$H_E = H_S + H_W - H_L - H_C - H_R \qquad (3\text{-}9)$$

where H_E = energy used for evaporation
  H_S = gain of energy in body of water from shortwave solar radiation
  H_W = net gain or loss of energy through inflow and outflow

FIGURE 3-1.   Class A evaporation pan.

H_L = net change of energy in water body from long-wave radiation

H_C = conduction and convection of energy from water surface to atmosphere

H_R = net change of energy storage within the body of water

All terms in the equation are expressed in calories per square inch, so the amount of evaporation in inches of depth may be calculated as

$$E = \frac{H_E}{L}$$
(3-10)

where L = latent heat of vaporization. The approach is too complicated to be of practical use in aquaculture, but it is instructive because it illustrates the energy relationships involved.

3.3.3 Water Budget

This technique relies on measuring the change in storage, all inflows, and all outflows, except evaporation, for a body of water. The measuring done, evaporation may be calculated readily:

$$E = I - O \pm \Delta S$$
(3-11)

where I = inflows
O = outflows
ΔS = change in storage

Most bodies of water have several outflow and inflow terms. The seepage term, which may have both inflow and outflow components, is virtually impossible to assess accurately independent of evaporation. Hence, the water budget technique rarely is useful.

3.4 MEASURING EVAPORATION WITH INSTRUMENTS

Instruments are widely used in estimating evaporation. Three types were recommended during the International Geophysical Year (Hounam 1973); they were the 3,000-cm^2 sunken tank, the 20-m^2 sunken tank, and the Class A evaporation pan, which is widely used in the United States. The Class A pan is made of stainless steel and is 4 ft in diameter by 10 in. deep (Fig. 3-1). The pan is mounted on an open wooden platform with

FIGURE 3-2.   Stilling well and hook gauge for evaporation pan.

the pan base 2–4 in. above ground. The pan is filled within 2 in. of the rim with clear water. Water depth in an evaporation pan usually is measured with a stilling well and hook gauge. The stilling well (Fig. 3-2) provides a smooth water surface and supports the hook gauge. The hook gauge (Fig. 3-2) consists of a pointed hook that can be moved up and down with a micrometer. With the point below the water surface, the micrometer is turned to move the hook upward until its point makes a pimple in the surface film. The hook is not permitted to break through the surface film. The micrometer is read to the nearest thousandths of an inch, and the value is recorded. After a period of time, usually one day, the procedure is repeated. The difference in the two micrometer readings is the water loss from evaporation. A rain gauge must be positioned beside the evaporation pan to correct for rain falling into the pan.

The evaporation rate from a pan is not the same as the evaporation rate from an adjacent body of water. A pan has a smaller volume than a lake or pond and a different exposure to the elements. There is usually a considerable discrepancy between water temperature patterns in a pan and in a lake or pond. The water surface in an evaporation pan is smaller and smoother than the surface of a lake or pond. Because of these differences, a pan coefficient is multiplied by pan evaporation data to afford an estimate of lake or pond evaporation. Pan coefficients have been developed by estimating evaporation from lakes by mass transfer, energy budget, or water budget techniques and relating these values to evaporation from an adjacent Class A pan as follows:

$$C_p = \frac{E_L}{E_p} \qquad (3\text{-}12)$$

where E_L = evaporation from lake
E_p = evaporation from pan
C_p = pan coefficient

Pan coefficients for Class A pans range from 0.60 to 0.81 (Hounam 1973); a factor of 0.70 is recommended for general use.

Pan coefficients are more reliable when applied to weekly or monthly totals for pan evaporation. Kohler et al. (1955) presented the following equation for estimating daily lake evaporation from Class A pan evaporation data:

$$E_{day} = C_p(0.37 + 0.0041L)(e_{s5} - e_{a5})^{0.88} \qquad (3\text{-}13)$$

where E_{day} = daily evaporation (in.)
e_{s5} = saturation vapor pressure at 5 ft above ground level (in. Hg)
e_{a5} = actual vapor pressure at 5 ft above ground level (in. Hg)
L = wind run 6 in. above rim of evaporation pan (mile/day)
C_p = pan coefficient

However, data for using this equation seldom are available, and lake evaporation usually is estimated as

$$E_L = 0.7E_p \qquad (3\text{-}14)$$

Because there was no assurance that 0.70 was a reliable coefficient for estimating pond evaporation, Boyd (1985a) made evaporation measurements at Auburn, Alabama, for a pond, lined to prevent seepage and fitted with a barrier to divert runoff, and an adjacent Class A evaporation pan. A chloride budget for the lined pond and a laboratory permeability test of the liner material revealed that the pond liner did not seep. Pan coefficients for estimating pond evaporation from a Class A pan averaged 0.81 over a year. Monthly coefficients were as shown in Table 3-1. Data

TABLE 3-1   Monthly Pan Coefficients for Estimating Pond Evaporation from Pan Evaporation at Auburn, Alabama

Month	Pan Coefficient	Month	Pan Coefficient
January	0.87	July	0.82
February	0.77	August	0.84
March	0.72	September	0.90
April	0.76	October	0.83
May	0.81	November	0.77
June	0.77	December	0.85

collected by Boyd (1985a) also revealed that pond evaporation may be estimated by the equation

$$Y = -0.108 + 0.848X \qquad (3\text{-}15)$$

where Y = monthly pond evaporation (in.) and X = monthly Class A pan evaporation (in.).

The coefficient of determination (r^2) for the relationship between monthly values for pan and pond evaporation was 0.995. Relationships between daily and weekly evaporation from the pan and from the pond had r^2 values of 0.668 and 0.902, respectively. Hence, it is best to make monthly estimates of pond evaporation from pan evaporation data. Lakes are much larger than ponds, and they store more heat and change in temperature much more slowly than do ponds. Therefore, the coefficients and equation should be more reliable for estimating pond evaporation than those developed from lake evaporation data. Climatic differences among regions will no doubt influence the pan coefficient, but climate is similar over the fish-farming area of the southeastern United States, and coefficients developed by Boyd (1985a) should be applicable in this region.

3.5 EVAPORATION RATES

In the United States, records of pan evaporation are summarized for each state by month and annual summary and published in Climatological Data by the National Climatic Data Center, Asheville, North Carolina. Normal, annual lake evaporation data are summarized for the United States in Figure 3-3. Pan evaporation data were multiplied by pan coefficients in order to estimate lake evaporation for the map (Kohler et al. 1959). The authors used a pan coefficient of 0.70 when air and water

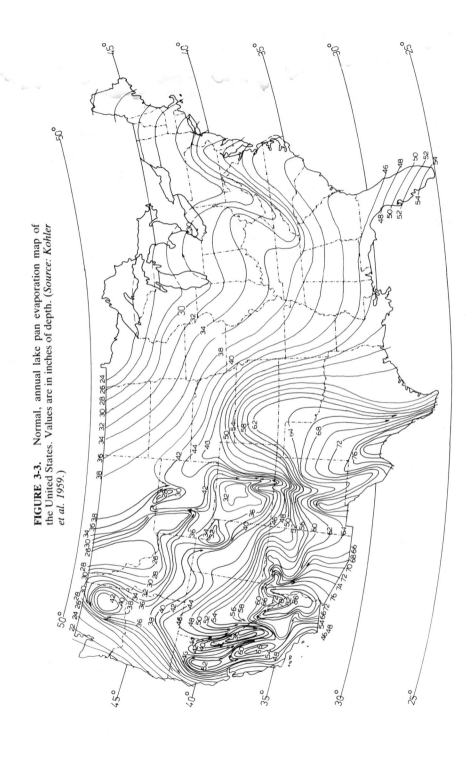

FIGURE 3-3. Normal, annual lake pan evaporation map of the United States. Values are in inches of depth. (*Source: Kohler et al. 1959.*)

temperatures were equal. A monograph procedure was developed for adjusting coefficients when air and water temperatures were unequal.

Rates of evaporation differ greatly from hour to hour and from day to day at a location. During at 24-hr period, air temperature is normally lowest at dawn and highest during the midafternoon (Fig. 3-4). Relative humidity usually is highest at dawn; it decreases as air temperature rises and often reaches its lowest point during afternoon (Fig. 3-4). Relative humidity rapidly increases as temperature decreases during the evening hours. The vapor pressure gradient has a trend similar to relative humidity. The evaporating power of air depends largely on the vapor pressure gradient, so evaporation rates usually are greater during the day. The rate of evaporation cannot be determined from vapor pressure gradient alone, because wind velocity, surface roughness, and air density and pressure also influence evaporation. There is considerable day-to-day variation in evaporation, because variables influencing evaporation rates differ on a day-to-day basis. Temperature, however, has a dominant influence on evaporation. Monthly pan evaporation rates follow the annual march of temperatures and solar radiation nicely (Fig. 3-5, Table 3-2).

There is a notable exception to the statement that evaporation is greatest during warm months. In large, deep lakes, there may be little or no evaporation during spring. Heat absorbed by the lake surface is stored in deep water. Surface water warms up slowly, so there is little movement of water vapor until summer. Heat from deep-water storage is given up

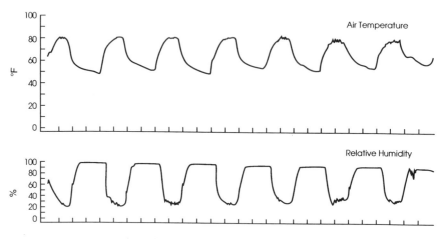

FIGURE 3-4.   A portion of a recorder chart from a hygrothermograph showing daily cycles in air temperature and relative humidity at Auburn, Alabama.

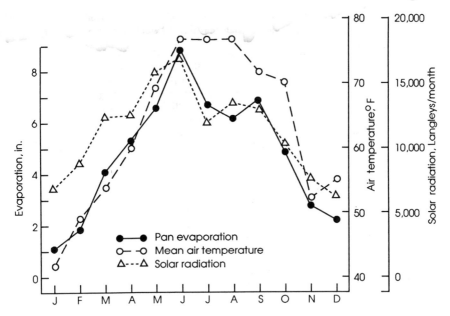

FIGURE 3-5.   Monthly Class A pan evaporation, solar radiation, and average monthly air temperatures at Auburn, Alabama, in 1984.

during fall and winter, and this heat favors high evaporation rates even though air temperatures are low (Miller 1977). Fish ponds are not large and deep enough to store appreciable heat, and evaporation rates follow the annual march of temperatures. In tropical regions, evaporation rates are appreciable throughout the year whether bodies of water are small and shallow or large and deep.

Monthly and annual pond evaporation totals are not as variable as annual rainfall totals. Annual pond evaporation over a 19-year span at Auburn, Alabama, was 44.24 in., with a range of 41.08 to 50.10 in. and a coefficient of variation of 5.7% (Boyd 1985a). Over the same years, the coefficient of variation for annual rainfall was 14.9%.

Applications of feeds and fertilizers to ponds to favor greater fish production lead to plankton blooms and increased turbidity. Idso and Foster (1974) reported that surface-water temperature was higher during afternoons when a plankton bloom was present in a pond than when the water was clear; this phenomenon favors greater evaporation. Nevertheless, studies of evaporation from waters with different levels of turbidity failed to reveal a significant influence of turbidity on evaporation (Stone 1988). Aeration of ponds with surface aerators splash water and increase

TABLE 3-2   Average, Minimum, and Maximum Pond Evaporation at Auburn, Alabama (based on data for 1965–1982)

Month	Average (in.)	Minimum (in.)	Maximum (in.)
January	0.95	0.10	2.24
February	1.46	0.61	2.48
March	2.95	1.97	3.99
April	4.25	3.69	5.35
May	5.12	3.77	6.89
June	5.98	4.79	7.09
July	5.87	5.09	7.78
August	5.59	4.85	6.48
September	4.88	4.41	5.55
October	3.74	3.30	5.01
November	2.20	1.16	3.45
December	1.26	0.69	2.37
Annual	44.42[a]	41.08[b]	50.10[b]

Source: Data from Southeast Agricultural Weather Service Center, Auburn University, Alabama.

[a] Annual pond evaporation is the sum of monthly estimates given in the average column.
[b] Minimum and maximum annual pond evaporation estimates are for years with lowest and highest totals. The minimum and maximum columns will not be added to the minimum and maximum annual values.

the surface area exposed to the air for evaporation. Stone (1988) found that 1 hp of aeration with an electric paddle wheel aerator increased daytime evaporation by 10%. Nighttime aeration did not increase the evaporation rate.

3.6   PRECIPITATION–EVAPORATION RELATIONSHIP

The relationship between precipitation and evaporation is especially important in aquaculture. It is instructive to plot monthly pond evaporation and rainfall data for a location (Fig. 3-6). The resulting graph clearly shows periods with excess precipitation or evaporation. At Auburn, Alabama, average annual precipitation is 11.02 in. greater than normal annual pond evaporation. However, evaporation exceeds precipitation from May to October (Fig. 3-6). Thus, even in a humid region, rain falling directly into a pond barely replaces evaporation losses on an annual basis, and it fails to compensate for evaporation during most warm months. In a warm, arid region, pond evaporation greatly exceeds

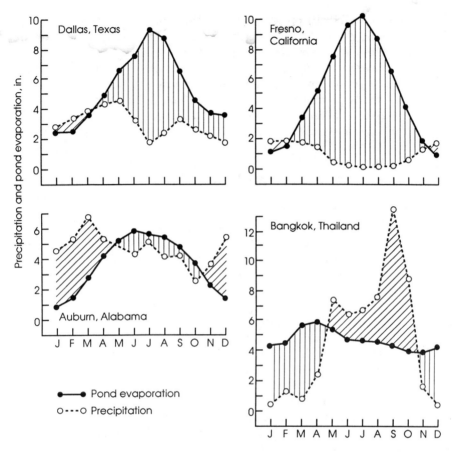

FIGURE 3-6.   Normal monthly precipitation and pond evaporation at four sites. (*Based on data from Wallis 1977; Thailand Ministry of Communications 1981; Farnsworth and Thompson 1982.*)

precipitation falling into a pond, as illustrated for Texas and California (Fig. 3-6). An interesting precipitation–evaporation pattern occurs in tropical areas such as Thailand where there is a pronounced dry season (Fig. 3-6). Of course, water also seeps out of ponds, and in areas with permeable soils, seepage losses may be as great or greater than losses to evaporation.

3.7 WATER LOSS BY PLANTS

Mechanisms involved in absorption, movement, and loss of water in plants are too complex for detailed analysis here, but a summary is in order. Kramer (1969) presents an excellent general treatment of plant–water relationships; any general plant physiology text also will provide details. Most water absorbed by a plant enters through its roots. Much water uptake results from diffusion into root hairs, but other epidermal cells of roots absorb water because of a diffusion pressure deficit gradient between fluids in root cells and soil solution. Solute concentrations in epidermal cells exceed solute concentrations in the soil solution, and water diffuses into cells to equalize solute concentrations. Water also may enter plants through mass flow; the suction force created by transpiration simply pulls water from the soil solution into the root. It seems that most water is absorbed without expenditure of energy by root cells, a passive process.

Total water loss by evaporation from vegetation, soil, and free-water surfaces at a place is called evapotranspiration. Transpiration is the process whereby water is absorbed by plant roots, translocated to leaves, and then lost through evaporation. Vessel elements and tracheids of the xylem are the cells associated with water movement in plants. These are perforated or open-ended cells that are dead when mature. Hence, other than friction, there is little to impede water movement through these cells. Cells are stacked end to end and form a tubelike structure called a xylem duct. Open cells form a network of ducts that extends from roots through stem or trunk to leaves; the network furnishes all living cells within the plant a supply of water.

Plants use large amounts of water, many times more than is contained in their tissues. For example, a single corn plant may consume 50 gal or more of water during a growing season. The corn plant contains up to 1 gal of water at maturity; the rest is lost by transpiration. Some of the water lost from plants merely evaporates directly from leaves, but a waxlike layer of cutin covers leaves and greatly restricts evaporation. Most water is lost through stomates, small pores located in the leaf epidermal surface (Fig. 3-7). Mesophyll cells lining the stomatal cavity present less resistance to evaporation than do epidermal cells on the leaf surface. Some plants have as many as 250,000 stomates per square inch, but when fully open the pores occupy less than 2% of the leaf surface. Stomates are most numerous on undersides of leaves.

The stomatal pore is bordered by two guard cells. Guard cells have thickened inner walls (Fig. 3-8), and they open and close the stomate in response to their turgor pressure. When turgor pressure rises, thickened inner walls of guard cells are pulled apart to open the stomate. A loss of turgor pressure causes the stomate to close. The opening process is illustrated in Figure 3-8. The closing process is simply the opposite of the opening process.

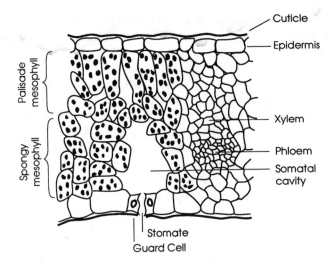

FIGURE 3-7.   Section of a leaf showing a stomate and stomatal cavity.

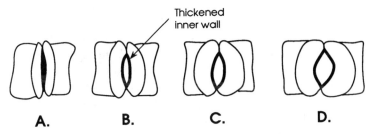

FIGURE 3-8.   Opening of a stomate. (A) Dark: the stomate is closed. (B) Sunrise: starch converted to sugar, and water diffuses into guard cells causing them to swell. (C, D) Morning hours: more starch converted to sugar, and water continues to enter and swell guard cells.

Turgor pressure in guard cells results from osmotic pressure, and the degree of turgor pressure within cells is a function of the diffusion pressure deficit. Through the action of the enzyme starch phosphorylase, starch reacts with phosphate to form sugar (glucose-1-phosphate). The sugar concentration in guard cells rises when pH increases above 7. The pH of leaf cells increases during daylight, because carbon dioxide, which is acidic, is rapidly used in photosynthesis and does not accumulate. Thus, the sugar concentration of guard cells increases during the day, and the diffusion pressure deficit increases as sugar accumulates. Water enters guard cells from mesophyll cells, and the resulting rise in turgor pressure pulls apart the inner walls of the guard cell to open the stomate. Photo-synthesis halts at night, and carbon dioxide accumulates in guard cells. The pH of cellular fluids declines, and starch concentration increases at the expense of sugar concentration. The diffusion pressure gradient decreases, and water moves from guard cells to mesophyll cells. Turgor pressure falls, and the thickened walls of guard cells relax to close the stomate. Hence, stomates generally open in daylight and close at night. When water is in short supply, the transpiration stream cannot deliver enough water to leaf cells to maintain high turgor pressure in guard cells even in daylight. Closing of stomates in response to water shortage is a water conservation measure in plants.

Many theories have been advanced to explain how water moves within the xylem from roots to leaves of plants. The cohesion–tension theory is the most generally accepted. Water evaporates from the leaf mesophyll (Fig. 3-7), and water vapor exits to the atmosphere through the open stomate. This causes a decrease in water content of mesophyll cells and an increase in diffusion pressure deficit of cells near the leaf surface. Water lost by evaporation is replaced by water that comes from cells deeper in the leaf interior. This sets up a situation, whereby, in an attempt to equalize diffusion pressure deficits, leaf cells finally draw water from xylem tissue. This puts water in xylem tissue under negative pressure (tension), and the tension is transmitted through the unbroken column of water in the xylem, which extends from the leaves to the root system. Water is pulled up through the xylem as a result of tension and cohesive properties of water, which prevent the column of water in the xylem from breaking. This event, called *transpiration*, will continue as long as water evaporates from leaves.

A physical model of transpiration is shown in Figure 3-9. As water evaporates from the moist sponge, water is drawn up the capillary tube. This model differs from transpiration mainly in that water passes through living tissue in plant roots and leaves.

3.8 DETERMINING EVAPOTRANSPIRATION

Transpiration is different from evaporation because escape of water vapor from plants is controlled to a considerable degree by leaf resistances, which are not involved in evaporation from a free-water surface. The soil water supply for plant use may be limited. Water for evaporation is never limiting for a free-water surface. All factors influencing evaporation from a free-water surface plus leaf characteristics and soil moisture supply influence the rate of evapotranspiration. The amount of leaf area exposed to the atmosphere is especially important. The leaf area index (LAI) is

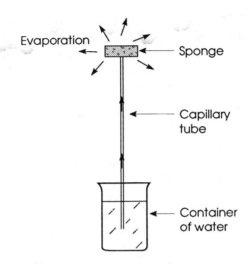

FIGURE 3-9.   A physical model of transpiration.

the area of leaf surface divided by the area of land or water surface over which the leaves stand. Other factors being equal, evapotranspiration increases with increasing LAI. In natural plant communities, LAI values range from less than 1 to more than 10. If living plant leaves behaved like a free-water surface, evapotranspiration rates of some plant communities would be many times greater than evaporation rates for a free-water surface. The leaf offers considerable resistance to evaporation, but evapotranspiration from a well-watered site still may exceed evaporation from a free-water surface under the same climatic conditions.

Most research on evapotranspiration has involved crop species, and it has been stimulated by a need to improve irrigation practices. Many procedures have been used for estimating evapotranspiration. The most popular have been moisture depletion studies on small plots, lysimeter techniques, mass-transfer procedures, energy budgets, water budgets, and empirical equations (Gray 1970; Sharma 1984). Three techniques—lysimeters, empirical equations, and water budgets—for estimating evapotranspiration are discussed next.

3.8.1 Lysimeters

According to McIlroy and Angus (1964), a lysimeter consists of a block of soil, together with vegetation, if any, enclosed in a suitable container and exposed in natural surroundings to permit determination of any term of the hydrologic equation when the others are known. Aboukhaled et al.

(1982) state that, in its simplest form, lysimetry involves the volumetric measurement of all incoming and outgoing water of a container that encloses an isolated soil mass with bare or vegetated surface. The incoming and outgoing flux of water can be represented by the following version of the hydrologic equation:

$$P + I \pm RO = ET + D \pm \Delta W \qquad (3\text{-}16)$$

where P = precipitation
         I = water added by irrigation
        ET = evapotranspiration
         D = deep percolation of water that drains from soil
        RO = overland flow into or out of the lysimeter
   ΔW = change in water content of soil in lysimeter

Most lysimeters are constructed to eliminate the overland flow term, so evapotranspiration may be estimated as follows:

$$ET = P + I - D \pm \Delta W \qquad (3\text{-}17)$$

A simple drainage lysimeter that may be constructed from an oil drum (Gilbert and van Bavel 1954) is illustrated in Figure 3-10. The amount of

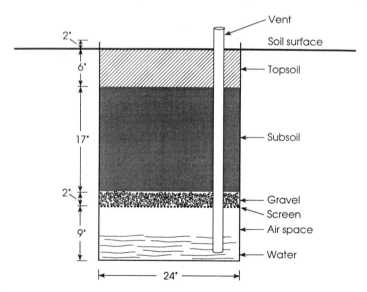

FIGURE 3-10.   Lysimeter constructed from drum. Dimensions are in inches. (*Source: Gilbert and van Bavel 1954.*)

water draining to the bottom of the lysimeter is measured with a dipstick. An alternative is to construct a sloping bottom of concrete in the barrel and pump water out for measurement. The moisture content of soil in the lysimeter is measured with tensiometers or by removing small soil cores for gravimetric measurement.

A discussion of all types of lysimeters is provided by Aboukhaled et al. (1982). The most sophisticated lysimeters provide a continuous record of lysimeter weight. The weight change can be related to water loss.

3.8.2 Empirical Equations

Several empirical equations are available for estimating evapotranspiration. The Penman method, which relies on an aerodynamic and energy balance approach, is probably the most theoretically sound and widely used. Because of their complexity, details of the Penman method are not given here. The reader may consult Penman (1948, 1956) or Gray (1970) for appropriate equations and nomographs. In essence, the Penman method uses air temperature, humidity, and wind speed data to estimate evaporation from a free-water surface. Potential evapotranspiration, which is the amount of the evapotranspiration expected with unlimited water supply, is estimated for a plant community by multiplying by a factor related to the type of crop, season, and length of day.

A fairly simple procedure presented by Thornthwaite (1948) permits estimation of potential evapotranspiration from temperature and day length data. Four steps are necessary for this computation. In step 1, the heat index is computed for the months over which an estimate of potential evapotranspiration is needed:

$$I = \sum_{m=1}^{12} \left(\frac{T_m}{5}\right)^{1.51} \tag{3-18}$$

where I = heat index for a 12-month period and T_m = mean monthly temperature (°C) for month m.

In step 2, the heat index (I) is used to compute an exponent a:

$$a = 67.5 \times 10^{-8}I^3 - 77.1 \times 10^{-6}I^2 + 0.0179I + 0.492 \tag{3-19}$$

The exponent a is used in step 3 to estimate potential evapotranspiration for each month as follows:

$$PET = 0.64\left(\frac{10T_m}{I}\right)^a \tag{3-20}$$

where PET = monthly potential evapotranspiration in inches.

Finally, in step 4, values for monthly PET must be adjusted for daytime hours with information from Table 3-3. The Thornthwaite equation assumes 12 hr of possible sunshine and a 30-day month. The evapotranspiration estimate for each month is adjusted by multiplying it by the possible number of daylight hours in a given month divided by 12. For example, at Auburn, Alabama (32°N), the average daytime hours for May is 13.7 hr by interpolation, and there are 31 days. Total daytime hours for May is 424.7 hr; for a 30-day month with 12-hr day length, the total daytime hours is 360 hr. The factor to multiply by the Thornthwaite estimate of PET to adjust for day length is 424.7 ÷ 360 = 1.18. Obviously, day length corrections will vary with latitude.

Example 3-1

To illustrate use of the Thornthwaite equation, we calculate potential evaporation from a vegetated area at 30°N latitude for a 12-month period. The average monthly temperatures are as follows: January, 8°C; February, 10°; March, 15°C; April, 18°C; May, 21°C; June, 23°C; July, 24°C; August, 25°C; September, 20°C; October, 16°C; November, 12°C; December, 10°C.

Solution

Step 1.

$$I = \left(\frac{8}{5}\right)^{1.51} + \left(\frac{10}{5}\right)^{1.51} + \left(\frac{15}{5}\right)^{1.51} + \left(\frac{18}{5}\right)^{1.51} + \left(\frac{21}{5}\right)^{1.51} + \left(\frac{23}{5}\right)^{1.51}$$
$$+ \left(\frac{24}{5}\right)^{1.51} + \left(\frac{25}{5}\right)^{1.51} + \left(\frac{20}{5}\right)^{1.51} + \left(\frac{16}{5}\right)^{1.51} + \left(\frac{12}{5}\right)^{1.51} + \left(\frac{10}{5}\right)^{1.51}$$

$I = 78.34$

Step 2.

$a = 67.5 \times 10^{-8}(78.34)^3 - 77.1 \times 10^{-6}(78.34)^2 + 0.0179(78.34) + 0.492$
$a = 1.745$

Step 3.

$$\text{PET}_{\text{January}} = 0.64\left(\frac{10(8)}{78.34}\right)^{1.745}$$

$$= 0.66 \text{ in. (value not adjusted for day length)}$$

TABLE 3-3 Average Monthly Day Lengths in Hours at Different Latitudes in the Northern Hemisphere

						Degrees North Latitude						
Month	0	5	10	15	20	25	30	35	40	45	50	55	60
January	12.00	11.76	11.50	11.22	10.95	10.64	10.45	10.10	9.75	9.25	8.65	7.95	6.90
February	12.00	11.84	11.69	11.52	11.36	11.17	11.15	10.95	10.65	10.35	9.05	9.65	9.10
March	12.00	11.96	11.94	11.90	11.88	11.85	11.95	11.95	11.95	11.90	11.85	11.80	11.75
April	12.00	12.11	12.22	12.34	12.45	12.59	12.90	13.00	13.20	13.45	13.70	14.05	14.50
May	12.00	12.22	12.45	12.68	12.93	13.19	13.60	13.95	14.35	14.75	15.30	16.05	17.05
June	12.00	12.28	12.57	12.86	13.17	13.49	14.00	14.40	14.90	15.50	16.15	17.10	18.50
July	12.00	12.26	12.50	12.77	13.06	13.34	13.85	14.20	14.65	15.20	15.80	16.65	17.85
August	12.00	12.15	12.31	12.47	12.65	12.84	13.20	13.45	13.75	14.05	14.45	14.95	15.55
September	12.00	12.02	12.04	12.08	12.10	12.14	12.35	12.40	12.50	12.55	12.70	12.80	12.95
October	12.00	11.89	11.78	11.66	11.54	11.42	11.45	11.35	11.20	11.05	10.80	10.55	10.50
November	12.00	11.78	11.55	11.32	11.07	10.80	10.70	10.40	10.05	9.65	9.15	8.55	7.70
December	12.00	11.72	11.43	11.13	10.81	10.48	10.30	9.90	9.50	8.95	8.25	7.40	6.25

Step 4.

January has 31 days, so calculate the day length conversion factor using day length data from Table 3-2:

$$\frac{10.45 \, \text{hr/day} \times 31 \, \text{days}}{12 \, \text{hr/day} \times 30 \, \text{days}} = 0.90$$

Step 5.

$\text{PET}_{\text{January}} = 0.66 \times 0.90 = 0.59 \, \text{in.}$

Step 6.

Evapotranspiration values in inches for all months follow:

January	0.59
February	0.85
March	2.04
April	3.03
May	4.19
June	5.05
July	5.38
August	5.51
September	3.49
October	2.19
November	1.24
December	0.87

The annual PET computed by the Thornthwaite equation is 34.43 in.

Turc (Gray 1970) presented an equation for estimating potential evapotranspiration from annual mean temperature and precipitation:

$$\text{PET} = \frac{P}{[0.90 + (P/I_T)^2]^{0.5} (25.4)} \tag{3-21}$$

where PET = annual evapotranspiration (in.)
 P = annual precipitation (mm)
 T = mean annual air temperature (°C)
 $I_T = 300 + 25T + 0.05T^3$

Example 3-2

The Turc equation will be used to calculate evapotranspiration for a region where rainfall is 1,000 mm (39.37 in.) annually and air temperature averages 18°C.

Solution   First, compute the value of I_T:

$$I_T = 300 + 25(18) + 0.05(18)^3 = 1{,}041.6$$

Then use equation (3-21) to estimate annual PET:

$$\text{PET} = \frac{1{,}000}{[0.90 + (1{,}000/1{,}041.6)^2]^{0.5}\,(25.4)} = \frac{1{,}000}{34.29} = 29.16\,\text{in.}$$

The equations by Thornthwaite and Turc are useful where few data are available. In locations where daily meteorological data are available, the PET model of Baier and Robertson (1965), which employs daily values of maximum–minimum air temperatures, solar radiation, and wind velocity, may be used to obtain reliable, daily estimates of PET. This method is not presented, for few aquaculturists have need for daily PET values.

Many workers have developed equations for estimating the actual evapotranspiration from potential evapotranspiration. Examples of such equations have been developed by Mustonen and McGuinness (1968) for grassland in Ohio:

$$ET_D = 0.70(\text{PET})(SM)^{0.25} \tag{3-22}$$

$$ET_M = 0.60(\text{PET})(SM)^{0.25}(P + 10\,\text{mm})^{0.125} \tag{3-23}$$

where ET_D, ET_M = daily and monthly evapotranspiration, respectively (mm)

SM = soil moisture (mm)
P = precipitation (mm)

Such equations usually are rather limited in application, because they were developed to fit data collected at specific locations.

3.8.3 Water Budgets

Actual evapotranspiration may be estimated from water budget studies of drainage basins. In the hydrologic equation of a basin, inflow equals precipitation, and outflow equals evapotranspiration and runoff. The water table depth is roughly at the same depth at the beginning of each year, indicating no net storage of water in aquifers from one year to the next. Therefore, the equation for ET is

$$ET = \text{precipitation} - \text{runoff} \tag{3-24}$$

Runoff can be calculated from equations in Chapter 4. Where stream discharge information is available for an area, it provides a direct measurement of runoff. If precipitation on a basin was 45 in. in a year and stream discharge was equal to a water depth of 15 in. over the basin, then ET was 30 in.

3.9   RATES OF EVAPOTRANSPIRATION

Some selected studies of evapotranspiration will be summarized. Harrold and Dreibelbis (1958) presented numerous evapotranspiration measurements made by lysimetry at Coshocton, Ohio. Lysimeters contained various types of vegetative cover as follows: pasture grasses, oats, wheat, meadow, and corn. Annual evapotranspiration rates ranged from 30.98 to 52.01 in.; values were greatest in lysimeters where crops were irrigated. Evapotranspiration rates usually exceeded pan evaporation; the difference was most marked during summer.

Gilbert and van Bavel (1954) maintained a cover of clover and weeds clipped at 3-in. height in lysimeters in North Carolina. Lysimeters were irrigated so that water supply did not limit evapotranspiration. Monthly rates of evapotranspiration closely followed air temperatures (Fig. 3-11); annual evapotranspiration was 32.00 in. Potential annual evapotranspiration by the Penman and Thornthwaite methods was 33.62 in. and 29.09 in., respectively. Monthly estimates by the Penman and Thornthwaite methods differed from measured monthly values by about the same proportions noted in annual comparisons.

According to Todd (1970), annual consumption of water by crops in the Sacramento–San Joaquin Valley in California ranged from 17.99 in. for celery to 42.13 in. for alfalfa. In the Central Valley of California, consumptive use values ranged from 27.91 in. for corn to 50.79 in. for pasture. Rice, an aquatic plant, used 55.98 in. Todd also reported that evapotranspiration rates for nonagricultural land in the western United States ranged from 12 in. for semiarid grass and shrubs to 30 in. for forests of Douglas fir, hemlock, and redwood.

Moran and O'Shaughnessy (1984) reported annual evapotranspiration rates of 26.61 to 44.21 in. for eucalyptus forest in Australia, where rainfall ranged from 45.52 to 73.86 in. Annual pan evaporation was 19.56 to 22.05 in. Annual evapotranspiration from grass in an irrigated lysimeter at Aspendale, Australia, was 51.57 in. and evaporation from a water-filled lysimeter was 43.11 in. (McIlroy and Angus 1964). In areas of western Australia, where rainfall is 24 in. or less, evapotranspiration losses roughly equal the annual rainfall (Stewart 1984).

Phreatophytes and hydrophytes (aquatic macrophytes) extend their

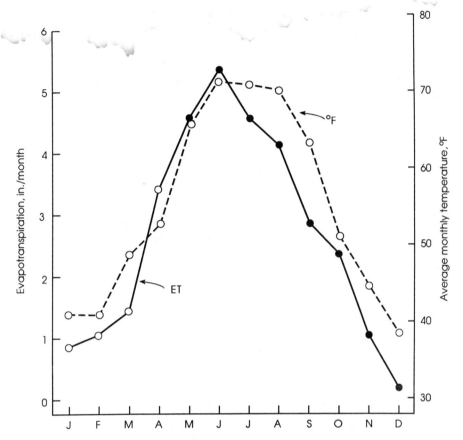

FIGURE 3-11.   Evapotranspiration from a grass-covered lysimeter and air temperature at a site in North Carolina. (*Source: Gilbert and van Bavel 1954.*)

roots below the water table and have a continuous supply of water. There has been much research on phreatophytes because they may grow along irrigation canals. Todd (1970) summarized data from the western United States showing that stands of phreatophytes may consume up to 100 in. of water per year; however, most estimates did not exceed 50 in. Davenport et al. (1982) found that evapotranspiration by a phreatophyte, saltcedar, varied with weather factors, stand density, and water availability. In the summer, daily evapotranspiration rates ranged from 0.08 to 0.63 in.

Relatively few estimates of evapotranspiration have been obtained for aquatic macrophytes. These studies have generally demonstrated that evapotranspiration from stands of aquatic macrophytes usually exceeds evaporation from a free-water surface. Ratios of evapotranspiration (ET) to evaporation (E) ranged from 0.9 to 3.7 (Brenzy et al. 1973; DeBusk et al. 1983; Snyder and Boyd 1987; Boyd 1975, 1987a). The smallest ET/E ratios were for duckweeds, which cover the water surface in essentially a single layer (LAI = 1), and the greatest values for ET/E were for large plants like water hyacinth and cattail, which have LAI values of 5 or more.

3.10 HYDROCLIMATE

The study of hydroclimate embraces the influences of climate on water availability. In some places more rain falls each month than is lost by evapotranspiration. The excess water either infiltrates or becomes stream flow. In other places, the monthly rainfall never meets the demands of evapotranspiration. Such regions have no permanent streams, and runoff is limited to unusually heavy rains. Most places have a hydroclimate between these two extremes. Some months experience excess rainfall; other months have a deficit of precipitation.

A common way of describing the hydroclimate of an area is to plot

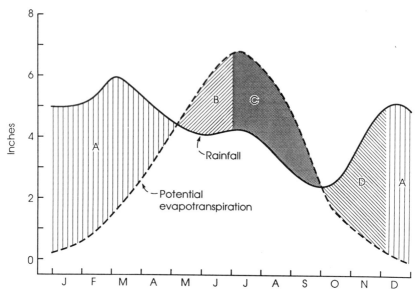

FIGURE 3-12.   Hydroclimate at a site in Alabama. Soil moisture status: (A) surplus, (B) utilization, (C) deficiency, (D) recharge. Runoff occurs primarily when there is a soil moisture surplus. (*Data from Southeast Weather Service Center, Auburn University, Alabama.*)

monthly rainfall totals and monthly potential evapotranspiration estimates over an entire year. A net gain in soil moisture occurs in any month in which precipitation exceeds potential evapotranspiration. If there is more rain than needed to bring the soil to field capacity with moisture, the remainder infiltrates deeper to become ground water, or else it becomes overland flow. A water surplus exists when there is more rain than needed to satisfy soil moisture storage. Soil moisture is utilized when rainfall is less than potential evapotranspiration. If the period of soil moisture utilization continues, the soil moisture will be exhausted. During periods of water deficiency, plant growth is limited by a shortage of water. Soil moisture recharge occurs when precipitation exceeds evapotranspiration and the soil moisture content is below field capacity. The annual hydroclimate of a locality in Alabama is given in Figure 3-12. The figure illustrates periods of water surplus, soil moisture utilization, water deficiency, and soil moisture recharge.

For the hydroclimate illustrated in Figure 3-12, it is clear that most overland flow and aquifer recharge occur between mid-December and May. The peak discharges of streams also occur between December and May. Between early May and mid-December, streams are sustained by base flow. Many small streams cease to flow during this period. Unusually heavy rains during the summer and fall generate some overland flow and possibly recharge aquifers slightly.

4

Runoff

4.1 INTRODUCTION

All of the water that flows over the surface of the land is referred to collectively as *runoff*. Runoff flows over watershed surfaces entering small streams that merge with larger streams that finally discharge into the ocean. Streams also receive the inflow of ground water from springs or from aquifers where the stream beds cut below the water table. In cold regions, water from melting snow and glaciers contributes to runoff. Stream flow consists of several components: rain falling directly into streams, water flowing over watershed surfaces, water flowing downslope within the upper layer of watershed soil, and ground water seeping into streams. For large watersheds or drainage basins, runoff can be computed directly from stream discharge. However, for watersheds too small to support streams or for ungauged watersheds or basins, runoff must be estimated.

This chapter explains the runoff process, provides simple procedures for estimating runoff, and discusses stream flow and fluvial processes.

4.2 THE PROCESS

Runoff is that portion of the rain falling on a watershed that is not retained on or beneath the watershed surface or lost from the watershed by evapotranspiration. It is the water that is ultimately transported from the watershed by flow over the land surface either as overland flow or stream flow.

Gravity is the force causing runoff, and runoff begins after there has been sufficient rainfall to exceed the capacity of watershed surfaces to detain water by absorption or in depressions and the rate of rainfall exceeds the rate of infiltration of water into the soil. Of course, evaporation extracts water during the runoff process. Features of the watershed, rainfall characteristics, season, and climate influence the amount of runoff generated by a watershed. Factors favoring large amounts of runoff are intense rainfall, heavy rainfall, impervious soil, frozen or moist soil, high proportion of paved surface, steep slope, little surface storage, sparse vegetative cover, soils with low moisture-holding capacity, and a shallow water table.

Consider a watershed with a small stream. Depending on the nature of the rainstorm, four different patterns of runoff are possible:

1. A brief, light rain may be detained on watershed surfaces and cause no downward infiltration of water into the aquifer or overland flow over the watershed surface.
2. A light rain of a longer duration may not be of sufficient intensity to exceed the infiltration rate and generate overland flow, but infiltration may raise the level of the water table, thereby increasing base flow into the stream and runoff from the watershed.
3. A brief, intense rain may quickly exceed the ability of a watershed surface to retain water and cause overland flow even though there is insufficient infiltration to raise the water table and increase base flow into the stream.
4. An intense rain of longer duration will fill the storage capacity of watershed surfaces to generate overland flow and produce sufficient infiltration to raise the water table and increase base flow into the stream.

4.3 METHODS FOR ESTIMATING RUNOFF

Several relatively simple procedures that incorporate important runoff-producing variables to allow estimates of amounts of water discharged from watersheds will be presented.

4.3.1 Peak Discharge

Estimates of peak discharge from watersheds are used in designing dams and water-conveying structures such as spillways and culverts. It is important that these structures not fail because of heavy rains and associated runoff. Of course, the probability of failure and the economic cost of this failure must be considered against the construction costs necessary to prevent failure. For example, if little damage would result from failure, it may not be economical to construct a structure that is not expected to fail but once every 50 years. On the other hand, if damage would be great or life put in jeopardy, a probability of a failure every 50 years probably would not be acceptable regardless of high construction costs.

The rational method is popular for estimating peak discharge. The rational peak discharge equation is

$$Q = CiA \tag{4-1}$$

where Q = peak discharge (cfs)
     C = runoff coefficient (dimensionless)
     i = intensity of rainfall for duration equal to concentration time of watershed for desired return period (in./hr)
     A = watershed area (acres)

The concentration time of a watershed is the time for overland flow to move from the most distant point on the watershed to the mouth of the watershed. It is found from the following equation:

TABLE 4-1   Runoff Coefficients for the Rational Method of Calculating Peak Discharge

Description of Area	Topography		
	Flat	Rolling	Hilly
Paved roads	0.70	0.80	0.90
Suburban residential areas	0.25	0.30	0.40
Parks and cemeteries	0.15	0.20	0.25
Lawns			
Sandy soil	0.10	0.15	0.20
Heavy soil	0.15	0.20	0.30
Woodland			
Sandy loam soil	0.10	0.25	0.30
Clay or silt loam soil	0.30	0.35	0.50
Clay soil	0.40	0.50	0.60
Pasture			
Sandy loam soil	0.10	0.15	0.22
Clay or silt loam soil	0.30	0.36	0.42
Clay soil	0.40	0.55	0.60
Row crops			
Sandy loam soil	0.30	0.40	0.52
Clay or silt loam soil	0.50	0.60	0.72
Clay soil	0.60	0.70	0.82

$$T_c = 0.0078 L^{0.77} S^{-0.385} \qquad\qquad (4\text{-}2)$$

where T_c = concentration time (min)
  L = maximum length of flow (ft)
  S = watershed gradient (ft vertical/ft horizontal)

Runoff coefficients (Table 4-1) are based on watershed features. Rainfall intensity must be obtained from a rainfall intensity–duration–frequency curve for the vicinity, such as Figure 2-13. It is assumed in the rational method that all parts of the watershed receive rainfall uniformly.

Example 4-1

To apply the rational method, consider a woodland watershed with tight clay soil that has an area of 250 acres, a slope of 0.2%, and a maximum length of flow of 2,000 ft. Peak discharge will be calculated for a 10-year return period; the rainfall intensity–duration–frequency curve for Montgomery, Alabama (Fig. 2-13), will be used.

Solution   First, estimate concentration time:

$$T_c = (0.0078)(2,000)^{0.77}(0.002)^{-0.385}$$
$$= (0.0078)(348.17)(10.94) = 30\,\text{min}$$

It will take water 30 min to flow from the farthest part of the watershed to the mouth. If overland flow starts as rain begins—an unusual situation, but the one that produces peak discharge—30 min will pass before all parts of the watershed contribute discharge at the mouth. Once this occurs, rainfall can continue indefinitely at the same intensity, but discharge will not increase. Therefore, solving the rational runoff equation for the maximum intensity of rainfall expected over a 30-min period for a 10-year return period gives the peak discharge of the watershed for a 10-year return period. The maximum intensity of rain expected every 10 years at Montgomery, Alabama, is 3.8 in./hr (Fig. 2-13); the runoff coefficient is 0.40 (Table 4-1). Peak discharge is

$$Q = (0.40)(3.8)(250) = 380\,\text{cfs}$$

The rational method gives peak discharge for the combination of maximum rainfall intensity and most favorable conditions for runoff on the watershed, which is soils saturated with water. This combination likely will not occur, and actual discharge will be less than calculated peak discharge.

Some readers may wonder how a dimensionless runoff coefficient times rainfall in inches per hour times area in acres gives discharge in cubic feet per second in the rational runoff equation. One acre is $43,560 \, \text{ft}^2$, 1 ft is 12 in., and 1 hr is 3,600 sec. Therefore, to get the answer in cubic feet per second, you must divide rainfall intensity in inches per hour by 43,200 (the product of 12 in./ft and 3,600 sec/min) and multiply area in acres by $43,560 \, \text{ft}^2/\text{acre}$. The quotient of $43,560 \div 43,200$ is essentially unity. Hence, by coincidence, the numerical answer is the same whether or not we adjust variables to proper dimensions.

The rational method is slightly more complicated for a nonuniform watershed.

Example 4-2

Suppose a watershed has 50 acres of hilly woodland with clay and silt loam soils, 25 acres of pasture with rolling sandy loam soils, and 90 acres of flat, cultivated, tight clay soil; a maximum length of flow of 1,800 ft; and an average slope of 4%. Peak discharge will be estimated for a 25-year return period.

Solution   Concentration time of the watershed is

$$T_c = (0.0078)(1,800)^{0.77}(0.04)^{-0.385}$$

$$= (0.0078)(321.04)(3.453) = 9 \, \text{min}$$

For a 25-year return period, maximum rainfall intensity at Montgomery, Alabama, for 9 min is 7 in./hr (Fig. 2-13). Rational runoff coefficients (Table 4-1) are as follows: woodland, 0.30; pasture, 0.16; cultivated land, 0.60. A weighted runoff coefficient (C_w) is calculated by summing the products of runoff coefficient and area for the different segments of the watershed and dividing the sum of the products by total watershed area:

$$C_w = \frac{(50)(0.30) + (25)(0.16) + (90)(0.60)}{50 + 25 + 90}$$

$$= \frac{73}{165} = 0.44$$

Using C_w in place of C in the rational runoff equation, we have

$$Q = (0.44)(7)(165) = 508 \, \text{cfs}$$

4.3.2   Overland Flow by Curve Number Technique

The U.S. Soil Conservation Service (1972) developed a technique for computing excess rainfall based on antecedent soil moisture, hydrologic soil group, land use, and hydrologic condition. A particular combination of these factors is associated with a specific curve that relates depth of overland flow to depth of rainfall produced by a given storm. This procedure, called the *curve number method*, is easy to use and requires no data that cannot be obtained by field observation. The curve number method is widely used, but it is based on a number of assumptions that may not be correct for a particular situation, and many hydrologists question its accuracy (Hewlett 1982). Nevertheless, the method is valuable because it is simple and often the only means of making approximate estimates of overland flow for ungauged watersheds. Antecedent moisture conditions are based on conditions observed in the field or the five-day antecedent rainfall. The three moisture condition categories are presented in Table 4-2.

Soils are grouped according to their abilities to infiltrate and transmit water. When runoff from an individual storm is of concern, hydrologic properties of soils can be represented by a single variable: the minimum rate of infiltration obtained for a bare soil after prolonged wetting. The U.S. Soil Conservation Service (1972) listed 4,000 soils by name and hydrologic soil group. In the United States, soil maps are available from the Soil Conservation Service for most counties. For most purposes, the definitions of hydrologic soil groups provided in Table 4-3 usually are sufficient to assign a soil to a particular group.

TABLE 4-2   Antecedent Moisture Conditions (AMC) for Estimating Runoff (overland flow) by Curve Number Technique

Condition	Description
AMC I	Watershed soils are dry enough to till. The total 5-day antecedent rainfall was less than 0.5 in. during the dormant season or less than 1.5 in. during the growing season. This condition has the lowest potential for overland flow.
AMC II	This is the average condition for overland flow. Total 5-day antecedent rainfall was 0.5–1.1 in. during the dormant season and 1.5–2.1 in. during the growing season.
AMC III	This condition has the greatest potential for overland flow because watershed soils are essentially saturated. The 5-day antecedent rainfall was greater than 1.1 in. during the dormant season and more than 2.1 in. during the growing season.

Source: U.S. Soil Conservation Service (1972).

TABLE 4-3   Hydrologic Soil Group Descriptions for Estimating Runoff (overland flow) by Curve Number Technique

Hydrologic Soil Group	Description
A	Soils having high infiltration rates even when thoroughly wetted. They consist chiefly of deep, well-drained sands and gravels. Infiltration rates are 0.30–0.45 in./hr. This group has the lowest potential for overland flow.
B	Soils having moderate infiltration rates when thoroughly wetted. They consist chiefly of moderately deep to deep, moderately well to well-drained soils with moderately fine to moderately coarse textures. Infiltration rates are 0.15–0.30 in./hr.
C	Soils having slow infiltration rates when thoroughly wetted. They consist primarily of soils with a layer that impedes downward movement of water, or soils with moderately fine to fine texture. Infiltration rates are 0.05–0.15 in./hr.
D	Soils having very slow infiltration rates when thoroughly wetted. They consist largely of clay particles with a high swelling potential, soils with a permanent high water table, soils with a clay layer near the surface, and shallow soils over nearly impervious material. Infiltration rates are 0–0.05 in./hr.

Source: U.S. Soil Conservation Service (1972).

Vegetative cover and land use also influence overland flow. The U.S. Soil Conservation Service (1972) developed curve numbers (CN) for combinations of hydrologic soil group (soil) and land use and vegetation (cover). The combination is termed a *hydrologic soil–cover complex*, and curve numbers for different hydrologic soil–cover complexes are provided in Table 4-4.

Reference is made to hydrologic condition in Table 4-4. Row crops, small grain, and close-seeded legumes or rotation meadow are referred to as being in either good or poor hydrologic condition. Farming practices that reduce overland flow, contouring, good vegetative cover, and rotation of crops produce a good hydrologic condition. Straight rows where land slopes, crops that do not cover the land surface, and continuous planting of a single crop species produce a poor hydrologic condition. The hydrologic condition for native pasture, range, or woodland may be found in Table 4-5.

Curve numbers for hydrologic soil–cover complexes are applicable only for soils that are not frozen. Impervious surfaces and water surfaces, which are not listed in Table 4-4, are always assigned a curve number of 100. Curve numbers listed in Table 4-4 are for antecedent moisture condition (AMC) II. The curve number for AMC-II is related to curve

TABLE 4-4   Curve Numbers for Hydrologic Soil-Cover Complexes and Antecedent Moisture Condition II

Land Use	Treatment or Practice	Hydrologic Condition	Hydrologic Soil Group			
			A	B	C	D
Fallow	Straight row	—	77	86	91	94
Row crops	Straight row	Poor	72	81	88	91
	Straight row	Good	67	78	85	89
	Contoured	Poor	70	79	84	88
	Contoured	Good	65	75	82	86
	Contoured and terraced	Poor	66	74	80	82
	Contoured and terraced	Good	62	71	78	81
Small grain	Straight row	Poor	65	76	84	88
	Straight row	Good	63	75	83	87
	Contoured	Poor	63	74	82	85
	Contoured	Good	61	73	81	84
	Contoured and terraced	Poor	61	72	79	82
	Contoured and terraced	Good	59	70	78	81
Close-seeded	Straight row	Poor	66	77	85	89
legumes or	Straight row	Good	58	72	81	85
rotation meadow	Contoured	Poor	64	75	83	85
	Contoured	Good	55	69	78	83
	Contoured and terraced	Poor	63	73	80	83
	Contoured and terraced	Good	51	67	76	80
Pasture or range		Poor	68	79	86	89
		Fair	49	69	79	84
		Good	39	61	74	80
	Contoured	Poor	47	67	81	88
	Contoured	Fair	25	59	75	83
	Contoured	Good	6	35	70	79
Meadow		Good	30	58	71	78
Woods		Poor	45	66	77	83
		Fair	36	60	73	79
		Good	25	55	70	77
Farmsteads		—	59	74	82	86
Roads/right-of-way						
dirt		—	72	82	87	89
hard surface		—	74	84	90	92

Source: U.S. Soil Conservation Service (1972).

numbers for AMC I and AMC III in Table 4-6. Table 4-2 shows five-day antecedent rainfall to determine the antecedent moisture conditions.

Potential for overland flow increases as curve number increases. Relationships among curve number, rainfall, and overland flow were employed to construct Figure 4-1. Once the curve number has been established,

TABLE 4-5   Hydrologic Condition for Native Pasture, Range, or Woodland for Estimating Overland Flow by Curve Number Technique

Watershed Type	Description	Hydrologic Condition
Native pasture	Heavily grazed. No mulch or less than 50% plant cover.	Poor
	Not heavily grazed. Plant cover of 50–75%.	Fair
Woodland	Heavily grazed or regularly burned. Litter, small trees, and brush absent.	Poor
	Grazed but not burned. Scant litter present.	Fair
	Protected from grazing. Litter and understory vegetation present.	Good

Source: U.S. Soil Conservation Service (1972).

TABLE 4-6   Conversion of Curve Numbers for Various Antecedent Moisture Conditions (AMC) I and III from Curve Number for Condition II

Curve Number for Condition II	Corresponding Curve Number for Condition	
	I	III
100	100	100
95	87	99
90	78	98
85	70	97
80	63	94
75	57	91
70	51	85
65	45	83
60	40	79
55	35	75
50	31	70
45	27	65
40	23	60
35	19	55
30	15	50
25	12	45
20	9	39
15	7	33
10	4	26
5	2	17
0	0	0

Source: U.S. Soil Conservation Service (1972).

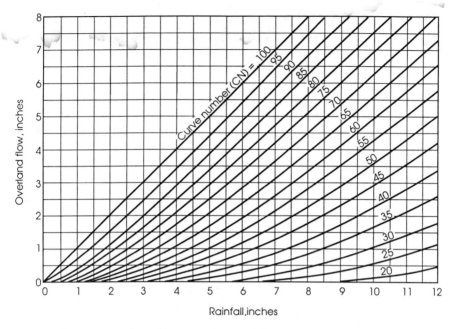

FIGURE 4-1.   Rainfall-runoff curve number relationships. Method for selecting curve number is provided in the text. (*Source: U.S. Soil Conservation Service 1972.*)

Figure 4-1 may be used to estimate overland flow after a storm. The relationships include

$$Q = \frac{(I - 0.2S)^2}{I + 0.8S} \tag{4-3}$$

and

$$S = \frac{1{,}000}{CN} - 10 \tag{4-4}$$

where Q = direct runoff or overland flow (in.)
I = storm rainfall (in.)
S = maximum potential difference between rainfall and runoff (in.)

Use of the curve number method will now be illustrated.

Example 4-3

Suppose the following situation exists for a pasture: contoured, good hydrologic condition, hydrologic soil group C, and AMC II. Overland flow for a 3-in. rain can be estimated as follows.

Solution   First, determine the curve number from Table 4-4: $CN = 70$. Next, find curve number 70 (Fig. 4-1) and move vertically from 3 in. of rainfall on the abscissa until $CN = 70$ is intersected. Finally, move horizontally from this intersection until the ordinate is reached. An estimated overland flow of 0.75 in. is read directly from the ordinate.

The curve number procedure is slightly more involved when the watershed is not uniform.

Example 4-4

Suppose a watershed consists of the following subunits: 62 acres of pasture—uncontoured, poor hydrologic condition, and group A soil; 124 acres of row crop—straight row, good hydrologic condition, group C soil; 50 acres of woods—fair hydrologic condition, and group D soil. The watershed has AMC III. Estimate overland flow for a 4-in. rainfall.

Solution   First, curve numbers for AMC II are obtained from Table 4-4 for each subunit. Values for CN are pasture, 68: row crop, 85; woods, 79. Values for AMC III can be estimated by interpolation from Table 4-6. They are pasture, 84; row crop, 97; woods, 93. Curve numbers for subunits must be weighted. Weighting may be done by either of two methods: (1) overland flow is estimated for each hydrologic soil–cover subunit and weighted to get the watershed estimate; or (2) curve numbers are weighted to get a watershed curve number that is used to estimate overland flow or runoff (Q). Both methods are used below:

Weighting runoff:

Subunit	A (acres)	Q (in.)	$A \times Q$ (acre-in.)
Pasture	62	2.3	142.6
Row crops	124	3.7	458.8
Woods	50	3.3	165.0
Sums	236		766.4

$$\text{Weighted } Q = \frac{766.4}{236} = 3.25 \text{ in.}$$

Weighting curve number:

Subunit	A (acres)	CN	A × CN
Pasture	62	84	5,208
Row crops	124	97	12,028
Woods	50	93	4,650
Sums	236		21,886

$$\text{Weighted } CN = \frac{21{,}886}{236} = 93$$

$$Q = 3.3 \text{ in.}$$

Direct runoff estimates by the two weighting techniques are similar here. However, weighting runoff is preferable for small rains and for two or more widely different CN values on the same watershed. For other situations, the weighted curve number method is recommended because it is less time-consuming.

The curve number method permits comparisons of different watershed conditions on the potential for runoff. Remember, as curve number increases there is greater potential for runoff. Different cover often exists over the same hydrologic soil group. Consider curve number values for the following cover complexes on soil group B for AMC II: row crop, straight row, poor hydrologic condition ($CN = 81$); pasture, uncontoured, good hydrologic condition ($CN = 61$); meadow ($CN = 58$); woods, good hydrologic condition ($CN = 55$); roads, dirt ($CN = 82$). The same cover may exist on different hydrologic soil groups. Woodland in good hydrologic condition has curve number values of 25, 55, 70, and 77 at AMC II for hydrologic soil group A, B, C, and D, respectively. These two examples illustrate how the curve number method can be used to determine the influence of watershed variables on the potential for runoff.

4.3.3 Runoff Equations

Data on runoff from small watersheds have been collected at many sites and used to develop equations for predicting runoff. For example, van Keuren et al. (1979) developed an equation for runoff from sloping pastures in Ohio. Its solution requires data on rainfall, rainfall intensity, and soil moisture. Such equations are site-specific, and data for solving them seldom are available; they are of little use to aquaculturists.

Hewlett et al. (1977) developed an equation for predicting runoff from small, forested drainage basins in the eastern United States; this approach was termed the R-index method. The equation is

$$Q = \frac{0.35RP^{1.5}}{2.54} \tag{4-5}$$

where Q = overland flow for a storm (in.)
        P = rainfall for storm (cm)
        R = hydrologic response (dimensionless)
        2.54 = conversion factor from centimeters to inches

The answers calculated by the equation must be adjusted for season by the following multipliers: January, 1.4; February, 1.5; March, 1.5; April, 1.4; May, 1.3; June, 1.2; July, 1.1; August, 1.0; September, 1.0; October, 1.1; November, 1.2; December, 1.3. The hydrologic response (R) of a watershed is the amount of direct runoff divided by the amount of precipitation. The magnitude of R varies among watersheds and with conditions on a given watershed. Hewlett (1982) provided a map of the eastern United States with R values indicated by physiographic region; values ranged from 0.04 to more than 0.4. The average was about 0.2. Although the R-index method is easier to use and possibly more accurate than the curve number method, lack of R values for individual watersheds and for physiographic regions outside the eastern United States make the technique unattractive for our purposes.

4.3.4   Water-Accounting Method for Estimating Runoff

The hydroclimate idea discussed in Chapter 3 has been used to develop water-accounting methods for estimating monthly or seasonal runoff from ungauged watersheds. A water-accounting procedure developed by the U.S. Soil Conservation Service (1964) involves a monthly accounting of rainfall, soil moisture, and evapotranspiration. Rainfall data may consist of information for a specific year or averages for several years. The available moisture-holding capacity of soils is provided in Table 4-7 for different soil textures. More specific information on the available moisture capacity often can be found in local soil survey records. Local soil data may give available moisture capacities for different depths of soil. Total available moisture capacity should be estimated for the depth of the intensive root zone. If the intensive root zone exists to a depth of more than 3 ft, estimate total available moisture for the surface 3-ft layer of soil. If soil features vary over the watershed, the total available moisture

TABLE 4-7   Available Moisture-Holding Capacity in Soils of Different Texture

Description	Inches of Water per Foot of Soil
Very coarse texture—very coarse sands	0.40–0.75
Coarse texture—coarse sands, fine sands, and loamy sands	0.75–1.25
Moderately coarse texture—sandy loams and fine sandy loams	1.25–1.75
Medium texture—very fine sandy loams, loams, and silt loams	1.50–2.30
Moderately fine texture—clay loams, silty clay loams, and sandy clay loams	1.75–2.50
Fine texture—sandy clays, silty clays, and clays	1.60–2.50
Peats and mucks	2.00–3.00

Source: U.S. Soil Conservation Service (1964).

capacity must be weighted according to the area of each different soil group. The U.S. Soil Conservation Service (1964) suggested using the method of Blaney and Criddle (1952) for obtaining potential evapotranspiration (PET) estimates for use in the water-accounting procedure for runoff. However, this method requires data on major crops and land uses, and the more general method of Thornthwaite (1948), given in Chapter 3, for computing PET is suitable for use in aquacultural considerations.

To use the water-accounting method for estimating runoff, obtain monthly rainfall data, use temperature data to estimate PET by the Thornthwaite method, use Table 4-7 to obtain the available moisture capacity of the soil, and begin with a month when the soil is either normally saturated or depleted of moisture. The example in Table 4-8 is for a place in the southern United States with coarse-textured soils. The moisture budget was calculated for the upper 3 ft of soil. Calculations begin on 1 April when the soil is normally saturated with moisture. A coarse-textured soil holds about 1.0 in. of moisture per foot of depth or 3.0 in. for the upper 3-ft layer. Rainfall in April was 5.31 in. Thus, the total available moisture for April was

$$\text{Total available moisture} = \text{initial soil moisture} + \text{rainfall}$$

$$= 3.00 + 5.31 = 8.31 \text{ in.}$$

For April, PET was 3.18 in. (calculations of PET are not presented). This is the quantity of water removed from the soil by evaporation, and it may be thought of as being removed from the pool of total available moisture. Therefore, the remaining available soil moisture at the end of the month is

Remaining available moisture = total available moisture − PET

$$= 8.31 - 3.18 = 5.13 \text{ in.}$$

However, the soil can only hold 3.00 in. of moisture, so the runoff may be computed by subtracting the final soil moisture from the remaining soil moisture:

Runoff = remaining available moisture − final soil moisture

$$= 5.13 - 3.00 = 2.13 \text{ in.}$$

These computations were repeated for May through November, and the results are tabulated in Table 4-8.

The water-accounting method does not include a term for the water that infiltrates downward and becomes ground water. However, this is not a serious error in the logic of the procedure. Over years, the amount of ground water stored beneath a watershed is relatively constant on a given date. For example, if water table depth beneath the ground surface was plotted for the first of April each year for a period of years, water table depth would be fairly constant. This suggests that there usually is little net storage of incoming water in aquifers from year to year, and water infiltrating into aquifers seeps into streams to become runoff.

4.3.5 Estimates Based on Stream Discharge

Methods for measuring stream discharge are provided in Chapter 10. However, if stream discharge data are available, they can be used as direct estimates of monthly or annual runoff. Data on watershed or basin area may be used to estimate runoff per unit area. To illustrate, suppose that stream discharge for a 500-acre watershed averages 2 cfs for March, and it is desired to compute runoff in inches of water. The computation follows:

$$2 \text{ cfs} \times 86{,}400 \text{ sec/day} \times 31 \text{ days} = 5{,}356{,}800 \text{ ft}^3 \text{ runoff}$$

$$500 \text{ acres} \times 43{,}560 \text{ ft}^2/\text{acre} = 21{,}780{,}000 \text{ ft}^2 \text{ watershed}$$

$$\text{Depth of runoff} = \frac{5{,}356{,}800 \text{ ft}^3}{21{,}780{,}000 \text{ ft}^2} = 0.25 \text{ ft} = 3 \text{ in.}$$

TABLE 4-8   Water Accounting Technique for Estimating Runoff (all data are in inches of water depth)

Item	April	May	June	July	August	September	October	November
Average rainfall	5.31	4.23	3.85	5.74	3.59	4.32	1.83	4.72
Initial soil moisture[a]	3.00	3.00	3.00	1.81	1.01	0.01	0.56	0.00
Total available moisture[b]	8.31	7.23	6.85	7.55	5.60	4.33	2.39	4.72
Potential evapotranspiration[c]	3.18	3.42	5.04	6.52	5.59	3.87	2.41	1.40
Remaining measurable moisture[d]	5.13	3.81	1.81	1.01	0.01	0.56	0.00	3.32
Final soil moisture[e]	3.00	3.00	1.81	1.01	0.01	0.56	0.00	3.00
Runoff[f]	2.13	0.81	0.00	0.00	0.00	0.00	0.00	0.32

[a] Available soil moisture when fully saturated is 1-in. water per foot of depth. The top 3-ft layer of soil is used in calculations. Soil normally is fully saturated in April.
[b] Rainfall plus initial soil moisture.
[c] Thornthwaite (1948) method.
[d] Total available moisture minus potential evapotranspiration.
[e] Final soil moisture cannot exceed maximum capacity of soil to hold water.
[f] Remaining available moisture minus final soil moisture.

4.3.6 Estimates Based on Precipitation and Evapotranspiration

If measured at 12-month intervals, the total amount of water stored on the surface, in the soil, and in aquifers beneath a watershed is usually almost constant. Therefore, the runoff may be computed as

$$RO = P - PET$$

where PET = potential evapotranspiration. Precipitation data usually are available, and PET may be computed by one of the methods in Chapter 3.

4.4 AMOUNTS OF RUNOFF

Stream discharge data are obtained for many streams on all continents, and it is possible to compute average runoff for continents or for various basins within continents from these data. The water balance for the continents is given in Table 4-9. Runoff ranges from 2.4 in./yr in Australia to 19.3 in./yr in South America. The percentage of the precipitation becoming runoff varies from 13% in Australia to 40% in Europe and North America. In the United States, precipitation averages 20 in., and runoff averages 8.5 in. or 28% of precipitation. A runoff map for the United States is shown in Figure 4-2.

An analysis of runoff on a smaller scale is provided in Table 4-10, with monthly runoff data for the major soil areas of Alabama. These data are for large streams draining 81 to 2,400 mile^2. Total runoff is about twice the average for the North American continent. Rainfall over the geographic area represented in Table 4-10 is about 50 in./yr, so roughly 35–45% of the annual precipitation was converted to runoff; this is in

TABLE 4-9   Water Balance of the Continents

Continent	Precipitation (in.)	Evaporation (in.)	Runoff	
			in.	% precipitation
Africa	26.4	20.1	6.3	24
Asia	24.0	15.4	8.7	36
Australia	18.5	16.1	2.4	13
Europe	23.6	14.2	9.5	40
North America	26.4	15.7	10.6	40
South America	53.1	33.9	19.3	36

Source: Data from van der Leeden et al. (1990).

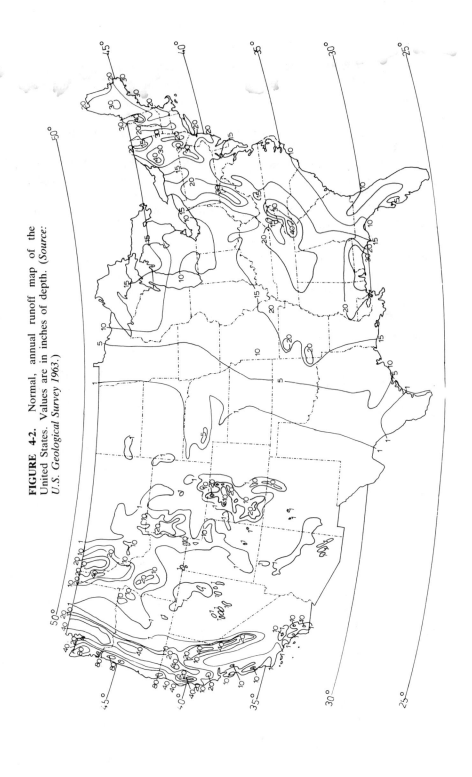

FIGURE 4-2. Normal, annual runoff map of the United States. Values are in inches of depth. (*Source: U.S. Geological Survey 1963.*)

close agreement with the average for North America (Table 4-9). Notice that runoff tends to be greatest in winter and spring, lowest in summer and fall. This occurs because rainfall is greater and temperature is lower in the winter and early spring than at other times. Lowrance et al. (1985) reported that 28–32% of the annual precipitation falling on basins in the Coastal Plain area of Georgia becomes stream flow.

However, one should not be misled into believing that 35–45% of the rainfall in well-watered regions becomes runoff. When many diverse basins are considered, annual runoff may vary from less than 1% to more than 75% of the annual rainfall. The proportion of rainfall becoming stream flow varies tremendously with watershed characteristics. Vegetative cover is especially important. Douglass and Swank (1975) demonstrated that conversion of hardwood cover to pine cover reduces runoff, and conversion of hardwood cover to grass cover enhances runoff. Bosch and Hewlett (1982) reviewed 92 watershed studies worldwide and showed that reducing commercial forest vegetation increases water yields, although an initial reduction in vegetation of less than 20% has no apparent effect. Cutting coniferous and hardwood forests increases annual water yield about 1.6 to 1.0 in., respectively, for each additional 10% removal of trees above the initial 20% reduction. Establishing a forest on sparsely vegetated land reduces water yield. Differences in water yield for different types of vegetation results because evapotranspiration varies with

TABLE 4-10   Average Monthly Runoff from Various Soil Areas of Alabama
(values are in inches)

Month	Limestone Valleys	Appalachian Plateau	Piedmont	Upper Coastal Plain	Black Belt	Lower Coastal Plain
January	3.0	4.1	2.4	3.2	2.3	2.6
February	3.7	4.9	2.4	3.6	2.9	2.5
March	3.6	4.2	3.5	3.9	4.1	4.1
April	2.6	2.8	2.6	3.2	2.2	3.0
May	1.7	1.0	1.6	1.8	1.1	1.9
June	1.0	0.7	1.1	1.0	0.9	1.5
July	0.8	0.9	1.6	1.2	1.0	1.7
August	0.7	0.6	1.1	1.4	1.2	1.9
September	0.5	0.3	0.6	0.8	0.3	1.4
October	0.6	0.3	0.5	0.6	0.1	0.9
November	1.0	0.9	1.4	1.7	0.6	1.9
December	1.8	1.8	1.5	1.9	1.5	2.1
Total	21.0	22.5	20.3	24.3	18.2	25.5

Source: Swingle (1955).

type of vegetation (Evans and Patric 1983). Grassland yields the most water because interception and transpiration rates of grassland are less than those of forests.

Often, ponds for aquaculture are constructed on small watersheds that do not maintain permanent streams. The major source of water for ponds on such watersheds is overland flow. The runoff yield from small watersheds often is smaller than that of larger basins, because ground water is exported from beneath these watersheds to enter streams at places removed from the watershed surface. To illustrate, overland flow from a 1,560-acre hydrologic unit consisting of nine different hydrologic soil–cover complexes was calculated by the curve number method for individual storms over a year (Boyd and Shelton 1984). This hydrologic unit is located on the Piedmont Plateau near Auburn, Alabama. It has an average slope of 4%, and average monthly temperatures ranged from 46°F in January to 80°F in July. Monthly rainfall and estimates of overland flow are presented in Table 4-11. Overland flow averaged only 14.5% of precipitation. Larger basins in the region usually yield about 20% of precipitation as runoff (Table 4-10). Monthly estimates for overland flow provided above are for a given year, and considerable variation in amounts occurs among years. However, the pattern exhibited appreciable overland flow during late fall, winter, and early spring with

TABLE 4-11   Estimates of Precipitation and Runoff (overland flow) from a 1,560-Acre Hydrologic Unit near Auburn, Alabama

Month	Rain (in.)	Overland Flow	
		in.	% rainfall
January	5.12	1.22	23.8
February	4.17	1.18	28.3
March	9.96	2.87	28.8
April	6.34	0.08	1.3
May	2.20	0.0	0.0
June	3.86	0.0	0.0
July	6.73	0.04	0.6
August	4.41	0.39	8.8
September	1.85	0.0	0.0
October	1.73	0.0	0.0
November	6.38	1.34	21.0
December	8.58	1.88	20.6
Total	61.33	8.89	14.5

Source: Boyd and Shelton (1984).

little during the rest of the year is typical for most sites in the southeastern United States.

4.5 STREAMS

Runoff flows over watersheds, collects in streams, and flows downslope toward the ocean. The energy of the flowing water erodes the land surface, and material cut from the watershed or the stream bed is transported and deposited in downstream reaches. Suspended soil particles in runoff often may be troublesome in ponds, where it produces turbidity and deposits sediment.

4.5.1 Stream Classification

A perennial stream normally flows year-round and has a well-defined channel. An intermittent stream flows only during wet seasons and after heavy rains and has a well-defined channel. An ephemeral stream only flows after rains and is often called a gully.

Area streams form a drainage pattern that can be seen on a topographic map. A tree-shaped pattern, called a *dendritic* pattern, is formed where the land erodes in fairly uniform manner so that streams randomly branch and advance upslope (Fig. 4-3). Where faulting is prevalent, streams tend to follow the faults. This gives rise to a *rectangular* pattern of streams (Fig. 4-3). Where the land surface is folded or is a broad, gently sloping plane, streams form a *trellis* pattern (Fig. 4-3).

A stream may be classified according to the number of tributaries it has. A stream with no tributaries is a first-order stream, and its basin (watershed) is a first-order basin. Two first-order streams combine to form a second-order stream. A stream that has a second-order branch is a third-order stream, and so on. This system of classification is illustrated in Figure 4-4; the stream system that contributes to the discharge at a specified point in a higher-order stream is called a drainage network. The number of streams decreases and stream length increases as stream order increases. Data presented by Leopold (1974) on the number and average lengths of streams of different orders in the basin of Seneca Creek above Dawsonville, Maryland, illustrate the rule (Table 4-12). The number of streams per unit area and the average length for a stream order differ from place to place. For Seneca Creek, the drainage basin is 100 square miles. There is one first-order stream per 387 acres. In the Appalachian Mountains, first-order perennial streams have watersheds of 40–60 acres, whereas on the coastal plain of the southeastern United States, first-order watersheds are often 750–1,000 acres (Hewlett 1982).

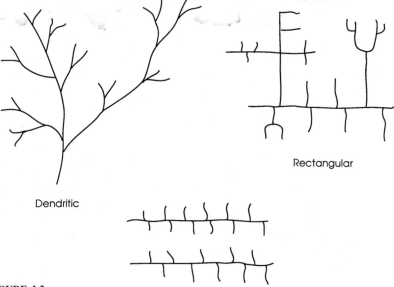

Dendritic

Rectangular

Trellis

FIGURE 4-3.
Three basic stream patterns.

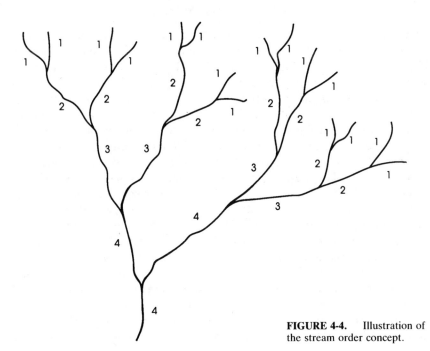

FIGURE 4-4.   Illustration of the stream order concept.

TABLE 4-12   Relationships Among Stream Order, Number of Streams in Creek Basin, and Average Stream Length

Order	Number	Average Length (mile)
1	165	0.73
2	36	0.83
3	8	2.74
4	2	7.96

Streams also are classified as young, mature, or old. Young streams flow rapidly and continually cut their channels. Their sediment loads are transported with essentially no deposition. In mature streams, slopes have been reduced and there is no downcutting of channels. Flows are adequate to transport most of the incoming sediment load. Old streams have gentle slopes and sluggish flows. They have broad floodplains, and their channels meander. Sediment deposition often leads to delta formation. Many streams may be classified as young near their sources, mature along middle reaches, and old near their mouths.

4.5.2 The Hydrograph

The discharge of a stream is the amount of water flowing through a cross section in a given time. The hydrograph is simply a plot of stream discharge versus time. The hydrograph in Figure 4-5 is typical; it shows the discharge of a small stream before, during, and after a storm. During dry weather, the hydrograph represents only base flow. During the initial phase of a rainstorm, only rain falling directly into the channel (channel precipitation) contributes to the hydrograph. Channel precipitation is seldom great enough to cause an appreciable increase in discharge. Once the rate of precipitation exceeds infiltration and depression and detention storage on the land surface are full, overland flow begins. For a fairly large rain, discharge rises sharply, the plot of which is called the *rising limb* of the hydrograph. Finally, peak discharge is reached; this is the crest segment of the hydrograph. Discharge then declines, and the plot of the declining discharge is termed the *falling limb* of the hydrograph. Subsurface storm drainage often is blocked from entering streams by high water levels until the overland flow component has passed downstream. Subsurface storm drainage cannot be separated from base flow on the hydrograph. After a heavy rain, the water table usually rises because of infiltration, and base flow increases. Because of subsurface storm drainage and greater base flow, some time may pass before discharge

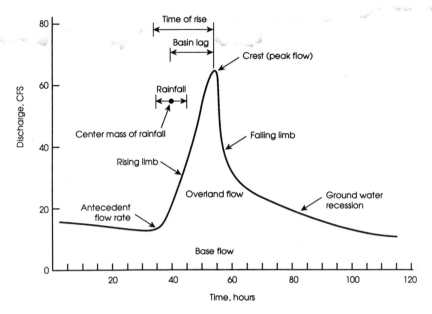

FIGURE 4-5.   A stream hydrograph.

declines to the prestorm rate. The plot of discharge during this time is known as the *ground-water recession* segment of the hydrograph.

Hydrograph shapes vary with storm characteristics; consider hydrographs for a small basin (Fig. 4-6). Hydrograph A is for a brief, intense storm. It rises, peaks, and falls quickly. For a long steady rain, hydrograph B results. The peak is reached more slowly and is maintained longer than that in hydrograph A. Hydrograph C is for a series of intense rains. Each rainfall event causes a peak, but before the stream discharge decreases to the prestorm level the next storm produces another peak. Several successive days of light rain will increase stream flow, yet there may not be a well-defined peak, as shown in hydrograph D.

Watershed characteristics also influence hydrograph shape. Watershed features that favor high rates of overland flow result in steep, triangular-shaped hydrographs. Streams from such watersheds are often said to be "flashy." Basins with permeable soils, good vegetative cover, and appreciable storage capacity tend to have less steep hydrographs that are often trapezoidal.

To estimate overland flow from data on stream discharge, we must separate out base flow. The simplest technique is to project a horizontal line (*AB*) from the point where overland flow begins to an intersection

Discharge in a 1st order stream

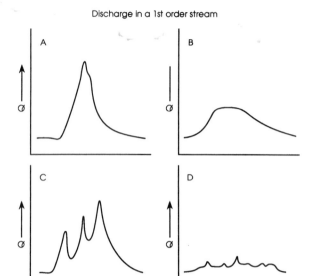

FIGURE 4-6.   Effects of storm characteristics on the hydrograph of a first-order stream. (A) Brief, intense storm. (B) Long, steady rain. (C) Intense, close-spaced rainfall events. (D) Several successive days of light rain.

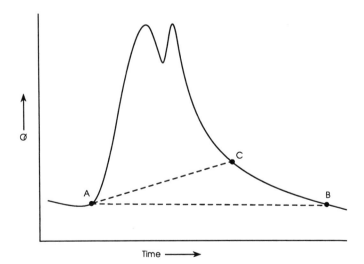

FIGURE 4-7.   Illustration of two techniques for separation of hydrograph into base flow and direct runoff. See text for explanation of how to draw either line AC or line AB.

with hydrograph recession (Fig. 4-7). Another method is to project a line (*AC*) from the point where overland flow begins to a point selected by judgment to represent the cessation of overland flow (Fig. 4-7). The second procedure is better because base flow usually is greater after the overland flow has passed downstream than before the storm. There are more complicated means of separating base flow, but all are arbitrary. Once the base flow section of the hydrograph has been separated by drawing a line, volumes of overland flow and base flow may be estimated by planimetry, by counting squares on graph paper, or by some other method of measuring the areas of the hydrograph that represent base flow and overland flow. Of course, the area must be equated to volume.

Hydrographs may be drawn to show discharge over a long time; Figure 4-8 shows a hydrograph of a first-order stream draining a 62-acre pasture watershed near Auburn, Alabama. Rainfall, measured at the center of the watershed also is presented in Figure 4-8. Notice that only the largest rains caused appreciable increase in discharge.

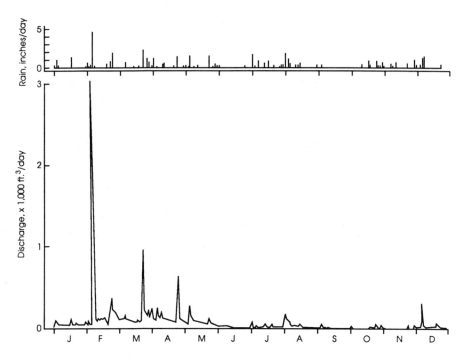

FIGURE 4-8.   Rainfall and stream discharge for a first-order stream near Auburn, Alabama, over a one-year period.

4.5.3 Erosion and Fluvial Processes

Three processes are involved in erosion: detachment, transportation, and deposition of soil particles. Energy for erosion is the kinetic energy of water, which may be expressed as

$$KE = M \frac{v^2}{2g} \tag{4-6}$$

where KE = kinetic energy
   M = mass
   v = velocity
   g = gravitational acceleration

As shown in equation (4-6), doubling velocity increases kinetic energy fourfold.

Raindrops are important in dislodging soil particles. The terminal velocities of different size raindrops (Fairbridge 1967) are shown in Table 4-13. The kinetic energy of a 0.16-in.-diameter drop is about 400 times as great as that of a 0.04-in.-diameter drop. Hence, the ability of raindrops to dislodge soil particles increases with drop size. Large drops are characteristic of high intensity precipitation events.

Soil particles detached by falling raindrops are moved along the land surface or channel bed and are carried in suspension by moving water. The ability of water to move or carry in suspension solid particles depends on its kinetic energy. The velocity of water flowing over the land or in streams usually is much less than the velocity of falling raindrops. Flowing-water velocity increases as the square root of the land slope and as the square of the water depth. Increasing slope has less effect on

TABLE 4-13   Relationship Between Raindrop Diameter and Terminal Velocities of Falling Drops

Drop Diameter (in.)	Terminal Velocity (ft/sec)
0.01	0.9
0.02	6.8
0.04	13.2
0.08	21.2
0.12	26.4
0.16	29.0
0.20	29.8

velocity than does increasing depth. Intense rains cause more erosion than light rains because they consist of more and larger raindrops, and they produce a greater depth of flow over the land surface and in rills and gullies. Thus, there is more kinetic energy to dislodge and transport soil particles. In fact, one or two of the most intense rains during a season will cause more erosion than all other rains combined. When land slope decreases and water moves more slowly, solid particles can no longer be moved along the land surface or streambed by flowing water. Finally, water velocity and turbulence are insufficient in gently sloping areas to maintain soil particles in suspension, and deposition occurs. You have seen where water has cut gullies in a denuded slope and left gravel deposits, sand, and silt at the base of the slope. Where the water continues to a stream, you also have noticed that overland flow entering the stream is turbid with suspended clay particles. Because of their small size, clay particles remain suspended at a lower turbulence than do larger particles.

In addition to the energy of rainfall and runoff, other factors influencing erosion are soil moisture content, soil chemical and physical prop-

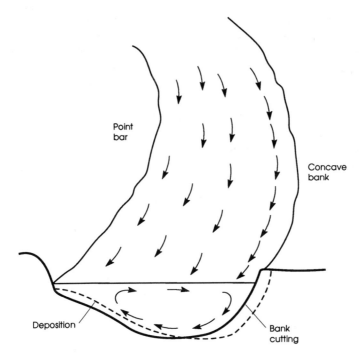

FIGURE 4-9.   Bank erosion and sedimentation in the bend of a stream.

erties, vegetative cover, slope, and length of flow. These same factors influence overland flow. Erosion is resisted by soils that are highly permeable, stable to dispersion and movement, or covered with vegetation.

In a stream, particles remain suspended as long as the slope of the channel is adequate to maintain the velocity necessary for suspension. However, sooner or later, depending on particle size, deposition occurs in the streambed. Most perennial streams flow in a channel bordered on both sides by a floodplain formed by fluvial processes. Deposition occurs in the streambed when overland flow crosses the floodplain, and when the stream overflows onto the floodplain.

A stream is seldom straight for any appreciable distance, and the course and shape of the channel change gradually over time. When water flows through a bend in the channel, the faster-moving surface water travels along the concave bank and the slower-moving water travels along the convex bank (Fig. 4-9). The faster-moving water cuts the concave bank. Further, some of the water moving along the concave bank flows downward and laterally across the channel bottom to emerge along the convex bank; this process is especially prevalent in narrow channels. Because water is moving downstream and laterally across the channel at the same time, water particles take a spiral or corkscrew path through bends in channels. Soil particles cut from the concave bank are deposited along the convex bank. Over time, the stream bend moves laterally across the floodplain; cutting of the concave bank causes lateral displacement of the bank. The channel tends to maintain the same width, for material cut from the concave bank is deposited along the convex bank. Water is deeper along the concave bank. Bends may occur to the right or to the left, so over the years the stream occupies most available places in its valley. The sideways movement of the stream along with sediment deposition in the valley by overland flow and flooding leaves a flat deposit—the floodplain.

5

Water Requirements for Aquacultural Ponds

5.1 INTRODUCTION

The two most common types of aquaculture ponds are levee ponds, constructed by building levees around areas where water is to be impounded, and watershed ponds, made by erecting dams across watersheds. Levee ponds are filled and maintained by water from wells, streams, or storage reservoirs. Watershed ponds usually are supplied by runoff; some depend entirely on overland flow, and others also receive inflow of intermittent or permanent streams. Pond design and construction is discussed in Chapter 13.

The quantity of water necessary for ponds depends on pond volume, soil characteristics, pond management, and climatic conditions. In this chapter, water budgets are developed for representative ponds, and factors influencing water supply requirements are discussed.

5.2 BASIC WATER BUDGET EQUATIONS

The water budget for a pond is prepared with the aid of the hydrologic equation, which states that inflows equal outflows plus or minus change in storage. Watersheds of levee ponds consist of inside slopes of levees; they are small compared with pond area, and runoff is a minor factor. The water budget for a levee pond is

$$P + I + RO = (S + E) \pm \Delta V \qquad (5\text{-}1)$$

where P = precipitation
   I = inflow from well, stream, or reservoir

RO = runoff
S = seepage
E = pond evaporation
ΔV = change in storage volume

Variables are expressed in volume or depth, usually depth.

When the water level is kept far enough below the overflow structure to conserve precipitation falling directly into a pond, inflow necessary to sustain the water level is

$$I = (S + E) - P - RO \qquad (5\text{-}2)$$

or

$$I = S - PX - RO \qquad (5\text{-}3)$$

where $PX = P - E$ or the precipitation excess. If runoff can be ignored, such as in levee ponds, the equation becomes

$$I = S - PX \qquad (5\text{-}4)$$

The sign for PX is negative when pond evaporation exceeds precipitation. No water needs to be applied when I is negative.

The water budget equation for a watershed pond is

$$P + RO = (S + E + OF) \pm \Delta V \qquad (5\text{-}5)$$

where RO = runoff and OF = spillway overflow. Water level fluctuates with amount of precipitation and runoff, and, depending on the pond, spillway discharge occurs continuously or only during wet months. Seepage into and out of ponds occurs simultaneously, but net seepage usually is outward. Seepage beneath and through dams normally is more important than seepage through pond bottoms.

5.3 MEASUREMENTS OF VARIABLES

Techniques for measuring precipitation and pond evaporation are covered in Chapters 2 and 3, respectively. Chapter 4 provides means of estimating runoff from ungauged watersheds. Procedures for measuring inflows and outflows of water in pipes, canals, or natural streams are discussed in Chapter 10. Only techniques for estimating pond volume and seepage are given here.

5.3.1   Pond Volume

To estimate the volume of a pond requires data on the pond bottom shape and area for different water depths. The relationship between stage and pond volume is constant for ponds whose bottoms are not rapidly changing because of erosion or sedimentation. The *stage* is simply the water level measured with respect to a reference elevation, and is often called *gauge height*. It can be measured with a staff gauge (Fig. 5-1) or recorded continuously with a water-level recorder, as discussed in Chapter 10. To use stage to estimate pond volume, we must tabulate stage-volume data. This is accomplished as follows.

Map the shoreline of a pond by traditional surveying techniques. Draw an outline map of the pond to scale and identify positions of survey stakes on the map. The survey stakes permit establishment of a grid over the pond surface (Fig. 5-2). Place range poles at opposite ends of the pond and traverse the length of the transect between the two range poles. Workers on each side of the pond must align themselves so that transects across the pond can be established. These workers ascertain when the boat reaches an intersection of two transects. Workers in the boat must make a sounding at each intersection with a sounding rod or line calibrated at 0.1-ft intervals. Insert traverse lines on the map of the pond and plot the depth at each intersection. Draw contours at 1-ft intervals. A finished contour map of a pond bottom is shown in Figure 5-3.

Determine the surface area of the pond and the area circumscribed by each submerged contour. Estimate the volume of the upper 1-ft stratum bounded by the surface and the plane circumscribed by the 1-ft contour from the formula

$$V = \frac{h}{3} (A_u + A_l + \sqrt{A_u + A_l}) \tag{5-6}$$

where V = volume between the upper (shoreline) and lower
            (1-ft) contour lines (acre-feet)
      A_u = area within upper contour (acres)
      A_l = area within lower contour (acres)
      h = vertical distance in ft between contours (1 ft in this case)

Continue to compute the volumes of succeeding contours in the same manner. For the stratum between 1 and 2 ft, A_u is the area within the 1-ft contour and A_l is the area within the 2-ft contour.

This procedure was used to calculate volumes for different depths for a pond with a maximum depth of 13 ft and a maximum surface area of 1.4

FIGURE 5-1.   A staff gauge.

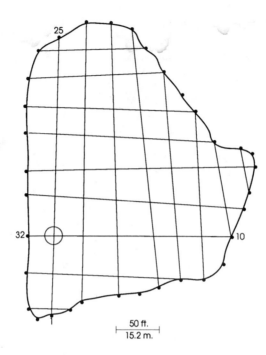

FIGURE 5-2.   Illustration of sounding for mapping ponds. Positions of stakes (dots) were established by transit readings and the outline drawn. Range poles were placed at stakes 1 and 25, and boat moved across pond while aligned between poles. Workers on shore aligned boat between stakes 10 and 32. Sounding was made at intersection in circle. Procedure was repeated until soundings were made at all intersections in grid. (*Source: Boyd and Shelton 1984.*)

acres (Fig. 5-3). Results are shown in Table 5-1. The deepest point in the pond was 13 ft. If a 13-ft-long gauge was positioned vertically at the deepest point, the water level would be at 13 ft on the gauge when the pond was full. At this gauge height or stage, the pond would contain 7.6 acre-ft of water. If the water level dropped until the stage was 8 ft, the volume would be 2.1 acre-ft. The relationship between stage and volume may be plotted on a graph to provide a rating curve if desired.

5.3.2 Seepage

Seepage from ponds is very difficult to measure accurately. The simplest approach for estimating seepage is to select a period during dry weather when no inflows are expected and when the only outflows are evaporation and seepage. The water-level decrease is seepage plus evaporation. Therefore, an estimate of pond evaporation subtracted from the measured

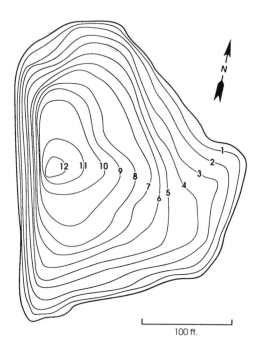

FIGURE 5-3.   Contour map of pond bottom. Contours are at 1-ft intervals.

TABLE 5-1   Water Surface Area and Volume of a Pond at Different Stage Heights

Contour (ft)	Stage (ft)	Water Area (acres)	Volume between Contours (acre-ft)	Volume at Stage (acre-ft)
0	13	1.4		7.6
			1.4	
1	12	1.3		6.2
			1.2	
2	11	1.2		5.0
			1.1	
3	10	1.0		3.9
			1.0	
4	9	0.9		2.9
			0.8	
5	8	0.7		2.1
			0.6	
6	7	0.6		1.5
			0.5	
7	6	0.5		1.0
			0.4	
8	5	0.3		0.6
			0.3	
9	4	0.2		0.3
			0.2	
10	3	0.1		0.1
			0.07	
11	2	0.04		0.03
			0.007	
12	1	0.01		0.003
			0.003	
13	0	0		0

stage change provides a value for seepage. Its accuracy depends primarily on the reliability of the pond evaporation estimate. For example, suppose the water level in a pond decreases by 1.12 in. over a 24-hr period when there was no inflow and outflow consisted of evaporation and seepage only. Further, suppose that Class A pan evaporation at the location was 0.30 in. and assume a pan coefficient of 0.81. Pond evaporation is 0.30 in. × 0.81 = 0.24 in. The total water loss is 1.12 in., so the seepage loss is 1.12 − 0.24 = 0.88 in.

A variation of this procedure employs the mass-transfer technique for handling the evaporation term (Eisenlohr 1966). Stage change is determined for one-day periods when there are no inflows or outflows other than seepage and evaporation. Wind speed and vapor pressure of air are measured, and daily averages are computed. Stage change (Y variable) and the product of wind speed and vapor pressure deficit (X variable) are plotted. The point where the regression line intercepts the Y axis represents water loss when there is no evaporation. Of course, this is seepage.

Seepage may be estimated as the water loss from a capped pipe inserted in the pond bottom. A modification of the capped-pipe technique is the seepage meter (Rohwer and Stout 1948). A seepage meter consists of a large inverted cup forced into the sediment. A tube is connected to the cup, and water is supplied to it to replace the water seeping out. The water input is measured, and this volume is equal to the seepage through the area under the cup.

Stone and Boyd (1989) compared different techniques for measuring seepage from ponds. The mass-transfer modification of the water budget technique takes too long for a measurement. Capped-pipe and seepage-meter procedures yielded highly variable estimates for different locations in the same pond, and average seepage values for a pond differed greatly from day to day. The water budget approach employing a pan coefficient to correct stage change for evaporation provided the most reliable seepage data.

A few records of pond seepage are available. Parsons (1949) determined seepage rates for a pond on coastal plain soils near Auburn, Alabama; the average rate of water loss to seepage averaged 0.15 in./day or 55 in./yr. Parsons also made seepage measurements during the summer for 12 ponds on soils of the Piedmont Plateau. Values ranged from 0.1 to 1.32 in./day with an average of 0.43 in./day. Seepage rates for a number of ponds on heavier clay soils of the Alabama Black Belt Prairie ranged from 0.02 to 0.18 in./day (average 0.06 in./day). Parsons noted that ponds with the greatest seepage rates always were located well upslope on watersheds. Most seepage apparently occurred under dams of ponds.

Manson et al. (1968) and Allred et al. (1971) obtained seepage estimates for small, natural bodies of water in Minnesota. Seepage out of ponds average 0.04 in./day; the maximum rate was 0.06 in./day.

The average seepage for 51 small (0.1–0.25 acre) embankment ponds without watersheds on the Fisheries Research Unit at Auburn University was 0.20 in./day, with a range of 0.10 to 0.49 in./day (Boyd 1982). Larger ponds (1–25 acres), constructed by building dams across watersheds, had an average seepage rate of 0.35 in./day (Boyd and Shelton 1984).

Stone and Boyd (1989) found that seepage was slightly greater during warm months than during cool months. This occurs because water viscosity decreases with increasing temperature, and seepage is favored by low viscosity.

Published information on pond seepage is limited to studies made in a few geologic regions. However, the results confirm that seepage varies greatly among ponds, and that some ponds may lose considerable water to seepage. Boyd (1986) classified pond seepage rates; a modified version of the classification is

Low   0–0.19 in./day
Moderate   0.20–0.39 in./day
High   0.40–0.59 in./day
Extreme   >0.60 in./day

Properly constructed ponds often have seepage rates below 0.1 in./day, and few seep more than 0.25 in./day.

5.3.3 Comments

Where one or two variables cannot be measured, they are estimated by solving the hydrologic equation. Seepage plus evaporation for a pond not completely full is

$$S + E = (WL_B + P + RO + I) - WL_E \tag{5-7}$$

where WL_B = water-level elevation at the beginning of the measurement period and WL_E = water-level elevation at the end of the measurement period. The equation (5-7) can be greatly simplified if the measurements are made for periods without inflow or outflow as mentioned earlier.

When it rains, a pond may overflow during the measurement period. If a weir is in place, overflow can be measured directly. Otherwise, overflow must be calculated as follows:

$$(P + RO + I) - \text{storage capacity} = \text{water loss} \tag{5-8}$$

TABLE 5-2   Differences Between Measured Gains and Losses of Water in Water Budgets for Four Ponds

Pond	Gains (in.)	Losses (in.)	Difference (in.)
67	55.4	57.3	1.9
70	119.5	118.1	1.4
72	136.1	133.8	2.3
73	88.8	87.8	1.0

Storage capacity in equation (5-8) is the distance (or volume) between the water level and the drain at the beginning of the measurement period. The loss term comprises evaporation, seepage, and overflow. Evaporation can be estimated from pan evaporation data, but seepage must be assumed equal to the seepage for a preceding period of one or more days without overflow:

$$OF = \text{water loss} - (S + E) \tag{5-9}$$

Normally, variables for water budgets are measured daily, but it is more convenient to summarize data on a monthly basis. When done carefully, water budgets are highly reliable. Boyd (1982a) prepared water budgets for four ponds over a seven-month period; gains and losses of water agreed closely (Table 5-2). Sometimes, it is desirable to make hypothetical water budgets for planning or design purposes, and values for hydrologic variables are taken from the literature rather than being measured.

5.4 EXAMPLES OF MEASURED WATER BUDGETS

Boyd (1982a) prepared water budgets for April 1 through October 31, 1981, for four, 0.1-acre experimental ponds at Auburn, Alabama. These were levee ponds, but, because of their small sizes, the area of the inside slopes of the levees was 50% of the water-surface area. Runoff was estimated for each rainfall event by the SCS curve number technique. Total inflows in inches of water were rainfall, 23.7; inflow from pipe, 30.7–111.4; and runoff, 0.9. Most rainfall was stored because water levels were kept below drain pipes. Outflows of water in inches over this period were pond evaporation, 30.3; seepage 23.1–102.4; overflow, 0.9–4.1. Seepage, the major variable among ponds, was the most important factor governing inflow necessary to maintain water levels. Water budgets for the four ponds are summarized by month in Table 5-3.

TABLE 5-3   Water Budgets for Four 0.1-Acre Ponds on the Auburn University Fisheries Research Unit, Auburn, Alabama (all values are in inches)

Variable and Pond	April	May	June	July	Aug	Sept	Oct	Total
Rainfall, all ponds	1.5	5.6	2.0	3.9	4.8	3.3	2.6	23.7
Runoff, all ponds	0	0.5	0	0	0.3	0.1	0	0.9
Water additions								
67	4.8	5.4	6.9	5.4	3.5	2.4	2.2	30.6
70	8.9	12.8	14.8	18.8	11.1	16.1	12.3	94.8
72	10.1	16.1	20.0	19.9	10.3	19.1	16.6	112.1
73	6.8	10.9	11.8	9.3	10.3	7.0	8.0	64.1
Evaporation, all ponds	4.3	4.5	5.6	5.9	3.5	3.8	2.7	30.3
Seepage loss								
67	4.2	3.0	3.1	3.1	3.9	3.0	3.6	23.9
70	8.6	10.3	10.7	12.8	13.7	13.8	14.9	84.8
72	14.3	12.8	14.7	15.4	15.6	14.6	15.1	102.5
73	7.0	7.8	8.4	7.5	8.1	7.4	7.2	53.4
Overflow								
67	0	2.4	0	0	1.5	0	0	3.9
70	0	2.0	0	0	1.1	0	0	3.1
72	0	1.0	0	0	0	0	0	1.0
73	0	1.8	0	0	2.1	0.2	0	4.1

Source: Boyd (1982a).

Boyd and Shelton (1984) measured the water budget for the 26.8-acre complex of small research ponds on the Fisheries Research Unit at Auburn University. Results are summarized in Table 5-4. The entire amount of water required from storage to keep ponds full was 95.6 acre-ft. Ponds usually are drained for fish harvest once each year and immediately refilled; this operation required an additional 80.4 acre-ft. Water necessary from storage to operate the ponds was 175 acre-ft. More water would be used if ponds were permitted to overflow or if ponds were drained and refilled more than once a year. The water supply for the small ponds is a 20.1-acre (140.6 acre-ft) reservoir. Small ponds normally are drained between October and December and refilled in late autumn or early winter. During winter, the reservoir usually is full or overflowing. Overflow normally ceases by May 1, but small ponds are full at this time. Between May 1 and October 31, an average of 77.4 acre-ft of water from the reservoir is comsumed in maintaining small ponds. Seepage and evaporation from the reservoir averages about 40.6 acre-ft between May and October. Natural losses and consumptive use account for about 118 acre-ft or 83.9% of reservoir volume. A small stream flows into the

TABLE 5-4   Water Budget for a 26.8-Acre Complex of Small Ponds on the Auburn University Fisheries Research Unit (values are in acre-feet of water)

Month	Rainfall	Runoff	Water Added	Evaporation	Seepage	Overflow
January	11.4	1.2	3.8	2.5	13.8	0.0
February	9.3	0.3	6.7	3.9	12.5	0.0
March	22.2	1.9	0.0	5.8	13.8	4.5
April	14.2	1.8	6.0	8.6	13.4	0.0
May	4.9	1.5	17.2	9.7	13.8	0.0
June	8.6	0.0	13.9	9.1	13.4	0.0
July	15.0	0.3	10.1	11.5	13.8	0.0
August	9.8	3.0	10.5	9.5	13.8	0.0
September	4.1	0.0	9.9	7.6	13.4	0.0
October	3.8	0.0	15.8	5.8	13.8	0.0
November	14.2	1.6	1.7	4.2	13.4	0.0
December	19.2	14.6	0.0	3.1	13.8	5.2
Total	136.7	26.2	95.6	81.3	162.7	9.7

Source: Boyd and Shelton (1984).

reservoir; still, the level of the reservoir typically declines during late summer and early autumn.

Parsons (1949) made a water budget for a 1.4-acre watershed pond near Auburn, Alabama. This pond had an average depth of 4.92 ft; it received overland flow from a 26.9-acre watershed. Ground water from a spring flowed into the pond during winter, spring, and early summer. Seepage out of the pond occurred through the dam; seepage water was collected by a tile field below the dam and measured. Results of this study, which extended over four years, are summarized in Table 5-5. Spillway discharge was 5.5 times pond volume; seepage was 1.6 times pond volume; evaporation was 0.8 times pond volume. Rainfall plus overland flow exceeded pond volume by a factor of 2.6, and seepage plus evaporation was 2.4 times pond volume. Spillway discharge resulted primarily from inflow of the spring. The pond had a greater seepage rate than normal for watershed ponds. Many watershed ponds depend more on overland flow than this pond does.

Shelton and Boyd (1993) measured water budgets for four watershed ponds near Auburn, Alabama, over a two-year period. On an annual basis, direct precipitation into three ponds was less than the loss of water in evaporation and seepage, and overflow from ponds averaged 60–90% of runoff entering ponds. During the drier part of the year (June–November), there was little overflow and water levels declined. Data for

TABLE 5-5   Average Water Budget over Four years for a Pond at Auburn, Alabama (values are in inches)

Month	Gains			Losses		
	Rainfall	Overland Flow	Spring Discharge	Evaporation	Seepage under Dam	Spillway Discharge
January	5.83	21.2	55.3	1.36	6.7	68.7
February	5.53	20.7	49.9	1.87	5.9	69.3
March	7.19	8.5	51.2	2.74	6.6	59.3
April	4.81	11.9	33.3	4.10	6.6	44.6
May	5.15	13.7	29.3	5.93	7.0	38.7
June	5.83	6.8	12.0	6.53	7.2	15.3
July	6.65	3.2	6.7	5.86	7.7	8.0
August	2.63	0.4	2.0	5.86	8.5	11.3
September	3.81	0.2	0.0	4.74	10.5	7.3
October	1.70	0.2	0.0	3.71	11.7	0.0
November	5.05	0.2	0.0	2.53	10.0	0.0
December	4.03	7.0	20.6	1.30	7.1	0.0
Total	58.21	94.0	260.3	46.53	95.5	322.5
		(Gains = 412.51)			(Losses = 464.53)	

Source: Parsons (1949).

one of these ponds is summarized in Table 5-6. From December to May, water flowed from ponds following periods of precipitation. The other pond (Table 5-7) received continuous inflow from a spring that exceeded the volume of evaporation and seepage out. On an annual basis, this pond had about 1.4 times as much overflow as entering runoff.

5.5   ESTIMATING WATER REQUIREMENTS

Water budgets will now be calculated for levee and watershed ponds. Note that several assumptions were made in solving these examples. Also, judgments had to be made regarding the magnitudes of some variables. Solutions of this type are not perfect, but they provide reasonable estimates of water budgets. It must be emphasized that, wherever possible, data from the site should be used. If data must be taken from a remote site, large discrepancies might occur. In practice, problems quite different from these examples will be encountered. However, if the reader studies the principles, techniques, and reasoning used in these examples, he or she can use a little ingenuity and modify the approach to solve other problems.

TABLE 5-6 Water Budget for Pond AE-2 on the Fisheries Research Unit of the Alabama Agricultural Experiment Station near Auburn, Alabama (all entries are in ten thousands of cubic feet of water)

| | Inputs (×10,000 ft³) | | | | Outputs (×10,000 ft³) | | | | | |
| | Rainfall | | Runoff | | Evaporation | | Seepage | | Overflow | |
Month	82–83	83–84	82–83	83–84	82–83	83–84	82–83	83–84	82–83	83–84
May	2.9	2.2	22.8	26.1	6.5	6.9	8.0	8.0	15.9	16.7
June	7.6	5.1	12.3	7.2	7.2	5.4	7.6	7.6	6.5	0.0
July	7.2	6.9	6.2	2.9	8.0	8.0	8.0	8.0	0.0	0.0
August	4.0	2.5	2.9	2.9	6.2	6.9	8.0	8.0	0.0	2.2
September	2.5	11.6	0.0	0.0	5.4	6.2	7.6	7.6	0.0	1.1
October	3.3	2.5	1.4	1.8	4.0	3.6	8.0	8.0	0.0	0.0
November	8.7	9.4	11.9	30.4	2.5	2.9	7.6	7.6	0.0	14.8
December	12.3	9.8	80.0	72.0	2.2	3.3	8.0	8.0	55.7	69.9
January	6.5	5.1	77.8	47.8	1.8	1.8	8.0	8.0	10.1	44.2
February	6.9	4.3	46.3	50.0	2.5	2.5	7.2	7.2	44.5	48.9
March	12.3	6.5	203.1	51.8	3.6	4.7	8.0	8.0	202.0	46.3
April	11.2	5.8	112.2	39.5	5.4	5.1	7.6	7.6	111.5	33.7
Annual total	85.4	71.7	577.0	332.3	55.4	57.2	93.4	93.4	511.5	277.7

TABLE 5-7 Water Budget for Pond S-28 on the Fisheries Research Unit of the Alabama Agricultural Experiment Station near Auburn, Alabama (all entries are in ten thousands of cubic feet of water)

Month	Inputs ($\times 10,000\ ft^3$)				Outputs ($\times 10,000\ ft^3$)							
	Rainfall		Runoff		Evaporation		Seepage		Overflow			
	82–83	83–84	82–83	83–84	82–83	83–84	82–83	83–84	82–83	83–84		
May	3.3	3.6	26.8	22.1	8.7	9.8	13.0	13.0	37.3	30.4		
June	11.2	6.2	22.1	11.2	10.1	7.6	12.7	12.7	32.6	21.4		
July	11.2	7.6	48.5	6.2	11.2	10.9	13.0	13.0	57.6	13.4		
August	5.4	4.0	4.7	0.0	8.7	9.4	13.0	13.0	18.8	12.3		
September	4.7	14.5	7.2	22.8	7.6	8.3	12.7	12.7	14.1	36.9		
October	4.0	3.6	0.0	0.0	5.4	5.1	13.0	13.0	13.0	8.3		
November	13.0	13.0	12.7	34.8	3.6	4.3	12.7	12.7	27.9	52.9		
December	15.6	13.8	73.8	68.4	2.9	4.3	13.0	13.0	101.4	89.4		
January	9.4	6.2	50.3	38.0	2.5	2.9	13.0	13.0	70.6	56.8		
February	9.8	6.5	40.9	16.3	3.3	3.6	11.6	11.6	56.8	48.9		
March	16.7	9.1	103.2	49.2	4.7	6.5	13.0	13.0	128.1	65.2		
April	14.5	8.3	81.8	40.9	7.6	6.9	12.7	12.7	102.8	55.4		
Annual total	118.7	96.3	472.0	328.0	76.4	79.6	153.5	153.5	661.0	491.2		

Note: Outputs exceeded inputs because of continuous spring discharge into the pond. Spring discharge could not be measured.

5.5.1 Levee Ponds

Three examples illustrate application of water budget procedures.

Example 5-1

Suppose levee ponds of 4-ft average depth are to be built where rainfall and pond evaporation patterns are known (Table 5-8). Soils are heavy clays, and seepage is not expected to exceed 0.1 in./day. Runoff from levees will be negligible. Rainfall will be stored during months when evaporation plus seepage exceeds rainfall. During other months, there will be overflow. Ponds will be drained every three or four years for repair work on levees, and enough water must be available to refill them within one month during winter. Estimate the minimum water supply per acre in gallons per minute and determine monthly water requirements for maintaining water levels.

Solution   Precipitation excess is calculated and subtracted from seepage to provide monthly estimates of required inflow (Table 5-8). Inflows in inches were converted to gallons per minute, as illustrated for May:

$$I_{May} = \frac{4.23\,\text{in.} \times 43,560\,\text{ft}^2/\text{acre} \times 7.481\,\text{gal/ft}^3}{12\,\text{in./ft} \times 31\,\text{day} \times 1,440\,\text{min/day}} = 2.57\,\text{gpm/acre}$$

TABLE 5-8   Data Used in Estimating Water Supply Requirements for Levee Ponds (see text for details)

| Month | Variables[a] (in.) | | | | | I (gpm/acre) |
	P	E	PX	S	I	
January	5.09	0.47	4.62	3.1	0.00	0.00
February	4.78	1.57	3.21	2.8	0.00	0.00
March	5.67	2.95	2.72	3.1	0.38	0.23
April	5.48	4.57	0.91	3.0	2.09	1.27
May	5.09	6.22	−1.13	3.1	4.23	2.57
June	3.72	6.50	−2.78	3.0	5.78	3.63
July	4.05	6.77	−2.72	3.1	5.82	3.54
August	2.37	6.26	−3.89	3.1	6.99	4.25
September	3.54	5.12	−1.58	3.0	4.58	2.88
October	2.46	3.74	−1.28	3.1	4.38	2.66
November	4.82	2.60	2.22	3.0	0.88	0.55
December	4.96	1.57	3.39	3.0	0.00	0.00
Total	52.03	48.34	3.69	36.4	35.13	21.58

[a] P = precipitation, E = pond evaporation, PX = precipitation excess, S = seepage, I = inflow.

A small, continuous inflow is needed to maintain water levels. Much greater inflow is necessary for filling ponds during January. The water requirement for filling ponds is pond depth plus seepage minus precipitation excess, or

$$(48\,\text{in.} + 3.1) - 4.62\,\text{in.} = 46.48\,\text{in.}$$

This depth is equal to 3.87 ft or 3.87 acre-ft because the pond area is 1 acre. The approximate required inflow is

$$\frac{3.87 \text{ acre-ft} \times 325{,}872\,\text{gal/acre-ft}}{31 \text{ days} \times 1{,}440\,\text{min/day}} = 28.25\,\text{gpm}$$

This computation ignores the side slopes of the ponds, so the calculated required inflow is slightly too large. Monthly inflows for sustaining water levels are greater for a dry year, but inflow necessary to fill ponds in one month always exceeds maintenance inflow.

Example 5-2

This example involves marine shrimp ponds. In pond culture of marine shrimp, water is flushed through ponds at about 10% of pond volume per day. A farm is to be built, and water from a tidal creek will be pumped into a supply canal from which it will flow by gravity through ponds and into a second canal for return to the creek. The farm will have 1,000 acres of ponds, and the supply canal will contain 50 acres. Ponds will be 3 ft in average depth, and canals will be 5 ft deep. Soils are heavy clays. No records of precipitation or pan evaporation are available. Determine the pump capacity for supplying water to the canal.

Solution   Because of tidal influence on water level in the creek, pumping will be limited to 16 hr/day. An equation for estimating pump capacity (inflow) may be written by modifying equation (5-1) as follows:

$$I + P + RO = S + E + OF \pm \Delta V \tag{5-10}$$

There is a distinct dry season at the place where the ponds will be built. Water supply calculations are based on this period, and $P + RO = 0$. Pond stage does not vary from day to day. Thus, the equation may be simplified to

$$I = E + S + OF \tag{5-11}$$

Pond evaporation rates in a hot dry climate may reach 0.35 in./day (Boyd 1986); this value of E will be assumed. Heavy clays do not seep appreciably, so let $S = 0.05$ in./day. Ponds are 3 ft deep, and 10% of the volume is 0.3 ft or 3.6 in./day. Seepage and evaporation losses for ponds and supply canal influence inflow requirements, but 10% exchange need be considered for ponds only. Calculations follow:

$$E = \frac{0.35 \text{ in./day}}{12 \text{ in./ft}} \times 1{,}050 \text{ acres} \times 325{,}872 \text{ gal/acre-ft} = 9{,}979{,}830 \text{ gpd}$$

$$S = \frac{0.05}{12} \times 1{,}050 \times 325{,}872 = 1{,}425{,}690 \text{ gpd}$$

$$OF = \frac{3.6}{12} \times 1{,}000 \times 325{,}872 = 97{,}761{,}600 \text{ gpd}$$

$$I = 9{,}979{,}830 + 1{,}425{,}690 + 97{,}761{,}600 = 109{,}167{,}120 \text{ gpd}$$

$$I = \frac{109{,}000{,}000 \text{ gpd}}{16 \text{ hr/day} \times 60 \text{ min/hr}} = 113{,}542 \text{ gpm}$$

Example 5-3

This example involves the water requirements for ponds in an arid region. A fish culture research station will be established at a site where the only source of water is a well that yields 500 gpm. It is desired to operate the pump on more than 14 hr/day. Pond evaporation exceeds precipitation by 60 in./yr, and seepage is expected to be 36 in./yr. There will be little seasonal variation in evaporation and seepage. Ponds will average 3 ft in depth. Throughout the year, individual ponds will be drained and refilled to accommodate research interests. Therefore, 60% of the water supply will be reserved for rapidly refilling ponds.

Solution   Permissible pond area for the research station is determined by the amount of water available for maintaining water levels divided by the quantity of water necessary per acre for this purpose. The amount of inflow necessary is the seepage plus the precipitation excess:

$$I = S - PX \tag{5-4}$$

Of course, precipitation is 60 in. (5 ft) less than evaporation, so $PX = -5$ ft/yr, and

$$I = 3 \text{ ft/yr} - (-5 \text{ ft/yr}) = 8 \text{ ft/yr}$$

Thus, 8 ft of water must be applied annually to each acre for maintenance of water levels. Of the 500-gpm water supply, 40%, or 200 gpm, may be used for sustaining ponds. With the pump operating 14 hr each day, the total volume of water for maintenance is

$$\frac{200\,\text{gpm} \times 60\,\text{min/hr} \times 14\,\text{hr/day} \times 365\,\text{days/yr}}{325{,}872\,\text{gal/acre-ft}} = 188\,\text{acre-ft/yr}$$

Permissible pond area is

$$\frac{188\,\text{acre-ft/yr}}{8\,\text{ft/yr}} = 23.5\,\text{acres}$$

Assuming ponds are 1 acre in size, the approximate filling time for a single pond using all 300 gpm assigned for this task is estimated below:

$$325{,}872\,\text{gal/acre-ft} \times 3\,\text{acre-ft/pond} = 977{,}616\,\text{gal/pond}$$

$$\frac{977{,}616\,\text{gal}}{300\,\text{gpm} \times 60\,\text{min/hr}} = 54.3\,\text{hr/pond}$$

$$\frac{54.3\,\text{hr/pond}}{14\,\text{hr pumping/day}} = 3.9\,\text{days}$$

Evaporation plus seepage is 8 ft/yr or about 0.022 ft/day. Over four days, 0.088 ft (1.06 in.) would be lost from the 3 acre-ft applied to a pond. Thus, filling time would be slightly greater than estimated here.

5.5.2 Watershed Ponds

Watershed ponds must be constructed so that they will fill from available runoff, and they must store enough water to ensure sufficient depth during dry weather. The task is to construct a 1-acre-by-6-ft average-depth watershed pond on the Piedmont Plateau near Auburn, Alabama. The watershed consists of 20 acres of woodland. The small watershed does not yield a permanent stream, but a wet-weather stream is present. An accounting of monthly inflows and outflows will be made to ascertain if runoff from the watershed will fill the pond and provide adequate water for dry months. The pond will be built in the fall, and the drain closed at the end of December.

Data on precipitation and pond evaporation at Auburn, Alabama, are provided in Tables 2-3 and 3-2, respectively. Seepage in watershed ponds

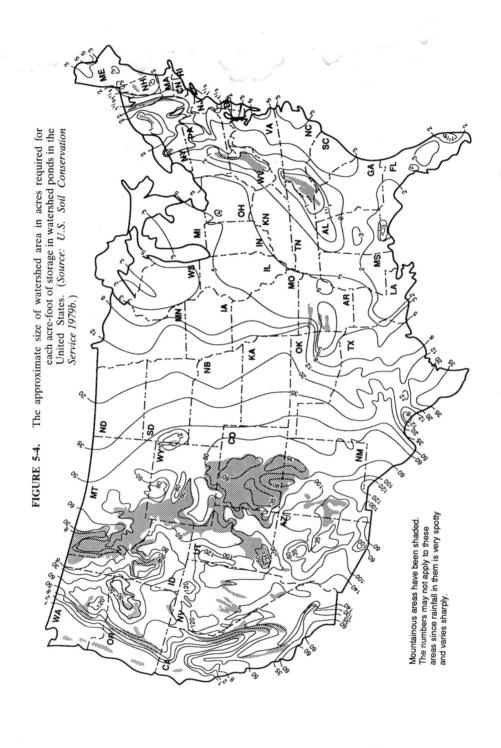

FIGURE 5-4. The approximate size of watershed area in acres required for each acre-foot of storage in watershed ponds in the United States. (*Source: U.S. Soil Conservation Service 1979b.*)

Mountainous areas have been shaded. The numbers may not apply to these areas since rainfall in them is very spotty and varies sharply.

on the Piedmont Plateau in Alabama averaged 0.13 in./day from May through November (Boyd and Shelton 1984). From Figure 5-4, it can be interpolated that for a pond in the Piedmont area of Alabama about 2 acres of watershed are usually required for each acre-foot of storage. Thus, a 1-acre pond will need a minimum watershed size of 12 acres. However, such data may not be available in all nations, so we will show how local data may be used to estimate watershed size. Average discharge of streams of the Alabama Piedmont (Table 4-10) permit an estimate of the percentage of rainfall becoming runoff. Data from Table 4-10 are for fairly large streams. Yield of runoff per acre would be somewhat less for small, upland areas because the base flow contribution to runoff would be less. Wet-weather streams from small watersheds in the Piedmont typically cease flowing in July or August, and flow does not resume until late November or early December. Therefore, runoff data (Table 4-10) for July through November were averaged. This average was subtracted from runoff values for all months in order to adjust runoff data so that they would be representative of small watersheds. Adjusted monthly runoff data were divided by normal monthly rainfall from Table 2-3 to estimate monthly runoff as a percentage of monthly rainfall. Results are listed here:

January	27.2%
February	26.2
March	36.5
April	30.1
May	14.2
June	2.6
July	10.5
August	2.8
September	0.0
October	0.0
November	8.8
December	9.1

The amount of runoff (RO) per month may be estimated in terms of pond depth from the equation

$$RO = P \times \frac{\%RO}{100} \times \frac{\text{watershed area}}{\text{pond area}} \qquad (5\text{-}12)$$

To illustrate for January, use normal January precipitation from Table 2-3 and the percentage runoff from the preceding list to get

$$RO = 5.14 \times \frac{27.2}{100} \times \frac{20}{1} = 27.96 \text{ in.}$$

Runoff in inches for other months follow: February, 27.98; March, 50.00; April, 31.97; May, 12.01; June, 2.00; July, 12.05; August, 2.01; September, 0.00; October, 0.00; November, 6.02; December, 9.97.

The water level at the end of each month can be estimated by the following modification of the basic hydrologic equation (inflow = outflow ± Δ storage) to fit existing conditions:

$$WL_E = WL_B + RO + PX - S \qquad (5\text{-}13)$$

where WL_E = water level at the end of the month and WL_B = water level at the beginning of the month.

Water will flow from the spillway when depth exceeds the maximum storage depth of 6 ft (72 in.). Therefore, when the water level is calculated to exceed the maximum storage depth, overflow (OF) can be calculated as

$$OF = WL_E - SD_{max} \qquad (5\text{-}14)$$

where SD_{max} = maximum storage depth in inches.

The equation for estimating water level at the end of a month was solved beginning with January, for which WL_B was assigned the value 0.00. Seepage was assumed to be 0.13 in./day (3.9 in. for a 30-day month), and PX was estimated by subtracting pond evaporation (Table 3-2) from precipitation (Table 2-3). The water level at the end of January would be

$$WL_E = 0.00 + 27.96 + 4.19 - 4.03 = 28.12 \text{ in.}$$

At the end of January, the pond would contain 28.12 in. of water; no overflow would occur during the month. Water-level and overflow data for a 12-month period follow:

	WL_E (in.)	OF (in.)
January	28.12	0.00
February	56.35	0.00
March	72.00	34.22
April	72.00	29.13
May	72.00	7.09
June	67.97	0.00
July	72.00	3.86
August	67.98	0.00
September	63.52	0.00
October	58.58	0.00
November	61.92	0.00
December	72.00	0.08

Remember, the pond was empty on January 1; had it been full, there would have been 28.12 in. of overflow in January, 28.33 in. in February, and 48.87 in. in March. Estimates for other months would be unchanged. Because of seasonal variation in precipitation, evaporation, and runoff, most ponds supplied by runoff have stable water levels only during wet months.

The reader should notice how the various equations for estimating different hydrologic variables were prepared by substituting the necessary inflow and outflow terms into the basic hydrologic equation, inflow = outflow ± Δ storage. The techniques for estimating seepage and runoff were educated guesses based on average situations. Seepage and runoff estimates may be increased or decreased depending on the site in question. In an area with sandy soils, seepage likely would be greater than 0.13 in./day used above, and the amount of runoff would be less. One might double the estimate for seepage and half the percentage runoff for a sandy watershed.

The number of acres of watershed necessary per acre of pond on the Alabama Piedmont will now be calculated. Different watershed sizes will be assigned to a 1-acre pond of 6-ft average depth. Seepage, evaporation, precipitation, and runoff data from the preceding example permit monthly water levels to be computed for different watershed sizes. Estimates in inches of depth are shown in Table 5-9. An 11-acre watershed would just barely fill the pond during a year with normal rainfall and runoff. A 15-acre watershed would be better, for it would supply more

TABLE 5-9   Water Levels at the End of Each Month for Ponds with Different Watershed Areas

	1 acre	5 acres	10 acres	11 acres
January	1.56	7.16	14.16	15.56
February	3.20	14.40	28.40	31.20
March	5.57	21.77	53.27	58.57
April	4.33	26.93	66.43	72.00
May	0.00	24.74	67.24	72.00
June	0.00	19.21	62.21	67.07
July	0.00	18.05	64.05	69.51
August	0.00	12.52	59.02	64.58
September	0.00	8.06	54.56	60.12
October	0.00	3.12	49.62	55.18
November	0.00	1.94	49.94	55.80
December	0.69	4.63	55.13	61.49

Note: Based on climatic, topographic, and edaphic conditions in the vicinity of Auburn, Alabama.

runoff on dry years. Notice that the answer obtained by this method is almost the same as that interpolated from Figure 5-4.

5.6 CLIMATIC FACTORS

The precipitation excess, precipitation minus pond evaporation, is an important factor governing the quantities of water it takes to stabilize water levels in levee ponds and the depths of water that must be stored in watershed ponds during rainy months to sustain satisfactory water levels during dry months. Boyd (1986) used equation (5-4) to estimate amounts of water required per year to sustain water levels in levee ponds at 43 locations in states where channel catfish are reared on farms. He obtained monthly Class A pan evaporation data from Farnsworth and Thompson (1982), and pond evaporation was taken as pan evaporation times 0.81. Precipitation data from Wallis (1977) were used. Seepage was assumed to be 0.1 in./day.

Estimates of water (I) required from wells, streams, or reservoirs to sustain water level are provided in Table 5-10. The annual water requirements ranged from 19 in. at Fairhope, Alabama, and Tallahassee, Florida, to 99 in. in Bakersfield, California. Differences in values among sites were related entirely to discrepancies between annual totals for precipitation and pond evaporation. Water requirements generally were between 30 and 40 in./yr in the southeastern United States, from 40 to 75 in./yr in Texas, Oklahoma, and Kansas, and more than 75 in./yr in California. Water requirements for other regions may be computed readily provided precipitation and evaporation data are on hand and a seepage estimate can be made.

Example 5-4

To illustrate the importance of the water supply requirement, consider fish farms of 300 acres each, one near Demopolis, Alabama, and one near Dallas, Texas. Water required to sustain ponds at Demopolis, Alabama, would equal a sustained inflow of 465 gpm:

$$\frac{(30\,in./yr \div 12\,in./ft) \times 300\ acres \times 325{,}782\,gal/acre\text{-}ft}{365\ day/yr \times 1{,}440\,min/day} = 465\,gpm$$

At Dallas, Texas, the water requirement increases to 1,054 gpm. Obviously, if all other factors were equal, the pumping cost for maintaining water levels in ponds would be more than twice as great at Dallas than at Demopolis.

TABLE 5-10   Amounts of Water Required to Sustain Levee Ponds at Selected Locations in the United States

Site	Inches/Year	Site	Inches/Year
Auburn, AL	25	Lexington, KY	31
Demopolis, AL	30	Baton Rouge, LA	31
Fairhope, AL	19	Calhoun, LA	34
Birmingham, AL	27	Lake Charles, LA	32
Little Rock, AR	35	Shreveport, LA	45
Russellville, AR	31	Jackson, MS	26
Stuttgart, AR	31	Meridian, MS	27
Bakersfield, CA	99	Starkville, MS	33
Fresno, CA	86	Stoneville, MS	33
Sacramento, CA	76	Raleigh, NC	38
San Diego, CA	76	Tulsa, OK	50
Belle Glade, FL	31	Charleston, SC	33
Gainesville, FL	37	Columbia, SC	35
Tampa, FL	49	Greenville, SC	32
Tallahassee, FL	19	Knoxville, TN	30
Athens, GA	36	Memphis, TN	30
Augusta, GA	38	Brownsville, TX	73
Macon, GA	42	Dallas, TX	68
Rome, GA	25	Houston, TX	43
Savannah, GA	42	San Antonio, TX	68
Tifton, GA	36	Wichita Falls, TX	73
Wichita, KS	58		

Source: Boyd (1986).

Additional water equal to pond volume must be supplied in years when ponds are drained and refilled. Ponds for catfish farming have average depths of 48 to 72 in.; 60 in. is a typical average pond depth (Boyd 1985b). A comparison of the amounts of water necessary to sustain water levels in annually drained and undrained ponds at four sites follows:

	Undrained Pond (in./yr)	Annually Drained Pond (in./yr)
Demopolis, AL	30	90
Houston, TX	43	103
Wichita, KS	58	118
Fresno, CA	86	146

Ponds filled by runoff are filled during rainy months, and the stored water must last through dry months. In the southeastern United States, ponds fill during winter and early spring, and water levels are fairly constant until June. During summer and autumn there is little or no runoff to sustain water levels, and seepage and evaporation take a toll. Earlier, it was calculated that the water level in a 1-acre pond with an 11-acre watershed on the Alabama Piedmont would decline by about 17 in. between June and October in a year with normal rainfall. During a dry summer and fall, the decline could be twice the normal drop. Hence, a pond of less than 3-ft depth might dry up completely. In previous calculations we assumed vertical sides for watershed ponds. This is often not feasible for real ponds, because the cost of excavating earth may be prohibitive. The usual practice is to deepen edges of watershed ponds so that all of the bottom is covered by water at least 2 ft deep. Thus, for a pond with an average depth of 6 ft, a fairly large area of the bottom must be covered with water deeper than 6 ft. Water-surface areas in watershed ponds decline as water levels decrease.

The procedures illustrated for estimating changes in water levels in ponds can be used to estimate the expected drop in water level for a watershed pond during dry months at a site. The most severe declines in level would come during periods of exceptionally low rainfall when little or none of the losses to seepage and evaporation are replaced by rainfall and runoff. For example, suppose it does not rain between July 1 and November 30 for a pond in the Alabama Piedmont. The seepage rate is 0.13 in./day; pond evaporation may be found in Table 3-2.

$$\text{Water-level decline} = S + E = 19.89 + 22.28\,\text{in.} = 42.17\,\text{in.}$$

The recommended minimum depths for watershed ponds in different climates (U.S. Soil Conservation Service 1982) are

Climate	Minimum Pond Depth (ft)
Wet	5
Humid	6–7
Moist subhumid	7–8
Dry subhumid	8–10
Semiarid	10–12
Arid	12–14

Deeper ponds are needed where seepage losses exceed 3 or 4 in. a month.

5.7   SEEPAGE EFFECTS ON WATER REQUIREMENTS

Special mention of seepage is essential, for it may sometimes be the largest outflow of water from a pond (Stone and Boyd 1989). If watershed ponds have high seepage rates, they will lose all of their water during dry months. Earlier, water levels were estimated each month for a 1-acre pond with a 20-acre watershed constructed in the fall and closed for filling on December 31. A seepage rate of 0.13 in./day was assumed. Calculations of water levels were repeated for seepage rates of 0.26, 0.39, and 0.52 in./day. Results are shown in Figure 5-5; water levels at the end of October would be 13.42, 29.45, and 50.31 in. below the full level for seepage rates of 0.13, 0.26, and 0.39 in./day, respectively. The pond would not fill if the seepage rate was 0.52 in./day, and it would be empty at the end of October. These calculations are based on a normal year, and water levels would drop lower on dry years. Also, calculations are for a 1-acre pond with a 20-acre watershed. This is almost twice the minimum watershed required for a 1-acre pond with a normal seepage rate on soils of the Piedmont Plateau of Alabama. Obviously, ponds that seep more than 0.2 in./day should be deeper than normal and have large watersheds.

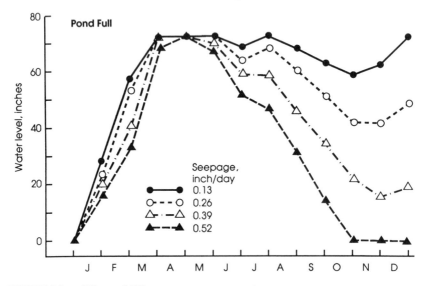

FIGURE 5-5.   Effects of different seepage rates on the water level of a hypothetical pond. See text for method of water level calculation.

Seepage from levee ponds may be offset by additions of water, but high seepage rates raise water requirements significantly. Consider the influence of different rates of seepage on amounts of water necessary to maintain the levels in a hypothetical pond at Demopolis, Alabama:

Seepage (in./day)	Inflow (in./yr)
0.1	30.0
0.2	66.5
0.3	103.0
0.4	139.5
0.5	176.0
0.6	212.5
0.7	259.0
0.8	285.5
0.9	322.0
1.0	358.5

Seepage rates above 0.2 in./day also are undesirable in levee ponds.

5.8 POND-FILLING TIME

A water supply sufficient to fill ponds in a reasonable time is essential. For watershed ponds, the control of refilling time usually is the size of the watershed. Therefore, it is essential to have a watershed large enough to generate adequate runoff. Inflow rates necessary to fill a levee pond of 5-ft average depth in different lengths of time when seepage plus evaporation is 0.2 in./day and there is no precipitation follow:

Refilling Time (days)	Inflow (gpm/acre)
1	1,148
5	232
10	118
15	80
30	42
45	29
60	23

6

Water Conservation

6.1 INTRODUCTION

Aquaculture in ponds requires a lot of water; hence, it often is desirable to implement techniques for conserving water. This chapter contains information on seepage and evaporation control, water harvesting, optimization of storage capacity, and water reuse. Seepage, one of the most common excessive losses of water from ponds, usually results from poor site selection or faulty construction. It is much easier to build a pond properly so that it does not seep excessively than to attempt to reduce seepage in a leaky pond. Most of the other topics covered in this chapter also can best be addressed during the construction phase. Water conservation not only lowers costs, but it minimizes the volume of effluents and reduces the pollution potential of aquaculture facilities. A little extra money spent during construction on water conservation measures will result in later savings.

6.2 CONTROL OF EVAPORATION

Efforts to control evaporation from free-water surfaces have focused on barriers to prevent movement of water molecules from the water surface to the atmosphere or to reduce the amount of solar radiation striking the surface (U.S. Nat. Acad. Sci. 1974). Certain aliphatic alcohols spread over water surfaces form a monomolecular layer that prevents gas exchange across the air–water interface. On still water, a few ounces of aliphatic alcohol per acre will form a film capable of greatly reducing the rate of evaporation. Unfortunately, wind and wave action make it impossible to maintain intact films of aliphatic alcohol over ponds or lakes, and little success in evaporation control has been achieved with monomolecular films. Nevertheless, research on this topic is continuing.

Blocks made of wax, styrofoam, or lightweight concrete have been floated on water surfaces and sheets of rubber or plastic have been spread over water surfaces to lower evaporation rates. This technique may reduce water loss by 85–90%, but it is not practical for large bodies of water. Shelters may be constructed over water storage tanks to lower evaporation rates.

In arid climates where water is stored in small ponds, evaporation can be reduced if the pond is filled with loose sand and the water stored in voids within the fill material. Water may be removed from the pond with drainage pipes or a well. Such a water storage structure is known as a *sand dam*.

These techniques are seldom valuable in conserving water in aquaculture ponds. Blocks or sheets on pond surfaces restrict light penetration, which then limits photosynthesis and might possibly result in low concentrations of dissolved oxygen. If research on monomolecular films leads to effective evaporation control techniques, further studies would be needed to decide if aquaculture can be conducted in ponds covered with such films. The alcohol layer readily transmits light, but it retards the escape of heat. Excessive heating might harm aquatic organisms. Sand dams possibly could be useful in storing runoff to use in some aquacultural applications.

When water is stored in reservoirs, the total evaporation loss increases as a function of increasing water-surface area. Suppose that pond evaporation at a location is 50 in./yr. Water loss by evaporation from a 25-acre pond with an average depth of 6 ft (assume vertical sides) would be 104.2 acre-ft, or 69.5% of storage volume. If, instead, a pond was 12.5 acres in area and 12 ft in average depth, it would have the same storage volume (150 acre-ft) as a 25-acre-by-6-ft-deep pond. However, the evaporation loss for the 12.5-acre pond would be only one-half that of the 25-acre pond because the smaller pond has only one-half as much surface exposed to the atmosphere as the larger pond. Total water loss by evaporation from the 12.5-acre pond would be 52.1 acre-ft, or 34.7% of storage volume.

Use of deep ponds for reducing the amount of evaporation per unit volume of storage is a useful concept where water for use in aquaculture is stored in one or more large ponds and conveyed by ditches or pipes to supply smaller production ponds. Unfortunately, the concept is not applicable to production ponds, for production of aquatic food organisms in ponds does not increase as a function of increasing depth. For example, production per acre would be about the same for a pond of 4-ft average depth as for a pond of 8-ft average depth. Deep ponds stratify, and waters below the thermocline often do not contain dissolved oxygen. Anoxic water and the sediment over which it lays is uninhabitable by most aquatic organisms. Most production ponds have average depths of no more than 4 or 5 ft. In some places, ponds may be put to multiple use; for example, fish may be produced in ponds used to store water for irrigation. Where aquaculture is not the only purpose of a pond, the deep pond concept may be a feasible means of conserving water. Sometimes, ponds for aquaculture must be filled only by runoff. In arid regions or in regions with long periods between rainstorms, such ponds must be deep to maintain sufficient water for survival of fish and other aquatic organisms.

Aquatic weeds growing in and around the edges of ponds enhance water loss through transpiration (Boyd 1987a). Pond edges should be deepened to a minimum of 2 ft to discourage the growth of rooted hydrophytes. Trees and shrubs should not be permitted to grow around pond edges, for their roots will withdraw water from the saturated zone created by seepage of pond water into pond banks. Grass cover around pond edges should be clipped periodically to reduce the leaf area for transpiration.

Plankton blooms usually exist in aquaculture ponds, and turbidity resulting from plankton restricts light penetration and discourages the growth of rooted hydrophytes. If plankton blooms are not present, they can be encouraged by applications of inorganic fertilizers (Boyd 1982b). Deep pond edges and plankton blooms usually prevent infestations of rooted hydrophytes within pond areas permanently covered with water. Floating plants such as water hyacinth, water lettuce, and duckweeds cannot be controlled through shading by plankton.

Grass carp eat large amounts of vegetation, and they can affect weed control in ponds. Grass carp do not reproduce in ponds, and they usually do not compete with other species. When used in combination with fertilizer applications, grass carp will eliminate infestations of floating and rooted hydrophytes from ponds (Whitwell and Bayne undated).

Rooted hydrophytes may be cut or dug from pond edges, and floating plants may be harvested. However, unless pond edges are deepened and turbidity increased through fertilizer applications, rooted hydrophytes will return. Of course, plankton blooms increase the organic matter content of water, and the organic matter absorbs heat. Hence, surface waters with plankton blooms are warmer than clear waters, and have slightly greater rates of evaporation when all other factors are equal. Mechanical control is the most effective means of controlling hydrophytes in water supply ditches.

Aquatic weed control may be effected with herbicides. Herbicides may have adverse ecological effects in aquaculture ponds, so they should be used as a last resort. Most weed infestations can be eliminated by biological and mechanical techniques.

6.3 SEEPAGE REDUCTION

Seepage can be a major loss of water from ponds, and seepage rates often exceed evaporation rates (Boyd 1982a). According to Boyd (1987b), the principle causes of seepage from ponds are

1. Permeable soils or strata of sand or gravel in the area over which water is impounded
2. Shallow soils underlain by fractured bedrock or solution cavities
3. Flocculated residual soils over cavernous and fractured limestone and calcareous shales
4. Soils with high gypsum content in which voids develop as gypsum dissolves
5. Improper methods of construction: failure to install a good core in dam, insufficient soil spread over exposed rock in bottom, poor keying of fill to residual soil, inadequate compaction of fill, no antiseep collars around drain structures

Seepage may occur through an earth-fill dam, beneath the dam, along drain structures extending through the dam, or through the pond bottom. Construction techniques discussed in Chapter 13 help reduce seepage through and beneath an earth-fill dam. Seepage through the pond bottom also may be reduced by initial construction techniques. However, many times, more attention is given to the construction of the dam than to preparation of the bottom. Later, a pond may be found to seep excessively through its bottom, and methods for reducing seepage are sought. Techniques for seepage control presented below are for reducing seepage through pond bottoms. Procedures were developed largely through efforts of the U.S. Soil Conservation Service, and they were initially presented by Holtan (1950a,b), Renfro (1952), and in various unpublished reports released over the past three decades. These techniques may be applied during initial construction of new ponds or used to reduce seepage from older ponds.

6.3.1 Compaction

Compaction alone is reliable only if soils of the bottom are well graded from small gravel or coarse sand to fine sand, silt, and clay. The pond must be drained and the bottom dried. All trees, stumps, vegetation, large rocks, and other objects should be removed. Holes left after removal of objects are filled with relatively impervious material. The bottom is pulverized to a depth of 8–10 in. with a disk harrow or other equipment. Pulverized soil is compacted with four to six passes of a sheep's-foot roller. For best results, compaction should be done under optimal moisture conditions (see Chapter 13). A thorough discussion of effects of moisture on compaction may be found in McCarty (1981). For ponds up to 10 ft deep, the thickness of the compacted seal should be 8–10 in. For deeper ponds, two or more 8- to 10-in. layers of soil should be compacted. The soil of the upper layers must be removed and stockpiled while the bottom layer is compacted.

6.3.2 Clay Layers

If soil of the pond bottom is not well graded and consists primarily of coarse particles, the bottom cannot be sealed by compaction alone. A layer of well-graded soil consisting of at least 20% by weight of clay particles may be installed over the bottom. The compacted layer of well-graded soil should be 1 ft thick for water depths up to 10 ft; allow 2 in. additional thickness of soil for each 1-ft increase in depth over 10 ft. The procedure for compaction is described earlier. The compacted area should be protected by rock riprap at points of concentrated inflow or outflow.

6.3.3 Bentonite

Incorporation of bentonite into well-graded but coarse soils will reduce seepage. Bentonite is a montmorillonitic clay; the mineral is a hydrated aluminum silicate. The crystalline mineral comprises three sheets or layers, and water can penetrate between the layers (Fig. 6-1). Absorption of water by the mineral causes it to swell. Sodium montmorillonite absorbs five times its dry weight of water, and the gel volume (volume of the swollen clay) is 10–15 times its dry volume. Calcium montmorillonite swells when wetted, but it does not swell as much as sodium montmoril-lonite. If bentonite is mixed into a coarse, highly permeable soil and wetted, it swells and fills the voids in the soil (Fig. 6-2). This process greatly reduces seepage. Bentonite shrinks when it dries and leaves cracks in the soil. Therefore, it is recommended only for sealing ponds where the water level does not decline and expose the bottom for drying.

Pond bottoms to be sealed with bentonite must be dried and cleared of all vegetation, stumps, large rocks, and like matter. Resulting holes

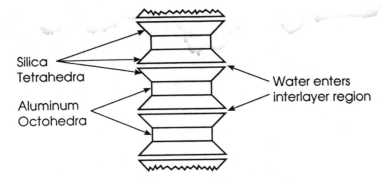

FIGURE 6-1.   Diagrammatic sketch of the montmorillonite structure.

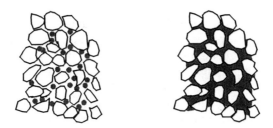

○ Coarse soil particles

∴ Dehydrated bentonite

● Hydrated bentonite

FIGURE 6-2.   Sealing of voids in coarse soil by bentonite.

should be filled, and the fill compacted with a roller compactor. The treatment rate for bentonite will vary from 1 to 7 lb/ft². The best estimates of treatment rate are based on laboratory permeability tests. However, a suitable alternative is to perforate the bottom of a large container, such as an oil drum, and put in 6–8 in. of pond bottom soil. Tamp the soil lightly and fill with water. Set the container so that water can drain out, cover it to prevent evaporation, and measure the water loss after a period of time. Repeat the test with a new batch of pond bottom soil into which bentonite has been mixed at the rate equivalent to 1 lb/ft². If this does not stop seepage, repeat the test using a greater amount of bentonite until a treatment rate is reached that stops seepage. Add 50–100% more bentonite to the pond bottom than the amount determined in the test. This safety factor is necessary because the pond water will be deeper and have more pressure than the water in the test container, and the mixing of bentonite into the pond soil is unlikely to be as thorough as that achieved for the test soil.

Bentonite can be spread best with an agricultural lime spreader. Spread half of the bentonite in one pass in one direction and the other half in a second pass made at right angles to the first pass. Other techniques that result in uniform coverage may be used to spread the bentonite. Mix the bentonite into the top 6-in. layer of soil with a rototiller or disk harrow. Compact the pond bottom with a roller compactor; do not use a sheep's-foot roller. For best results, the bentonite-treated layer should be covered with 1 ft of soil, and the slopes 2 ft above and below the water level should be covered with stone riprap. The pond should be filled immediately, or the bentonite-treated layer must be protected from drying. When filling the pond, do not allow the force of the inflow to erode the soil layer above the bentonite-treated soil.

Ponds with average depths greater than 10 ft require two or more 6-in. layers of bentonite-treated soil. The upper layer must be removed and stockpiled while the lower layer is treated.

6.3.4 Dispersing Agents

Particles of clays may be associated in several different manners. In a flocculated clay the particles, which may be thought of as tiny plates, are associated by edge-to-edge or edge-to-face attractions. The particles in a dispersed clay are associated by face-to-face attractions (Fig. 6-3). Van Olphan (1977) provides an excellent discussion of flocculated and dispersed clays. During deposition, sand, silt, and clay particles may form a loose or honeycomb structure (Fig. 6-3). The particles can develop a particle-to-particle contact that results in large voids, yet the structure is capable of supporting overlaying material. A flocculated structure in soils favors seepage. The application of chemicals known as dispersing agents to flocculated soils causes the honeycomb arrangement of particles to collapse or disperse. A dispersed soil will seep less than a flocculated soil.

Common dispersing agents are tetrasodium pyrophosphate (TSPP), sodium tripolyphosphate (STPP), salt (sodium chloride), and soda ash (sodium carbonate). Tetrasodium pyrophosphate is slightly more effective than sodium tripolyphosphate, but both are good general soil dispersants. Salt and soda ash are effective only for soils with high cation-exchange

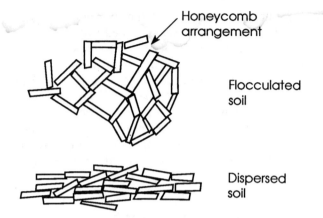

FIGURE 6-3.   Flocculated and dispersed soil particles. Note honeycomb arrangement in flocculated soil.

capacities, such as those soils containing montmorillonite clay. The effect of dispersants on the liquid limit of soils provides a field test for determining if a dispersant will likely reduce seepage. A pinch of the dispersant is added to a handful of soil, and the soil is thoroughly kneaded. If the soil becomes noticeably wetter, it has been dispersed.

Treatment with dispersing agents, like treatment with bentonite, is expensive and time-consuming. Therefore, it is advisable to make standard laboratory permeability tests to determine the best dispersant and to ascertain the optimal treatment rate for that dispersant. If this is not possible, a crude method for conducting permeability tests was described in Section 6.3.3. In absence of test data for dispersants, use salt or soda ash only where soils are known to have high cation-exchange capacities. Salt should be applied at $0.33\,lb/ft^2$; soda ash should be used at $0.2\,lb/ft^2$. Tetrasodium pyrophosphate and trisodium polyphosphate may be used at $0.1\,lb/ft^2$. These treatment rates are for a 6-in. layer of soil. Remember, dispersants are only effective in fine-grained soils that seep badly because they have a flocculated structure. As a rule, soils for treatment with dispersing agents should contain 50% or more of particles with diameters less than 0.003 in. and at least 15% of particles with diameters less than 0.00001 in.

The pond bottom must be dried and cleared of vegetation, stumps, and large rocks. Rock outcrops and other highly permeable areas are covered with a thick layer of fine-grained soil. Apply the soil dispersant uniformly over the bottom soil, and mix it into the upper 6-in. layer with a rototiller or disk harrow. With the soil at optimal moisture content, compact with a sheep's-foot or rubber-tired roller. If the average water depth exceeds 10 ft, two or more 6-in. layers must be treated with dispersing agents. Of course, upper layers must be removed and stockpiled while each lower layer is treated. When finished, the area near the high-water line and at any place where flow is concentrated should be protected from erosion with rock riprap or a layer of soil. The pond should be filled slowly to allow the dispersing agent plenty of time to react with the soil. Ponds often are turbid with fine clay particles following treatment of bottom soils with dispersing agents. This turbidity will gradually disappear; do not treat the pond with a coagulating agent such as alum (aluminum sulfate).

6.3.5 Flexible Membranes

The ideal membrane for lining ponds to reduce seepage would be water-tight and reasonably resistant to deterioration from weathering and mechanical damage. The membrane should be slightly flexible to allow it to adjust to small settlements in the subgrade without damage to the membrane. Polyethylene films are inexpensive and fairly resistant to weathering. They are quite susceptible to mechanical damage, and sheets of polyethylene can be joined or patched only by heat-sealing. Vinyl plastic and butyl rubber are fairly resistant to mechanical damage, and both can be joined and patched with a special cement. Woven fabrics (Marafi and Propex, Amoco Fabrics) sprayed with a substance of undivulged composition called Soil Crete (American Marketing and Research) provide a very tough and watertight membrane (Boyd 1983); however, this membrane quickly deteriorates when exposed to sunlight. Sheets of fabric membranes must be joined and patched by sewing (double stitch). For obvious reasons, the membranes are thin and must be handled carefully to prevent puncture. Polyethylene and vinyl membranes are 0.008–0.015 in. thick. Butyl rubber and woven fabric membranes are about twice as thick.

The pond bottom is dried and cleared of all vegetation, roots, rocks, and other sharp objects. If the surface is covered with coarse material (gravel and stones), a layer of fine-grained material should be provided as a cushion for the membrane. All banks and side slopes to be lined should be sloped no more than 1:1 for exposed membranes or 1:3 for covered membranes. The layer of soil covering the membrane may slide on slopes steeper than 1:3.

Plants can penetrate polyethylene, vinyl, and woven fabric membranes. Plants can be discouraged by applying herbicide to the soils of the side slope before the membrane is installed.

A trench about 1 ft deep and 1 ft wide is excavated around the pond just above the high-water line, and membrane edges are buried. Poly-ethylene, vinyl, and butyl membranes are laid in strips with about 6 in. of overlap for seaming. Woven fabric membranes must be sewn before placing them on the pond bottom. Vinyl, butyl, and woven membranes are laid smooth but loosely. At least 10% slack must be provided for polyethylene membranes. For best results, the membrane should be covered with a 4- to 6-in. layer of soil. This soil must be free of objects that could puncture the membrane.

Membranes are expensive, but they can reduce seepage essentially to zero. For example, the annual water loss from a pond was reduced from 202 to 52 in./yr by a woven fabric membrane treated with Soil Crete (Boyd 1983). Pan evaporation was 54 in. during the 12-month period when water loss totaled 52 in. from the lined pond, so seepage was negligible.

6.3.6   Organic Matter

A less expensive, but less reliable, means of reducing seepage is to treat pond bottoms with organic matter. Manure, grass, paper, cardboard, or other organic matter can be spread over the dry pond bottom at 0.5–1.0 lb/ft^2 and mixed with the soil by disk harrowing. Best results will be achieved if the soil is allowed to stand for a few weeks and then com-pacted before flooding. An alternative is to drain the leaky pond, fence around it, and hold and feed livestock in the dry pond. Manure from the animals will be compacted with the soil as the animals walk about. Swine are especially effective for sealing ponds.

6.4   STORAGE CAPABILITIES

Ponds usually have overflow structures. If water levels are maintained at the same elevation as the intake of the overflow structures, all rain and runoff entering ponds will be lost as overflow. Therefore, water levels should be maintained 6 or 8 in. below overflow structures so that most rainfall and runoff may be conserved. In arid climates, the saving of water may be small, but in the southeastern United States, rain falling directly into ponds is almost enough to replace losses to evaporation (Boyd 1982a, 1985a).

Some ponds are drained each year and refilled by overland flow. At many places, most of the overland flow occurs during cool, wet months, and ponds must be filled during this time. It was shown in Chapter 4 that most overland flow for a watershed near Auburn, Alabama, occurred during December, January, February, and March. During the rest of the year, rain falling directly into watershed ponds may be the only significant source of water. Ponds should be sufficiently deep to maintain water levels through drier months. However, pond volumes must conform to watershed areas, so that there is adequate overland flow to fill ponds. In the vicinity of Auburn, Alabama, 10–20% of annual rainfall becomes overland flow, and normal annual rainfall is about 54 in. Hence, overland flow normally is 5.4–10.8 in./year, and each acre of watershed will yield roughly 0.5–1.0 acre-ft of overland flow. Each acre-foot of pond volume will require at least 1–2 acres of watershed, and to be conservative the larger estimate should be used. Watershed ponds near Auburn, Alabama, usually have average depths of 5–6 ft, and there should be at least 10–12 acres of water per acre of pond surface.

Boyd and Shelton (1984) studied the hydrology of watershed ponds on the large pond research area of the Fisheries Research Unit at Auburn University, where soils have fairly high infiltration rates. Many of the ponds are constructed so that one pond backs water up to the base of the dam of another pond (Fig. 6-4). Seepage through the dam of the upper pond enters the lower ponds. Because of this arrangement, lower ponds have more stable water levels than upper ponds. This principle is illustrated by stage changes in ponds S-11, S-12, S-13, and S-19 (Fig. 6-5). Seepage through the dams of S-13 and S-19 enters S-12, and seepage through the dam of S-12 enters S-11. Between April 5 and November 9, the water level decreased by 2.49 ft in S-13 and by 2.80 ft in S-19. Water levels in S-11 and S-12 remained virtually constant over this period. Ponds S-13 and S-19 are of marginal value for aquaculture, because of water-level fluctuations, but ponds S-11 and S-12 are excellent. Without ponds S-13 and S-19, water levels would fluctuate greatly in ponds S-11 and S-12.

An example of the importance of storage capacity is afforded by the operation of the small pond research area on the Fisheries Research Unit. There are no good aquifers or large streams in the vicinity, so surface water must be used for filling ponds. There are 194 ponds ranging from 0.044 to 1.0 acre in area, for a total area of 26.8 acres. The water budget

FIGURE 6-4.   Cross section of an upper and a lower pond on the same watershed. Water seeps under the dam of higher pond and enters the pond below.

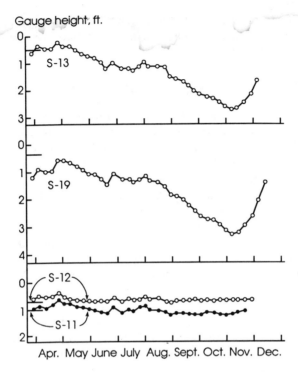

FIGURE 6-5.   Water-level changes during 1982 in four ponds (S-11, S-12, S-13, and S-19) on the Auburn University Fisheries Research Unit. Ponds S-13 and S-19 are higher than S-12 and seep into it. Pond S-12 is above S-11 and seeps into it. Horizontal lines between 0 and 1-ft gauge height represent elevation of spillways.

for this complex was discussed in Chapter 5. Ponds average 3 ft in depth, so the total storage volume is 80.4 acre-ft. The ponds are filled by gravity flow through a pipeline from a 20.1-acre (140.6-acre-ft) reservoir. A small stream feeds the primary water supply reservoir, and additional water can be released into the stream from secondary water supply reservoirs. During the winter, the primary water supply reservoir fills and overflows. On a normal year, overland flow is negligible after May, and the discharge of the small stream declines drastically. The water level in the primary reservoir declines as water is withdrawn for use in the small research ponds. Water also is lost from the primary reservoir through seepage and evaporation. During drier than normal years, water must be released from the secondary reservoirs to maintain adequate water for the research ponds.

These examples demonstrate that storage capacity is an important variable in ponds or in complexes of ponds. Interactions among local hydrologic factors, soil characteristics, and water supply requirements must be evaluated to determine the necessary storage capacity for an application.

6.5 WATER HARVESTING

The term *water harvesting* is often used in a limited sense to describe the process of collecting and storing water from an area that has been treated to increase runoff. The catchment area is the surface that has been treated to increase runoff. The primary use of water harvesting has been to collect and store water for consumption by humans, livestock, and wildlife. For this application, the catchment area usually consists of land that has been cleared of vegetation and either treated with a chemical to reduce infiltration or covered with an impermeable sheet of rubber, plastic, asphalt, asphalt-treated fabric, or metal (Frasier and Myers 1983). Water flowing from the catchment is stored in a tank and protected from evaporation until it is used. Water-harvesting systems usually are found in arid regions, but they have not been used for aquaculture. A typical water-harvesting system is illustrated in Figure 6-6.

The cost of constructing large catchments by chemical treatment or impermeable covers is probably prohibitive in aquaculture. However,

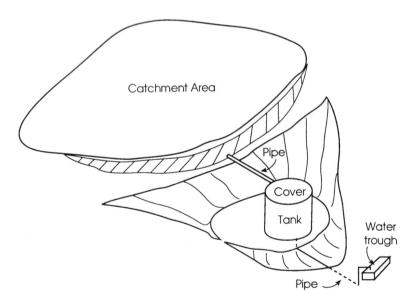

FIGURE 6-6.   A water-harvesting system for supplying water to livestock. (*Source: Frasier and Myers 1983.*)

there are relatively inexpensive ways of enhancing runoff that might be useful. Remember that the term *runoff* is synonymous with stream discharge, and that sheet flow is that part of the runoff that flows downslope over the land surface to enter streams. Small watersheds often do not have streams, and all runoff is overland flow. However, for watersheds or drainage basins that sustain streams for all or most of the year, runoff may be considerably greater than overland flow.

For practical purposes, runoff may be estimated as precipitation minus evapotranspiration. This generalization is correct because ground-water and soil-water storage for a watershed remains fairly constant over time. For example, soil moisture content and water table elevation usually would be about the same in March of this year as it was in March of last year. In other words, because of evaporation of intercepted water and water detained on the land surface and transpiration by vegetation (because of evapotranspiration), the amount of water taken into storage on the land surface and in the soil each year is roughly equal to the annual loss of water through evapotranspiration. Furthermore, the annual amount of water taken into storage by an aquifer is equal approximately to the base flow of streams whose bottoms intercept the aquifer. Therefore, watershed alterations that decrease evapotranspiration will increase runoff.

If a watershed has a rough surface with many depressions, water will be detained in the depressions and lost through evaporation. Small depressions may be filled; drainage ditches may be supplied for larger depressions. Of course, vegetation must be established where the land was cut or filled to prevent erosion.

Reduction in the density of forest vegetation increases runoff (Evans and Patric 1983). Douglass and Swank (1975) demonstrated in North Carolina that the runoff increase after cutting was greater on north-facing watersheds than on south-facing watersheds in steep terrain, because the south-facing watersheds receive more insolation and have greater rates of evapotranspiration. Coniferous forests have greater interception capacities and evapotranspiration rates than do hardwood forests, so converting watershed cover from conifers to hardwood will increase runoff. Grass cover yields more runoff than forest cover because of reduced evapotranspiration and interception. Short grass has a smaller interception capacity than tall grass, so short grass cover is more favorable to runoff than tall grass cover.

The benefits of grass cover on runoff may be visualized by using the SCS curve number technique (Chapter 4) to calculate overland flow for a watershed covered by uncontoured pasture and a wooded watershed. Assuming antecedent moisture condition II, hydrologic soil group C, and

3 in. of rain, overland flow would be 0.7 in. (2,540 ft^3/acre) and 1.65 in. (5,990 ft^3/acre) for wooded and pasture watersheds, respectively.

Even in humid climates, increasing runoff for ponds is beneficial. The ratio of watershed area to pond area can be reduced to permit more or larger ponds. During drier months, greater overland flow from storms would help maintain water levels in watershed ponds. However, if changes are made in watershed surfaces and cover, care must be taken to prevent erosion. In logging operations, the primary source of erosion leading to excessive turbidity in streams is poorly designed and improperly located logging roads. This observation reveals that relatively small denuded areas can contribute large amounts of suspended soil particles to runoff water.

6.6 WATER REUSE

For many years, channel catfish farmers drained ponds for fish harvest. Most channel catfish ponds are levee ponds supplied by well water. Refilling the ponds by pumping is a major expenditure. Furthermore, water in catfish ponds contains high concentrations of nutrients, organic matter, and suspended solids, and effluents from catfish ponds are a potential source of pollution to receiving streams (Boyd 1978). Catfish farmers have developed techniques for harvesting fish without draining ponds. Hollerman and Boyd (1985) demonstrated that over a three-year period water quality was no worse in undrained catfish ponds than in annually drained ponds, and the undrained ponds were just as productive of fish as annually drained ponds. Catfish ponds must be drained every three or four years to repair wave damage to the insides of levees. The practice of harvesting fish without draining ponds conserves water, reduces pumping costs, and reduces the water pollution potential of catfish farms.

The greatest concentration of bait minnow farms in the United States is in Lonoke County, Arkansas. For effective harvest of minnows, ponds usually are drained. However, on one large farm a method for water reuse has been implemented. Pipes were installed at normal water level between adjacent ponds. When minnows are to be harvested from a pond, a tractor-powered pump is used to transfer the water from the pond to two or more adjacent ponds. An elbow and short vertical extension pipe is placed on the pipes between adjacent ponds to provide storage capacity. After minnows are harvested, water is permitted to drain back into the ponds that were drained, through removal of the elbows and extension pipes. Water is saved and pumping costs are reduced. The pumping level in wells that supply the ponds is about 50 ft below the surface, but water has to be lifted no more than 6 or 8 ft to transfer it to an adjacent pond. Water is returned by gravity flow. Regardless of whether ponds are drained into a stream or water is transferred to adjacent ponds, the same volume of water must be pumped. However, the pumping head is much less for the water reuse scheme.

There is another benefit of water reuse on a minnow farm. Minnows are fed a commercial ration, but they also depend on natural food organisms. Fertilizers are applied to encourage plankton blooms. Plankton and inorganic nutrients are lost when ponds are drained into a stream, and well water used to refill the pond has low concentrations of nutrients. Water reuse reduces the amount of fertilizer that must be applied to minnow ponds.

The Fisheries Research Station at Auburn University has numerous research ponds that are filled from a reservoir supplied by a stream. There is plenty of water during winter, spring, and early summer, but stream flow decreases and the water level in the reservoir falls during late summer and autumn. The storage reservoir has some seepage under its dam, and the research ponds seep. The seepage water collects in a ditch and forms a small, permanent stream. A concrete sill was constructed across the stream to elevate the surface level, and water from the stream is pumped into the pipeline that delivers water from the reservoir to the research ponds.

A raceway system for fish production was constructed on the University of Georgia Coastal Plains Experiment Station at Tifton, Georgia. Water from the tail of the raceway flowed into a large pond where biological and chemical process removed nutrients and organic matter from the water and reoxygenation was effected by photosynthesis. Water from the pond was then pumped to the head of the raceway from where it flowed by gravity through the successive segments of the raceway and back to the pond again.

The four examples given show how water can be reused in aquaculture systems. Water reuse not only conserves water, but it often lowers cost and reduces the potential for stream pollution. Whenever possible, water reuse should be implemented on farms, hatcheries, and research stations.

7

Morphometric and Edaphic Factors Affecting Pond Design

7.1 INTRODUCTION

Ponds vary greatly in morphometry, and there is no consensus about the best combination of morphometric features for an aquaculture pond. For watershed ponds, morphometry is determined largely by terrain. Levee and excavated ponds can be built to design morphometry, but some designs are more feasible than others. Edaphic factors are important in construction, for soils at some places are much more suitable than at others. Soil characteristics also can influence physical, chemical, and biological processes, affect production, and be a factor in pond management. The possibility of turbidity, sedimentation, and organically stained waters should be considered, and control measures planned if necessary. This chapter provides guidelines on pond morphometry, soils, and sediment to aid in pond design, construction, and management. Details of engineering design and construction of aquaculture ponds are discussed in Chapter 13. Attention to these guidelines will help ensure that ponds will be easy to maintain and manage.

7.2 MORPHOMETRY

7.2.1 Surface Area

The size of a pond depends on species to be cultured, intensity of culture, availability of land, nature of the site, construction costs, and personal preference. Research ponds seldom are larger than 1 acre, and most are smaller than 0.25 acre. Small ponds are used because several ponds are needed for replication of experimental treatments, and the cost of operating several large ponds for research purposes usually is excessive. Fortunately, production levels in ponds provided similar management inputs do not differ appreciably with pond size, and information obtained in small research ponds may be applied to larger production ponds. Large ponds normally are less expensive to construct per unit surface area than are small ponds, because less earth work is required for embankments. In sparsely inhabited areas with plenty of land and water, commercial aquaculture ponds are typically several acres in area and managed for extensive or semi-intensive production. For example, channel catfish ponds in the southeastern United States usually are 5–20 acres in area, and brackish-water shrimp ponds in Central and South America normally contain 20–50 acres. In heavily populated areas where land is scarce or in areas without abundant water, ponds are typically small and managed for intensive production. Shrimp ponds in heavily populated Asian nations are seldom larger than 2.5 acres.

Practically any size pond can be used for aquaculture, but the difficulty of managing the pond increases in proportion to its size. Ponds greater than 5 acres are extremely difficult to manage for intensive production, and management for semi-intensive production becomes problematic in ponds larger than 20–30 acres. We feel that 0.05- and 0.1-acre ponds are desirable for most types of research, 0.5- to 2.5-acre ponds are most suitable for intensive aquaculture (10,000 lb/acre and above), 5- to 10-acre ponds are best for semi-intensive aquaculture (2,000–10,000 lb/acre), and even in extensive aquaculture (<2,000 lb/acre), ponds should not be larger than 50 acres.

7.2.2 Shape and Orientation

The shape of a watershed pond conforms to the contour of the watershed at the elevation of the pond water level. There is great variation in shape of watershed ponds. Levee and excavated ponds usually are rectangular or square, and some round ponds have been constructed. The major cost of levee ponds usually is levee construction, so it is important to minimize total levee length as much as practical. Total levee lengths for ponds of different sizes and shapes are given in Table 7-1. The amount of levee needed per acre of pond surface decreases drastically with increasing surface area. For example 0.1-, 1-, 5-, and 20-acre square ponds require 2,640, 835, 373, and 187 ft of levee per acre, respectively. Total levee length for ponds of the same surface area do not differ greatly among the three basic shapes.

Some researchers are touting the value of circular ponds. They claim that mechanically induced water circulation is more efficient in round

TABLE 7-1   Total Levee Lengths for Ponds of Different Sizes and Shapes

Area (acres)	Length of Levee (ft)		
	Square	2:1 Rectangle	Circle
0.1	264	280	233
0.5	590	626	523
1	835	885	740
5	1,867	1,980	1,654
10	2,640	2,800	2,340
20	3,734	3,960	3,309

ponds than in ponds of other shapes. In Figure 7-1, a circle, square, and 4:1 rectangle of the same area have been centered over a point. It is obvious from this figure that it would be easier to mechanically induce circular water currents in a round or square pond than in a rectangular pond. However, we doubt that mechanically induced circulation is much more efficient in a round pond than in a square one. Also, round ponds are much more difficult and expensive to build than square ponds.

In some places, there is a prevailing wind direction most of the year. If it is desired to take maximum advantage of this wind to circulate pond water, ponds can be oriented with their long axes parallel to the prevailing wind direction. Obviously, a rectangular pond provides a greater wind fetch than a round or square pond (Fig. 7-1). Aerators or other mechanical devices often are used to mechanically induce water circulation. A square or rectangular pond with a length-to-width ratio smaller than 2:1 provides less resistance to circular water movement than do rectangular ponds with length-to-width ratios of 2:1 or larger. Although aerators often are placed in corners of ponds to induce circular water movement, there is no experimental evidence that circular water movement is superior to other water movement patterns. Where ponds are harvested by seining, it is necessary to have a width no greater than the seine length. In channel catfish farming, mechanically operated seines can be used best in ponds no wider than 600 ft.

Natural bodies of water with long, irregular shorelines are thought to have greater productivity than water bodies with shorter and regular shorelines. A high degree of contact with land favors input of nutrients and organic matter to a lake, and a large littoral zone enhances productivity within the lake. The shore development index refers to the ratio of the actual length of shoreline of a body of water to the circumference of a circle the area of which is equal to that of the water body (Welch 1948):

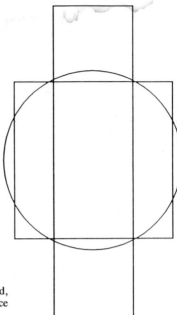

FIGURE 7-1. Graphic representation of round, square, and 4:1 rectangular ponds of equal surface area.

$$\text{Shore development} = \frac{s}{2\sqrt{a\pi}} \qquad (7\text{-}1)$$

where s = length of shoreline (ft) and a = area of lake (ft^2).

Round, square, and 2:1 rectangular ponds with areas of 5 acres have shore development indices of 1.00, 1.13, and 1.20, respectively. Thirty-six watershed ponds on the Auburn University Fisheries Research Unit in Alabama averaged 5.4 acres in area and had shore development indices of 1.05–1.86 with an average of 1.30 (Boyd and Shelton 1984). Natural lakes and large man-made reservoirs often have shore development indices of 5 or more. The shore development index is of no value in aquaculture, because fertilizers, manures, and feeds are applied to ponds to ensure high rates of productivity.

7.2.3 Depth

Most aquaculture ponds have average depths of 4–6 ft. Levee ponds have sloped levees along with gradual slopes and cross slopes of the bottom to facilitate draining, but most of the pond bottom is at least 4–5 ft deep. Excavated ponds have sloped sides, but as with levee ponds most of the

TABLE 7-2   Selected Morphometric Features for 36 Watershed Ponds on the Auburn University Fisheries Research Unit

Feature	Average	Min.	Max.
Surface area, acres	5.39	0.72	26.0
Water area less than 2 ft deep, %	17	7	36
Water area more than 5 ft deep, %	24.6	0.0	46.5
Average depth, ft	5.03	2.50	8.63
Maximum depth, ft	11.9	6.1	21.5

Source: Boyd and Shelton (1984).

bottom is 4 or 5 ft deep. Watershed ponds have greater variation in water depth. They usually have a maximum depth two or three times greater than average depth, and maximum depth tends to increase as pond area increases. This results because watershed surfaces slope upward, and when a pond is built on a watershed of similar slope, a large pond has a greater depth than a small pond. Watersheds differ in slope, and some small ponds are deeper than some large ponds. Thirty-six levee ponds on the Auburn University Fisheries Research Unit with areas of 0.72–26.0 acres averaged 5.03 ft deep with an average maximum depth of 11.9 ft (Table 7-2). This observation is consistent with the widely used rule-of-thumb that average depth is about four-tenths of maximum depth.

Thermal stratification is common in lakes and ponds. Part of the solar radiation striking the water surface penetrates, and the rest is reflected. Solar radiation changes in spectral quality and decreases in intensity as it passes through the water because of scattering and differential absorption by water and substances dissolved or suspended in it. Some light energy is used by plants for photosynthesis, some is converted to heat, and some is scattered. The amount of light decreases rapidly with depth, and heating of water occurs primarily within the 0- to 3-ft layer. Heat in surface layers is transferred to deeper strata mainly by wind mixing of waters. Warm water is less dense (lighter) than cool water, and warm surface waters may become so much lighter than deeper, cooler layers of water that wind-induced circulation cannot mix them. This causes thermal stratification (Fig. 7-2). The layer of warm surface water is called the *epilimnion*, and the bottom layer of cooler water is known as the *hypolimnion*. The temperature changes rapidly with increasing depth in the transition zone between the epilimnion and hypolimnion. This transition zone is termed the *thermocline*.

The duration of thermal stratification depends on pond depth and weather conditions. Ponds not over 6 ft deep usually stratify on warm

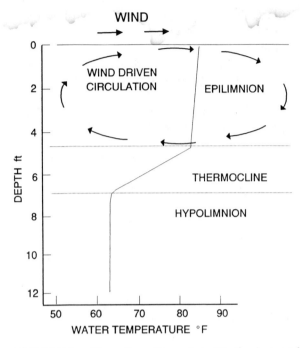

FIGURE 7-2.   Illustration of thermal stratification in a pond.

days, but at night surface waters cool by convection, and even light winds can cause the entire body of water to circulate. In deeper ponds, strong patterns of thermal stratification develop during warm months. Nightly cooling and wind mixing will not overcome the density difference between the epilimnion and hypolimnion, and ponds remained stratified until seasons change and the epilimnion cools or until high winds or intense rainfall induce mixing of the two layers.

Dissolved oxygen production by phytoplankton photosynthesis occurs mainly in the epilimnion, because there is not enough light for photosynthesis in the hypolimnion. Organic matter originating from photosynthesis, metabolic wastes from aquatic animals, and other sources settles into the hypolimnion, and its decomposition can result in oxygen depletion and the accumulation of toxic reduced substances (e.g., nitrite, ferrous iron, manganous manganese, hydrogen sulfide, and methane) in the hypolimnion. Differences in concentrations of dissolved substances with depth in a body of water is known as *chemical stratification*. Thermal and chemical stratification are undesirable in aquaculture ponds because they can make large parts of the pond bottom and lower strata of water unfavorable or unfit habitat for aquatic animals.

Because of thermal stratification, it is best to restrict the depth of aquaculture ponds to about 6 ft. This practice often is not practical for watershed ponds because relatively deep areas are necessary in order to have the desired surface area. Watershed ponds in Alabama typically stratify at about 5 ft deep. An average of 24.6% of the volume of water in 36 watershed ponds on the Auburn University Fisheries Research Unit was below the 5-ft contour (Table 7-2). The bottom profile of watershed ponds can be altered during construction by cutting soil from high areas and using it to fill the low areas. This procedure substantially increases construction costs. Aerators and water circulators (Boyd 1990) can prevent stratification in deep aquaculture ponds.

Excessively shallow ponds may become infested with underwater weeds, and waters may become too warm during summer. Weed infestations are usually not a problem in water deeper than 2 ft. Watershed ponds on the Auburn University Fisheries Research Unit had an average of 17% of pond areas in water less than 2 ft deep (Table 7-2). Levee and excavated ponds only have water less than 2 ft deep on side slopes. Edges of watershed ponds must be excavated to avoid water less than 2 ft deep.

In large brackish-water ponds for shrimp production, drainage ditches several yards wide are cut around the periphery of bottoms just beyond the bases of levees. Ditches usually are 2–3 ft deeper than the elevation of the original bottom. They facilitate draining and shrimp harvest, and they are also thought to be a haven of cool water for shrimp on hot days.

Forms of lake basins are sometimes expressed by an index figure known as the volume development (Welch 1948). Volume development represents the ratio of the total volume of a lake to the volume of a cone whose base area is equal to the surface area of the lake and whose height is equal to the maximum depth of the lake. A simplified formula is used to compute this index:

$$\text{Volume development} = 3 \times \frac{h_{\text{ave}}}{h_{\text{max}}} \qquad (7\text{-}2)$$

where h_{ave} = average depth and h_{max} = maximum depth. The value of volume development near unity indicates that the form of the lake bottom is close to that of a cone. Values less than unity suggest that the lake bottom walls are convex toward the water; the lake bottom walls are concave toward the water when the value is greater than unity. Levee ponds and excavated ponds typically have volume development indices of 1.5 and above. Watershed ponds on the Auburn University Fisheries

Research Unit had volume development indices of 0.92–1.61 with an average of 1.30 (Boyd and Shelton 1984). Volume development is presented because many aquaculturists are aware of it, but it apparently has no significance in pond aquaculture.

7.2.4   Baffle Levees

Many aquaculturists think that water circulation, especially at the pond bottom, is critical to maintaining good water quality. Mechanical aeration and water circulation devices (Boyd 1990) are widely used to promote water circulation in ponds. Water exchange also causes water circulation in ponds; water is typically drained from the pond bottom at one end and introduced at the surface on the other end. Water exchange is most common in brackish-water aquaculture. The daily exchange rate is usually 5–10% in semi-intensive ponds and 20–30% in intensive ponds.

Lawson and Wheaton (1983) considered the problem of water circulation in large, shallow, crawfish ponds. When water is simply pumped into the pond on one end and allowed to flow to the exit on the other end, there were considerable "dead" areas of poor water circulation (Fig. 7-3). They demonstrated that dead areas could be eliminated by using aerators to augment flow and baffle levees to direct the flow. The baffle levee concept is used widely in intensive, brackish-water shrimp ponds in Thailand and Indonesia. Rectangular ponds have a single baffle levee extending down the middle of the long axis of the pond with gaps at each end (Fig. 7-4). This design facilitates a circular movement of water when paddle-wheel aerators are operated. This system no doubt has merit in many types of aquaculture.

7.3   SOILS

7.3.1   Rating Site Soils

Soil properties most common in reference to pond construction are slope of terrain, soil texture, organic matter content, and sulfide content. Soil texture is probably considered most important, because soil must contain enough clay for constructing embankments and bottoms that do not seep excessively. Many other properties are important and should be considered in site evaluation, design, and construction.

Hajek and Boyd (1990) tabulated soil properties important in pond construction and developed a rating system for classifying site soil properties as having slight, moderate, or severe limitations for construction of excavated ponds (Table 7-3) and pond embankments (Table 7-4).

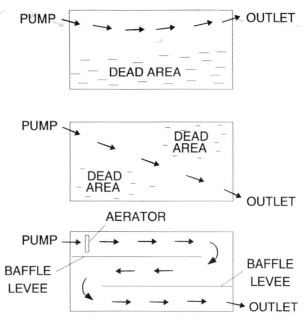

FIGURE 7-3. Problems of water circulation and improved water circulation by use of aerator and baffle levees. (*Source: Lawson and Wheaton 1983.*)

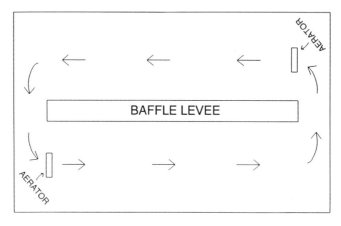

FIGURE 7-4.   Improved water circulation by a single-baffle levee and aerators.

TABLE 7-3   Soil Limitation Ratings for Excavated Fish Ponds

Property	Limitation Rating			Restrictive Feature
	Slight	Moderate	Severe	
Depth to sulfidic or sulfuric layer, ft	>3	1.5–3	<1.5	Potential acidity or toxicity
Thickness of organic soil material, ft	<1.5	1.5–2.5	>2.5	Seepage; hard to compact
Lime requirement, tons/acre	<2	2–10	>10	Mineral acidity
pH of 1.5- to 2-ft layer of pond bottom	>5.5	4.5–5.5	<4.5	Too acid
Clay content, %	>35	18–35	<18	Too sandy/silty; excessive seepage
	Clayey	Loamy	Sandy/silty	
Slope of terrain, %	<2	2–5	>5	Slope
Depth to water table, ft	>2.5	1–2.5	<2.5	Hard to drain; dilution
Frequency of flooding[a]	None	Rare/occasional	Common/frequent	Flooding
Small stones, %	<50	50–75	>75	Small stones
Large stones, %	<25	25–50	>50	Large stones
Organic matter, %				
Low clay content soil (<60% clay)	<4	4–12	>12	Excessive humus
High clay content soil (>60% clay)	<8	8–18	>18	Reducing environment
Depth to rock, ft	>5	3–5	<5	Shallow; seepage

Source: Hajek and Boyd (1990).

[a] None: no possibility of flooding. Rare: 0–5 times in 100 years. Occasional: 5–50 times in 100 years. Frequent: >50 times in 100 years. Common: occasional and frequent combined.

TABLE 7-4 Limitation Ratings for Fish Pond Embankments, Dikes, and Levees

Property	Limitation Rating			Restrictive Feature
	Slight	Moderate	Severe	
Clay content[a], %	>35	18–35	<18	Too sandy
	Clayey	Loamy	Sandy	
Depth to sulfidic or sulfuric material, ft	>3	1.5–3	<1.5	Toxicity; potential acidity
Lime requirement, tons/acre	<2	2–10	>10	Mineral acidity
pH of 1.5- to 2-ft layer of pond bottom	>5.5	4.5–5.5	<4.5	Too acid
Engineering classes	—	—	All G & S classes	Too stony
				Too sandy
Slope, %	<8	8–15	>15	Slope
Thickness of organic material, in.	<6	6–18	<18	Subsides; excess humus; difficult to compact
Depth to water table, ft	>3	1.5–3	<1.5	Wetness
Fraction >3 in. diameter, %	<25	25–50	>50	Large stones
Depth to bedrock, ft	>3	1.5–3	<1.5	Depth to rock
Shrink–swell potential	Low	Medium to high	Very high	Shrink–swell
Erodibility, K	<0.1	0.1–0.3	>0.3	Erosion

Source: Hajek and Boyd (1990).

[a]Weighted mean from surface to 3.3 ft depth.

A slight limitation indicates that a property is favorable for use, and no unusual construction, design, management, or maintenance is necessary. A moderate limitation means that a property requires special attention, but the limitation can be overcome through special design, construction, management, or maintenance. A severe limitation for a property renders soil unfit for use. However, it does not necessarily mean that the site must be abandoned. Usually, severe limitations can be overcome through special techniques, but in most cases, significant cost and effort are required to compensate for a severe limitation.

Evaluation of soil properties by use of the rating system outlined in Tables 7-3 and 7-4 requires data on several properties of site soils. Some of these data may be obtained from previous soil surveys and soil maps, but a thorough site investigation to obtain specific on-site soil data is essential.

Some explanation of properties listed in Tables 7-3 and 7-4 are necessary. The sulfidic or sulfuric layer refers to soils that may become highly acidic or are already highly acidic because of the oxidation of iron pyrite. This property will be discussed in detail in Section 7.3.2. Soil texture has to do with particle size distribution in soil. Methods for measuring soil texture, including clay content and percentage stone, are provided in Chapter 13. Boyd and Tucker (1992) give techniques for determining soil pH, exchangeable acidity, and lime requirement, but measurements of these variables usually can be obtained by sending samples to a nearby soil-testing laboratory. Shallow water tables and rock layers near the soil surface can be detected by boring a hole into the soil with a hand-operated soil auger. Slopes may be determined with a surveyor's level or a hand-held clinometer.

A rough estimate of soil organic matter content may be obtained as follows: dry soil sample at 105°C (220°F), weigh 5.0 g into a tared crucible, heat at 350–400°C (650–750°F) in a muffle furnace for 8 hr, cool in a desiccator, weigh, and calculate percentage weight loss. Wet oxidation methods presented by Nelson and Sommers (1982) may be used to provide more accurate data on soil organic matter concentrations.

Soil can vary in organic matter concentration from near 0% to almost 100%. Organic matter in soil contains 48–58% carbon. A factor of 1.9 times organic carbon concentration is thought to provide a reasonably reliable estimate of organic matter in surface soil (Nelson and Sommers 1982). Soil material that is not saturated with water for long periods each year is an organic soil material if it has an organic carbon concentration above 20% of the dry weight (Soil Survey Staff 1990). Soil material saturated with water most of the time is considered to be organic soil material at slightly lower organic carbon concentrations (Fig. 7-5). In

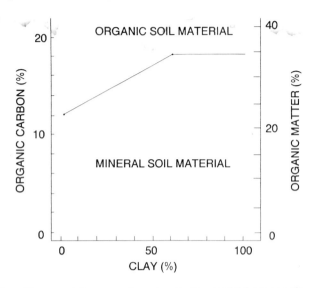

FIGURE 7-5.   Nomograph for separating mineral soil material from organic soil material.

general, a soil is classified as organic if more than half of the upper 80-cm layer can be considered organic soil material.

There are three basic types of organic soil materials (Soil Survey Staff 1990). Fibric materials consist largely of plant fibers. A fiber is a fragment of plant tissue, excluding live roots, large enough to be retained on a 0.15-mm sieve. Fibric soil materials must have a fiber content, after rubbing with the hand, of three fourths or more of the soil volume. Coarse fragments of wood and mineral layers are excluded when determining the volume of fiber. Fibric soil materials are commonly known as peat soil materials. When drained and exposed to the air, fibric soil materials may decrease greatly in volume because of decomposition. Sapric materials contain highly decomposed organic matter. After rubbing, the fiber content is less than one sixth of the soil volume. Sapric soil materials are dark gray to black and relatively stable with time when drained and exposed to the air. Colloquially, these materials are called *muck* soils. Soil materials with fiber contents between one sixth and three fourths of their volumes are called *hemic* soil materials.

The shrink–swell potential can be estimated from the change in length of an unconfined soil core. Take a 1-in.-diameter core of wet soil, remove the intact soil core from the sampling tube, set the wet soil core upright and measure its height, let the core remain in upright position and dry, and measure its height when dry. The linear coefficient of expansion (LE) is

TABLE 7-5   Soil Erodibility Factor (K) by Soil Texture in Tons/Acre

Textural Class	Organic Matter Content (%)		
	0.5	2	4
Fine sand	0.16	0.14	0.10
Very fine sand	0.42	0.36	0.28
Loamy sand	0.12	0.10	0.08
Loamy very fine sand	0.44	0.38	0.30
Sandy loam	0.27	0.24	0.19
Very fine sandy loam	0.47	0.41	0.33
Silt loam	0.48	0.42	0.33
Clay loam	0.28	0.25	0.21
Silty clay loam	0.37	0.32	0.26
Silty clay	0.25	0.23	0.19

Source: U.S. Department of Agriculture and Environmental Protection Agency (1975).

$$LE = \frac{\text{wet height} - \text{dry height}}{\text{dry height}} \times 100 \qquad (7\text{-}3)$$

Engineering classes G and S can be obtained from soil texture analysis. Class G is called *gravelly*, and more than 50% of the soil is composed of particles more than 0.2 in. in diameter. Class S is called *sandy* and includes all sands and loamy sands.

The erodibility factor, K, can be estimated from an equation provided by the U.S. Department of Agriculture, Agricultural Research Service (1978), but a K factor for most soils can be obtained from Table 7-5.

7.3.2   Acid-Sulfate Soils

In coastal areas, some soils developed in brackish-water marshes contain iron pyrite and may become extremely acidic upon exposure to air. Such soils are called *potential acid-sulfate* soils. Rickard (1973) described the process of potential acid-sulfate soil formation. When rivers and runoff water carrying heavy loads of suspended solids emptied into estuaries, sediment was deposited near shore. After the deposits rose above mean low-water level, vegetation became established. As deposition continued, the coast slowly accreted, and a swamp forest developed. In the swamp forest, tree roots trapped organic and inorganic debris, decomposition in dense masses of organic debris resulted in anaerobic conditions, and sulfur-reducing bacteria became abundant. Sulfides produced by bacteria accumulated in pore spaces of muds as hydrogen sulfide and combined with ferrous iron to form precipitates of iron sulfide. The iron sulfide underwent further chemical reaction to form iron disulfide that crystallized to form iron pyrite. The reactions may be summarized as

$$2CH_2O \text{ (organic matter)} + SO_4^{2-} \rightarrow H_2S + 2HCO_3^-$$
$$Fe(OH)_2 + H_2S \rightarrow FeS + 2H_2O$$
$$FeS + S \rightarrow FeS_2 \text{ (iron pyrite)}$$

As long as soils containing iron pyrite are submerged and anaerobic, they remain reduced and change little. However, if they are drained and exposed to the air, oxidation occurs according to the following equations:

$$FeS_2 + H_2O + 3.5O_2 \rightarrow FeSO_4 + H_2SO_4$$
$$2FeSO_4 + 0.5O_2 + H_2SO_4 \rightarrow Fe_2(SO_4)_3 + H_2O$$
$$FeS_2 + 7Fe_2(SO_4)_3 + 8H_2O \rightarrow 15FeSO_4 + 8H_2SO_4$$

According to Sorensen et al. (1980), the production of ferric sulfate from ferrous sulfate is greatly accelerated by the activity of bacteria of the genus *Thiobacillus*, and under acidic conditions the oxidation of pyrite by ferric sulfate is very rapid. In addition, ferric sulfate may hydrolyze according to the reactions

$$Fe_2(SO_4)_3 + 6H_2O \rightarrow 2Fe(OH)_3 + 3H_2SO_4$$
$$Fe_2(SO_4)_3 + 2H_2O \rightarrow 2Fe(OH)(SO_4) + H_2SO_4$$

Ferric sulfate also may react with iron pyrite to form elemental sulfur, and the sulfur may be oxidized to sulfuric acid by microorganisms:

$$Fe_2(SO_4)_3 + FeS_2 \rightarrow 3FeSO_4 + 2S$$
$$S + 1.5O_2 + H_2O \rightarrow H_2SO_4$$

Soils are termed *acid-sulfate* once oxidation of pyrite occurs.

Ferric hydroxide can react with adsorbed bases, such as potassium in acid-sulfate soils to form jarosite, a basic iron sulfate (Gaviria et al. 1986):

$$3Fe(OH)_3 + 2SO_4^{2-} + K^+ + 3H^+ \rightarrow KFe_3(SO_4)_2(OH)_6 \cdot 2H_2O + H_2O$$

Jarosite is relatively stable, but in older acid-sulfate soils where acidity has been neutralized, jarosite tends to hydrolyze:

$$KFe_3(SO_4)_2(OH)_6 \cdot 2H_2O + 3H_2O \rightarrow$$
$$3Fe(OH)_3 + K^+ + 2SO_4^{2-} + 3H^+ + 2H_2O$$

Sulfuric acid dissolves aluminum, manganese, zinc, and copper from soil, and runoff from acid-sulfate soils not only is highly acidic but it may contain potentially toxic metallic ions. The acid from acid-sulfate soil depends largely on the amount and particle size of the iron pyrite, the presence or absence of exchangeable bases and carbonates within the iron-pyrite-bearing material, the exchange of oxygen and solutes with the iron pyrite, and the abundance of *Thiobacillus*. Because the exchange of oxygen and solutes and the abundance of *Thiobacillus* are restricted with depth, acid-sulfate acidity in soils is essentially a surface problem.

Waterlogging of acid-sulfate soils restricts the availability of oxygen, and sulfuric acid production ceases when the soil becomes anaerobic. In fact, sulfate is reduced to sulfide under anaerobic conditions by bacteria of the genus *Desulfovibrio*. In ponds, the mud–water interface often is aerobic, so sulfuric acid production can occur on pond bottoms.

An acid-sulfate soil will have a pH less than 3.5 (1:1 mixture of soil and distilled water). The yellowish mottling of the mineral jarosite can sometimes be seen in surface layers. In strongly acid-sulfate soils, yellow deposits of elemental sulfur are often visible and the odor of sulfur is apparent. In Australia and Indonesia, we have often seen 0.25- to 0.5-in.-diameter by 1- to 2-in.-long cylindrical, rust-colored particles of iron oxides and hydroxides on levee surfaces where soils were highly acidic.

A potential acid-sulfate soil has a high total sulfur content (usually more than 0.75%). If it is dried, its pH often drops to 3.5 or less. Fleming and Alexander (1961) reported that dry soil-distilled water mixtures made from soils from a tidal marsh area in South Carolina had pH values of 2 to 3. These soils contained up to 5.5% total sulfur. Soils from some areas in Southeast Asia (Thailand and Philippines) contained between 2% and 5%. Another test for potential acid-sulfate soil is to take some of the soil and mix it thoroughly with 30% hydrogen peroxide. After 10 min, the pH of the mixture should be 2.5 or less. Universal pH paper with a pH range of 0–6 is suitable for field use to identify potential acid-sulfate soils.

Singh (1980) provided information on acid-sulfate soils at the Brackishwater Aquaculture Center in Iloilo, Philippines (Table 7-6). Drying of the soil resulted in decreased soil pH because of oxidation of iron pyrite upon exposure to the air. The lime requirements calculated from the total potential acidity values in Table 7-6 ranged from 38,000 to

TABLE 7-6   Total Acidity and pH of Acid-Sulfate Soils, Brackishwater Aquaculture Center, Leganes, Iloilo, Philippines

Profile Depth (in.)	Wet Soil pH	Dry Soil pH	Total Potential Acidity (meq H^+/100 g soil)
0–6	3.6	3.4	70
6–12	4.4	3.6	61
12–18	4.4	3.4	87
18–24	5.7	3.5	93
24–30	6.3	3.5	95

Source: Based on data in Singh (1980).

64,000 lb/acre of $CaCO_3$. This calculation assumes that the weight of soil in a 6-in. layer weighs about 1,400,000 lb/acre. The acid-sulfate soils at the Center resulted in low pH in pond water. The problem was especially severe after rains. Iron pyrite in the levees would oxidize during dry weather, and highly acidic runoff would enter ponds after rains. This would further lower pH in pond water and kill or stress fish and cause low production. Similar observations have been made in many areas where acid-sulfate soils occur in ponds.

There are several ways of dealing with potential acid-sulfate soils: (1) avoid such areas for pond construction; (2) drain soils and wait until natural oxidation and leaching removes the acidity; (3) apply large amounts of lime to neutralize the acidity; (4) build ponds and use techniques for rapidly oxidizing and removing the acidity; (5) prevent oxidation of the iron pyrite so that acidity is not expressed.

In some places where it is desired to grow shrimp, sites without potential acid-sulfate soils simply cannot be located. It takes many years for the acidity to be leached from acid-sulfate soil, so it is not economically feasible to build ponds and wait several years to use them. Neutralization of acidity through liming is seldom feasible because lime requirements normally range from 25 to 150 tons/acre.

Brinkman and Singh (1982) developed a method for rapid reclamation of brackish-water ponds with acid-sulfate soils. The procedure is as follows:

1. In the early part of the dry season, dry the pond bottom and harrow thoroughly.
2. Fill with brackish water. Frequently measure the pH of the water. The pH will drop from that of seawater (7–9) to below 4. Once the pH has stabilized, drain the pond. Repeat this procedure until the pH stabilizes

at a pH above 5. Often, three or more drying and filling cycles are required.

3. At the same time the pond bottom is being reclaimed, acid must be removed from the surrounding levees. To achieve this, level the levee tops and build small bunds along each side of the levee tops to produce shallow basins. Fill the basins with brackish water. When the pond is drained for drying, also drain the small basins on the levee tops for drying. Repeat if necessary. Finally, remove the bunds and broadcast agricultural limestone over the tops and sides of levees at $0.1-0.2 \, lb/ft^2$.

4. Once the last drying and refilling cycle is complete, broadcast agricultural limestone over the pond bottom at 500 lb/acre. Fertilize the pond with manures or chemical fertilizers as necessary to promote phytoplankton blooms.

5. To prevent fish kills by seepage of acid from levees, check pH frequently and apply agricultural limestone if necessary.

Gaviria et al. (1986) suggested that most problems with acidity can be solved through reducing oxidation rates of iron pyrite. They suggested that ponds always be full of water. After harvest, ponds should be refilled immediately to prevent the bottoms from drying. Ponds should be no deeper than necessary, and levees should have the smallest possible surface area. Grass cover should be established on levees to provide a barrier between the air and the iron pyrite and to minimize the contact of runoff and raw soil. Levees should be treated with agricultural limestone at $0.1-0.2 \, lb/ft^2$ and planted with native grass. Some experimentation may be required to determine the best species of grass. The grass should be fertilized and, during the dry season watered. Even with these measures, some acidity will be produced in pond bottoms and runoff will carry some acidity into ponds. Total alkalinity should be monitored, and if alkalinity drops below 40 or 50 ppm, agricultural limestone should be applied at 1 ton/acre.

7.4 SEDIMENT

Sedimentation is often a problem in brackish-water aquaculture ponds, because erosion on drainage basins of rivers brings large amounts of suspended soil particles into estuaries that serve as pond water supplies. Tidal action in estuaries maintains solids in suspension, but when turbid water is placed in ponds its motion is reduced and soil particles rapidly settle. Reducing water motion allows sand and silt particles to settle. Clay particles are electronegatively charged and may remain suspended even in

TABLE 7-7   Depth of Sediment Deposited over a Four-Month Period in Two Shrimp Ponds in Thailand

	Depth (in.)	
Month	Pond 1	Pond 2
1	0.28	0.43
2	0.67	0.35
3	1.02	0.63
4	1.38	1.10
Total	3.35	2.51

still water because of mutual repulsion. Dissolved cations can neutralize the electric field around clay particles and cause them to settle when turbulence is reduced. Salinity values of even 2–3 parts per thousand (ppt) greatly accelerate sedimentation rates of clay over the rate that clay settles in freshwater. Thus, brackish-water ponds are excellent settling basins.

In places where water supplies contain high concentrations of suspended soil particles, considerable sedimentation may occur if water is pumped directly into ponds. For example, data supplied by Yont Musig, Kasetsart University, Bangkok, Thailand (Table 7-7), showed that the average depth of loosely consolidated sediment in two shrimp ponds at Samut Songkram, Thailand, averaged 2.6 in. and 3.3 in. over a four-month period. The rate of sediment accumulation increased with time during the crop because water-exchange rates were increased as shrimp grew and feeding rates were increased. The sediment was 3.1% organic matter, 33.2% clay, 21.0% silt, and 45.8% sand on a dry-weight basis.

In Ecuador, shrimp farms in the Guayas River estuary receive heavy sediment loads. Most farms use the water supply canal as a settling basin (Fig. 7-6) and dredge sediment from the canal at regular intervals. Based on data from one farm, if all of the sediment was allowed to settle uniformly over the pond bottoms, it would form a 3-in.-deep layer within a year. Ponds have average depths of about 3 ft, so 3 in./year of sediment is an appreciable reduction in pond volume. This farm contains 1,100 acres of ponds, and sediment yield in the water supply canal is about 250,000 tons/year.

A better way of removing sediment is to provide a separate settling pond before the supply canal (Fig. 7-7). Regardless of the sedimentation system, in order to affect sedimentation water velocity must be greatly reduced, and a retention time of at least 2–4 hr is needed. Unless sedi-

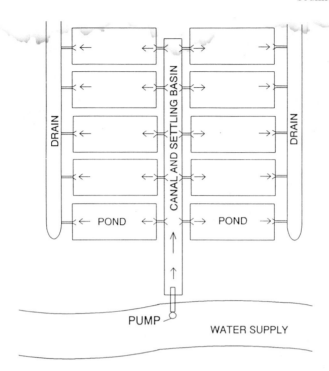

FIGURE 7-6.   Water supply canal used as a settling basin.

ment is removed from settling basins at regular intervals, they will fill in, have shorter retention times, and be less effective. Dredges or draglines may be used to remove sediment.

Heavy aeration of ponds can create sediment problems in the ponds where incoming water is free of settleable solids (Boyd 1992). Water currents created by aerators may erode the bottom in one part of a pond, and the suspended bottom material may settle in another part of the pond where currents are slower. It is popular in intensive shrimp ponds to position aerators in the corners of ponds to create a circular motion of water (Fig. 7-8). If excessive aeration is used, bottom material is eroded from inside slopes of the levees and around the periphery of the pond bottom. The eroded soil material and organic matter then settles to the bottom in the central part of the pond. This reshapes the pond bottom (Fig. 7-8) and, because water flows most easily in the deeper parts of a pond, further reduces water velocity over the central area of the pond. Anaerobic conditions typically develop in the central area, making it undesirable habitat for shrimp. Water currents in the peripheral area

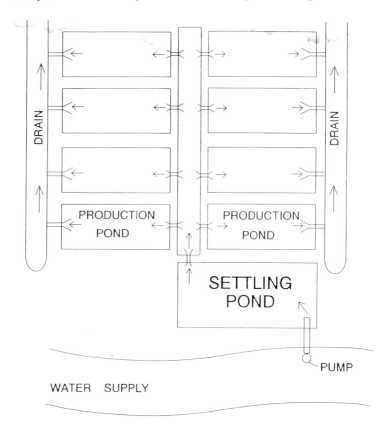

FIGURE 7-7.   Separate settling basin for removing settleable solids from water before it enters the supply canal.

often are so high that aerators must be turned off so that feed will sink when shrimp are fed. Such strong water currents obviously are undesirable for shrimp, for they must expend considerable energy to maintain their positions or to move against the currents.

Small, circular basins have been used for intensive culture of shrimp. Wyban et al. (1988) and Wyban and Sweeney (1989) used a 0.086-acre experimental pond built with 4.3-ft-high concrete walls and an earthen bottom with a 2% slope from the perimeter to the 8-in.-diameter center drain and a 0.5-acre commercial prototype round pond with center drain for intensive shrimp production. In the experimental pond, sedimentation rates in the pond center were 81% greater than at mid-radius and 172% greater than at the pond perimeter. They concluded that the round pond

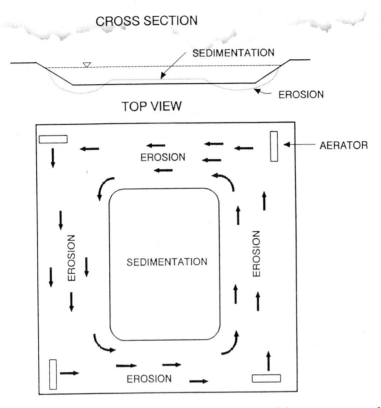

FIGURE 7-8. Erosion and sedimentation in a pond caused by water currents from aerators.

design was an effective technique for concentrating sediment so that it could be flushed through the center drain. Center drains work well in small basins, but not in larger ones. In ponds of area 0.5 acre or more, it is not economically practical to produce strong enough water currents to move solid wastes to the center, and even if it were soil eroded by these strong currents would settle in the center of the basin and bury the drain.

7.5 TURBIDITY AND COLOR

Sedimentation may also be a problem in freshwater ponds, but turbidity is normally of more concern than actual sediment accumulation. Turbidity from suspended clay particles restricts light penetration and reduces photosynthesis rates. The source of suspended soil particles in freshwater ponds usually is erosion on the watershed. If vegetative cover is established over the entire watershed, runoff entering the pond is less turbid and turbidity diminishes. If it is not possible to eliminate the source of turbidity, it may be possible to divert turbid runoff from a pond or to construct a settling basin above it to remove settleable material before it enters the pond. Alum, gypsum, or organic matter applications may be made to clear turbid water (Boyd 1990), but these techniques should only be used where clay turbidity does not settle from ponds after elimination of turbid runoff. Internal sources of pond turbidity also exist, including excessive wave action, heavy aeration, and bottom disturbances created by certain species of fish and other aquatic animals.

Some pond waters may be highly stained by dissolved and particulate organic matter. The source of this organic stain usually is vegetative remains that enter ponds from watersheds. Highly colored waters usually are acidic, and therefore have low rates of bacterial activity. Liming such ponds increases pH, and bacteria usually respond by decomposing the accumulated organic matter.

II

Design of Water Supply and Pond Systems

Aerial photograph showing expansion of the Small Pond Research Area at the Auburn University Fisheries Research Unit. (*Courtesy of the International Center for Aquaculture and Aquatic Environments.*)

8

Open-Channel Flows

8.1 INTRODUCTION

Streams and man-made channels have a free-water surface exposed to the atmosphere, so they are known as open channels. The cross-sectional area of flow can vary at a given place in a channel, and the velocity of flow depends on several variables. Pressure is atmospheric, and the energy causing the flow is elevational energy (gravity); gravity drives water in a channel from higher to lower elevations. This chapter discusses principles of open-channel flow and hydraulics; the design of unlined, lined, and vegetated channels are presented with some examples.

8.2 CONSERVATION OF ENERGY IN OPEN-CHANNEL FLOWS

In considerations of open-channel flows, we are often interested in the total available energy at a point relative to a reference plane of known elevation. The total energy of flow in any streamline passing through a channel section usually is expressed as the hydraulic head in feet of water. Figure 8-1 shows that the total energy head of an open-channel flow, h, consists of three terms: kinetic energy or velocity head, depth of flow in the channel, and elevation head. The elevation head is the vertical distance from the reference plane to the channel bottom. The total energy head (h) is written as

$$h = \frac{v^2}{2g} + y + Z \qquad (8\text{-}1)$$

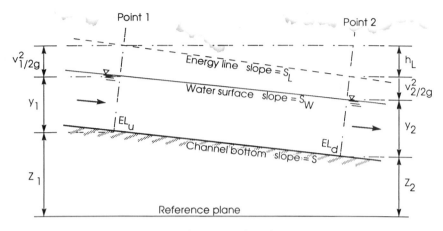

FIGURE 8-1.   Energy heads and the conservation of energy in an open-channel flow.

where $v^2/2g$ = kinetic energy or velocity head (ft)
v = flow velocity (ft/sec)
g = gravitational acceleration (g = 32.2 ft/sec^2)
y = depth of flow (ft)
Z = elevation head (ft)

Because total energy stays constant in a flow, the total energy at an upstream point in a channel must equal total energy at a downstream point:

$$h_1 = h_2 \qquad\qquad (8\text{-}2a)$$

or

$$\frac{v_1^2}{2g} + y_1 + Z_1 = \frac{v_2^2}{2g} + y_2 + Z_2 \qquad\qquad (8\text{-}2b)$$

where subscripts 1 and 2 represent upstream and downstream points, respectively, and all other terms are defined in equation (8-1). This is the well-known Bernoulli equation. However, elevation energy is converted to kinetic energy as water passes downstream, and some kinetic energy is converted to heat energy as the flowing water overcomes resistance to flow offered by the channel bottom. Conversion of kinetic energy to heat energy is known as *head loss* (h_L). According to the principle of conservation of energy, the total energy at the upstream section 1 should be equal to the total energy at the downstream section 2 plus the head loss between the two sections:

$$h_1 = h_2 + h_L \qquad (8\text{-}3a)$$

or

$$\frac{v_1^2}{2g} + y_1 + Z_1 = \frac{v_2^2}{2g} + y_2 + Z_2 + h_L \qquad (8\text{-}3b)$$

where h_L = head loss in feet, and all other terms are found in equation (8-1).

Example 8-1: Determining Total Energy at a Point in an Open-Channel Flow

A point on the bottom of an open channel with 3-ft average depth of flow is located 10 ft above a reference plane. The flow velocity in the channel is 5 ft/sec. What is the total energy of the flow?

Solution

Velocity head v_h is

$$v_h = \frac{v^2}{2g} = \frac{(5\,\text{ft/sec})^2}{2(32.2\,\text{ft/sec}^2)} = 0.39\,\text{ft}$$

Flow depth: $y = 3\,\text{ft}$
Elevation head: $Z = 10\,\text{ft}$
Total energy head: $h = 10 + 3 + 0.39 = 13.39\,\text{ft}$

As illustrated in Example 8-1, the velocity head term usually is small compared with other energy terms in open-channel flows. Furthermore, in calculation of open-channel flows, the distance above a reference plane is often inconsequential. We usually are interested in the elevation difference on the channel bottom between upstream and downstream points, for this elevation difference represents the primary source of energy available to drive water downstream in open-channel flows.

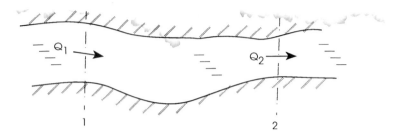

FIGURE 8-2.   The continuity of flow in an open channel.

8.3 CONTINUITY EQUATION

If there are no gains or losses in a channel reach, the discharge at one point must equal the discharge at a second point in the reach (Fig. 8-2), or

$$Q_1 = Q_2 \tag{8-4}$$

Because discharge is a product of flow cross-sectional area normal to the flow direction, A, and the mean velocity of the flow, v, the continuity equation may be written as

$$A_1 v_1 = A_2 v_2 \tag{8-5}$$

The continuity equation is invalid where water flows in or out along the channel section under consideration.

Example 8-2: Application of Continuity Equation

The flow depth in a rectangular, concrete-lined channel changes from 2 ft to 1.5 ft due to increased channel slope. The channel is 2 ft wide, and the velocity at the 2-ft-depth section is 4.5 ft/sec. What is the velocity at the downstream section of the channel?

Solution

Cross-sectional area at the upstream point: $A_1 = b_1 y_1 = 2 \times 2 = 4\,\text{ft}^2$
Cross-sectional area at the downstream point: $A_2 = b_2 y_2 = 2 \times 1.5 = 3\,\text{ft}^2$

From the continuity equation, the flow velocity at the downstream point is

$$v_2 = \frac{A_1 v_1}{A_2} = \frac{4 \times 4.5}{3} = 6.0\,\text{ft/sec}$$

8.4 THE MANNING EQUATION

For practical purposes open-channel flows in an aquaculture system are considered to be uniform flows that have the following features: (1) the depth of flow, cross-sectional area of flow, flow velocity, and discharge at every section of the channel reach are constant; and (2) the slopes of the energy line (S_L), water surface (S_W), and channel bottom (S) are all parallel, or $S_L = S_W = S$. Under these conditions the mean velocity of an open-channel flow may be expressed in the general form

$$v = CR^x S^y \tag{8-6}$$

where v = mean velocity (ft/sec)
C = flow resistance coefficient (dimensionless)
R = hydraulic radius (ft)
S = slope of channel (ft/ft)

Using this equation is very difficult without values for the flow resistance coefficient (C) and the exponents x and y. These terms have been determined experimentally by several researchers. One of the resulting equations is the well-known Manning equation

$$v = \frac{1.49}{n} R^{2/3} S^{1/2} \tag{8-7}$$

where n is Manning's surface roughness coefficient of the channel (dimensionless).

The Manning equation shows that channel flow is related to three factors: the nature of the channel bed (roughness coefficient, n), the ratio of the cross-sectional area of flow to the wetted perimeter of the inundated channel bed (hydraulic radius, R), and the slope of channel (S). The first two terms provide an estimate of the frictional resistance to flow, and the last term affords an estimate of kinetic energy available to cause flow. In applying the Manning equation the most difficulty comes in determining the roughness coefficient n, which is highly variable and depends on several factors. Sound judgment in selecting the roughness coefficients comes from experience, and individuals will select different values. Figure 8-3 shows a nomograph for solving the Manning equation. To find a velocity from the nomograph, the values for hydraulic radius

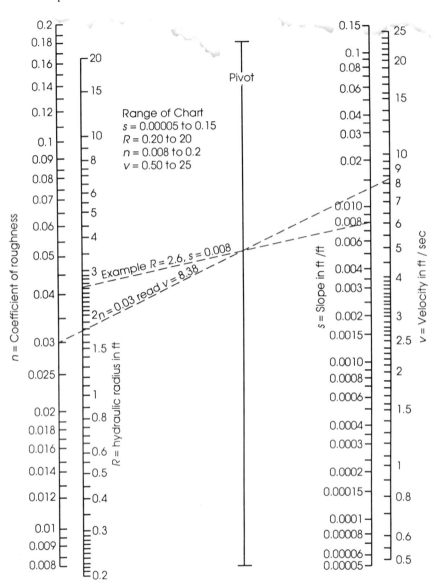

FIGURE 8-3. Nomograph for solving Manning's equation. (*Source: U.S. Soil Conservation Service 1979a.*)

TABLE 8-1   Surface Roughness Coefficient (*n*) of the Manning Equation for Open Channels

Type of Channel	Roughness Coefficient (*n*)
Excavated earth, straight and uniform	
Clean	0.018–0.020
Gravel	0.025–0.030
Short grass	0.027–0.030
Excavated earth, winding and sluggish	
Clean	0.025–0.030
Weedy	0.030–0.033
Dense weed	0.035–0.040
Excavated earth	
Asphalt-lined	0.013–0.016
Smooth-concrete-lined	0.015–0.020
Riprap	0.030–0.045
Brick	0.013–0.015
Masonry	0.025–0.030
Corrugated metal	0.025–0.030
Excavated earth, not maintained or lined,	
weedy, and sediment accumulation	0.060–0.075
Natural streams, straight	
Clean	0.030–0.033
Stones, weeds	0.035–0.040
Natural streams, winding	
Clean	0.040–0.045
Stones, weeds	0.045–0.050
Natural streams, sluggish, weedy, deep pools	0.070–0.080

Source: V.T. Chow, *Open-Channel Hydraulics*, Copyright © 1959 by McGraw-Hill, Inc., New York, selected from pp. 111–113. Reprinted by permission of McGraw-Hill, Inc.

and channel slope are connected to find a pivot point on the middle line. The pivot point is then connected to the value for roughness coefficient and extended to the velocity line. The figure shows an example solution of this method.

Sometimes, the Manning equation is presented in a slightly different form that permits direct calculation of discharge. The discharge of an open-channel flow is the product of cross-sectional area of flow and flow velocity:

$$Q = Av = \frac{1.49}{n} A R^{2/3} S^{1/2} \qquad (8\text{-}8)$$

The roughness coefficient (*n*) represents the resistance to flow caused by friction between flowing water and the channel bottom. Table 8-1 lists

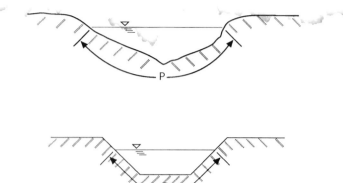

FIGURE 8-4.   Definition of wetted perimeter of open channels.

roughness coefficients for the Manning equation. Reference to Table 8-1 reveals that roughness coefficients for different types of channels differ greatly. The smaller values of n are recommended for channels with good maintenance. The values should be increased for channels in which poor maintenance is expected in the future.

The hydraulic radius is computed as the cross-sectional area of flow divided by the wetted perimeter:

$$R = \frac{A}{P} \tag{8-9}$$

where A = cross-sectional area of flow (ft^2) and P = wetted perimeter (ft). The definition of wetted perimeter in open channels is illustrated in Figure 8-4. It is the distance along the channel bottom below the water surface. Table 8-2 lists formulas to calculate geometric elements of common channel shapes for man-made channels.

Flow in an open channel is due to the slope of the energy line, as shown in Figure 8-1. In using the Manning equation, we assume that the energy line, water surface, and channel bottom are parallel. Therefore, the slope of the energy line can be estimated as the average slope of channel bottom (S) and is calculated as

$$S_L = S = \frac{EL_u - EL_d}{L} \tag{8-10}$$

where EL_u = elevation of channel bottom at upstream point (ft)
EL_d = elevation of channel bottom at downstream point (ft)
L = distance between the two points (ft)

TABLE 8-2 Geometric Parameters of Rectangular, Trapezoidal, Triangular, and Parabolic Channels

	Cross-sectional Area A	Wetted Perimeter P	Hydraulic Radius R	Top Width
Rectangular shape channel	by	$b + 2y$	$\dfrac{by}{b + 2y}$	$t = T = b$
Trapezoidal shape channel	$by + zy^2$	$b + 2y\sqrt{z^2 + 1}$	$\dfrac{by + zy^2}{b + 2y\sqrt{z^2 + 1}}$	$t = b + 2yz$ $T = b + 2Yz$
Triangular shape channel	zy^2	$2y\sqrt{z^2 + 1}$	Approx. $y/2$	$t = 2yz$ $T = Yt/y$
Parabolic shape channel	$0.67ty$	$t + 2.67y^2/t$	Approx. $0.67y$	$t = A/0.67y$ $T = t\sqrt{Y/y}$

Example 8-3: Application of the Manning Equation to Determine Flow Rate

The Manning equation will be used to estimate the mean flow velocity in a concrete-lined channel that originates at a spring of 600-ft elevation and travels 200 ft to a pond at 598 ft elevation. For simplicity, assume that the flow cross section is roughly rectangular and uniform, 0.5 ft deep and 2 ft wide.

Solution

Slope of channel bed:

$$S = \frac{600\,\text{ft} - 598\,\text{ft}}{200\,\text{ft}} = 0.01$$

Wetted perimeter: $P = b + 2y = 2\,\text{ft} + 2 \times 0.5\,\text{ft} = 3\,\text{ft}$
Cross-sectional area of flow: $A = by = 2\,\text{ft} \times 0.5\,\text{ft} = 1.0\,\text{ft}^2$
Hydraulic radius: $R = A \div P = 1.0\,\text{ft}^2 \div 3.0\,\text{ft} = 0.33\,\text{ft}$
Roughness coefficient: $n = 0.015$
Mean flow velocity: $v = \dfrac{1.49}{0.015} \times 0.33^{2/3} \times 0.01^{1/2}$
$$= 99.3 \times 0.48 \times 0.1 = 4.77\,\text{ft/sec}$$

Once mean velocity is found, the discharge (Q) can be readily calculated as

$$Q = Av = 1.0\,\text{ft}^2 \times 4.77\,\text{ft/sec} = 4.77\,\text{ft}^3/\text{sec}$$

Nevertheless, note that it is more accurate to measure channel flows with the methods discussed in Chapter 10. The Manning equation is used primarily in channel design and is rarely used to determine flow rate in the field.

8.5 BEST HYDRAULIC SECTION

Channel cross-sectional shapes can vary, but regardless of the shape selected it is usually desirable to maximize the hydraulic radius. The hydraulic radius is an index of the efficiency of the flow cross section of a channel. The amount of channel surface present to create frictional resistance to flow increases with increasing wetted perimeter. The larger that the flow area is compared with the wetted perimeter, the easier water moves through a channel. The channel section having the least wetted perimeter for a given cross-sectional area has the maximum hydraulic radius, which is known as the best hydraulic section. Table 8-3 lists geometric elements of best hydraulic sections commonly used in channel design. However, structural and economic considerations related to the nature of the soil and materials available for lining channels may impose limitations on the use of best hydraulic sections. The best hydraulic section should be applied only to the design of lined nonerodible channels (Chow 1959).

In aquacultural applications, most channels are trapezoidal. The best hydraulic section for a trapezoidal channel is defined when the channel bottom width is

$$b = 2y \tan \frac{\theta}{2} \tag{8-11}$$

where θ is the angle of side slope. Half of a hexagon, which has $\theta = 60°$, is the best hydraulic section for trapezoidal-shape channels. However, the

TABLE 8-3   Dimensions for the Best Hydraulic Sections

Cross Section	Area A	Wetted Perimeter P	Hydraulic Radius R	Top Width t
Trapezoid, half of a hexagon	$1.73y^2$	$3.46y$	$0.5y$	$2.31y$
Rectangle, half of a square	$2y^2$	$4y$	$0.5y$	$2y$
Triangle, half of a square	y^2	$2.83y$	$0.35y$	$2y$
Parabola	$1.88y^2$	$3.77y$	$0.5y$	$2.83y$

Note: y = depth of flow.

angle of the side slope varies, depending on the material with which the channel is built; a channel built in a sandy soil requires a milder side slope than one built in a clay soil. The best hydraulic section for a rectangular channel results when the channel width is twice the flow depth, which is half of a square.

8.6 STABILITY OF OPEN CHANNELS

The relationship between water velocity and erosion is a primary consideration in channel design. The discharge of a channel of given cross-sectional area increases with velocity, so high velocity is desirable because it favors a smaller channel cross section. High velocity also prevents sedimentation and filling of channels with silt. However, if the velocity is too great, the channel's sides and bottom will be degraded by erosion.

A number of factors favor channel erosion. A constriction in a channel causes velocity to increase and leads to erosion of the channel's sides and bottom. Impinging flow on the concave bank in a bend of a channel causes erosion in the same manner that it does in natural streams. Steep channel slopes increase velocity and favor erosion. Sudden changes of channel width may produce eddies, which increase the potential for erosion. At a channel junction, flow from one channel may impinge on the opposite bank of the other channel and cause erosion. Steep side slopes for channels cut through unstable soils favor rapid erosion. Increased flow in a channel because of heavy rains, changes in watershed management, or diversion of adjacent watercourses may also increase flow velocity and erosion.

Erosion may be reduced by sloping the channel sides in accordance with the soil stability and protecting the sides with stone or concrete. Sudden constrictions or increases in channel width should be avoided, but if they are necessary the channel should gradually change in width. Sharp bends in channels should be avoided. Drop spillways, chutes, and flumes should be used to control channel erosion where rapid elevation changes occur (Schwab et al. 1981). Channels may be lined throughout or in vulnerable reaches to control erosion. However, these erosion control measures increase already high construction costs.

8.6.1 Channel Side and Bottom Slopes

Channel side slopes are often measured as the ratio of horizontal distance to vertical rise; for example, a 2:1 side slope indicates a displacement of 2 ft horizontally for a 1-ft rise in elevation (Fig. 8-5). In some countries side slopes are measured as the ratio of vertical rise to horizontal dis-

FIGURE 8-5.   Definition of side slope and side-slope factor (z) for open channels.

tance. Aquaculturists studying channel design should be extremely careful to notice how the side slopes are given in a particular case. Side slope also may be given as a side-slope factor (z), which represents the horizontal displacement in feet necessary to provide 1 ft of rise. A side slope of 2:1 is the same as a slope with a z value of 2. The slope can also be represented as an angle. For 2:1 side slopes shown in Figure 8-5, the slope angle is

$$\tan \theta = \frac{1}{z} = \frac{1}{2} \tag{8-12a}$$

Then

$$\theta = \tan^{-1}\left(\frac{1}{2}\right) = 26.4° \tag{8-12b}$$

Allowable side slopes and corresponding slope angles to prevent erosion are presented in Table 8-4 for different channel materials. Notice that a z value of 1.5 is permissible for a stiff clay soil, but the z value should be increased to 3 for a loose clay or sandy loam.

Excessive flow velocity erodes the sides and scours the bottom even in a channel with proper side slopes. The maximum permissible velocity is the greatest mean velocity that will not cause channel erosion. This velocity is usually uncertain, depending on the condition of the channel flow for a given channel material. Old channels sustain higher velocity than newly constructed channels, because they are usually well stabilized (Chow 1959). If other conditions are the same, a deep channel can convey water at higher velocity than a shallower one without causing erosion. One of the purposes of lining a channel is to prevent erosion. However, if the flow velocity is extremely high, the lining material may be damaged.

The maximum permissible velocity for unlined earthen channels may be calculated from an equation of Simon (1976):

TABLE 8-4   Allowable Side-Slope Factors and Erosion Coefficients for Various Channel Materials

Soil Type	Side Slope z	θ (deg)	G
Sandy loam	3.0	18.4	2.0
Silty clay	3.0	18.4	2.5
Silty sand	2.0	26.5	2.5
Soft shale	2.0	26.5	3.2
Stiff clay	1.5	33.7	3.5
Soft sandstone	1.5	33.7	3.5
Riprap lining of excavation	1.0	45.0	—
Concrete lining of excavation	0.5–1.0	63–45	—
Peat	0.25	76.0	—
Rock	0.0	90.0	—

Source: U.S. Bureau of Reclamation (1952).

$$v_{max} = Gy^{0.2} \tag{8-13}$$

where v_{max} = maximum permissible velocity (ft/sec)
G = erosion coefficient (Table 8-4)
y = normal flow depth (ft)

The maximum permissible velocities given in Table 8-5 may be also used in designing erodible channels. These values are for aged channels with small slopes and flow depths less than 3 ft.

The flow velocity in a channel must also be maintained above a minimum value or siltation will occur in the channel. The minimum permissible velocity (v_{min}) to prevent siltation may be calculated as follows (Simon 1976):

$$v_{min} = 0.63y^{0.64} \tag{8-14}$$

Example 8-4: Determining v_{max} and v_{min}

A trapezoidal-shape channel ($z = 3.0$) carrying 50-cfs flow will be built into a sandy-loam soil having a channel slope of 1.2%. The channel will be 12 ft wide on the bottom and 0.65 ft deep. Will erosion or siltation occur in the channel?

TABLE 8-5   Maximum Permissible Velocities for Various Channel Materials (for straight channels of small slopes, after aging)

		v_{max} (ft/sec)	
Channel Material	Manning's n	Clear Water	Water Carrying Colloidal Silts
Fine sand, colloidal	0.02	1.50	2.50
Sandy loam, noncolloidal	0.02	1.75	2.50
Silt loam, noncolloidal	0.02	2.00	3.00
Alluvial silts, noncolloidal	0.02	2.00	3.50
Ordinary firm loam	0.02	2.50	3.50
Volcanic ash	0.02	2.50	3.50
Stiff clay, very colloidal	0.025	3.75	5.00
Alluvial silts, colloidal	0.025	3.75	5.00
Shales and hardpans	0.025	6.00	6.00
Fine gravel	0.02	2.50	5.00
Graded loam to cobbles when noncolloidal	0.030	3.75	5.00
Graded silts to cobbles when colloidal	0.030	4.00	5.50
Coarse gravel, noncolloidal	0.025	4.00	6.00
Cobbles and shingles	0.025	5.00	5.50

Source: U.S. Bureau of Reclamation (1952).

Solution

Cross-sectional area:

$$A = by + zy^2 = 12 \times 0.65 + 3.0 \times 0.65^2 = 9.1 \text{ ft}^2$$

Expected flow velocity:

$$v = \frac{Q}{A} = \frac{50 \text{ cfs}}{9.1 \text{ ft}^2} = 5.5 \text{ ft/sec}$$

Erosion coefficient: $G = 2.0$ for sandy loam soil
Maximum permissible velocity: $v_{max} = 2.0 \times 0.65^{0.2} = 1.83 \text{ ft/sec}$
Minimum permissible velocity: $v_{min} = 0.63 \times 0.65^{0.64} = 0.48 \text{ ft/sec}$

The expected velocity exceeds the maximum permissible velocity, but it is higher than the minimum velocity. Siltation is not expected, but the channel will have a scouring problem due to the high velocity.

If design velocity exceeds maximum permissible velocity, the channel bottom slope must be limited because excessive slope leads to scouring,

due to high velocity. In some cases, excessive slope can be reduced as the channel slope. Common hydraulic structures used are straight drop structures, such as shown in Figure 8-6. They are concrete weirs where the elevation difference between two structures is concentrated at the downstream-side structure. The required distance, L, between two drop structures may be calculated as (Kraatz and Mahajan 1975)
structure. The required distance, L, between two drop structures may be calculated as (Kraatz and Mahajan 1975)

$$L = \frac{h - 0.3}{S_o} \tag{8-15}$$

where L = required distance between two drop structures (ft) (Fig. 8-7)
h = drop height (ft)
S_o = original slope of channel (ft/ft)

A minimum of 0.3 ft is recommended for smooth flow and to facilitate operation of the structures. Structure sizes are determined by the amount of flow carried in the channel. The flow capacity over the structure may be calculated by weir equations, which are discussed in Chapter 10. When drop structures are installed the downstream side must be well protected to prevent scouring by the concentrated hydraulic energy. Total length of the protected floor along the channel should be

$2 \times$ (water depth above the crest + 3.6 ft) + drop height

(Kraatz and Mahajan 1975).

Example 8-5: Determining the Required Distance between Drop Structures in a Sloping Channel

A channel described in Example 8-4 will be used. Drop structures that afford a 3-ft elevation drop will be installed.

Solution: The distance between two drop structures is

$$L = \frac{3 \text{ ft} - 0.3 \text{ ft}}{0.012} = 225 \text{ ft}$$

The channel should be installed with drop structures placed at 225-ft or shorter intervals to prevent scouring of the channel bottom.

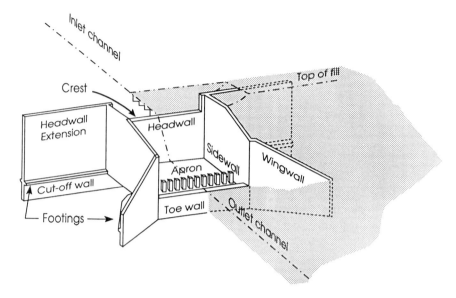

FIGURE 8-6.   A straight drop structure and its components.

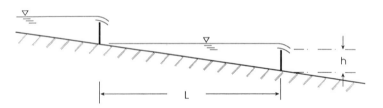

FIGURE 8-7.   Application of a drop structure to reduce flow velocity and protect channel from erosion.

8.6.2 Freeboard of Channel

In channel construction it must be remembered that discharge may exceed the design discharge after an unexpected high flow. This may damage the embankment by overflow or wave effects. Therefore, the channel must be deeper than the depth computed for normal discharge to protect the embankment from damage. The vertical distance from normal water level to the top of the channel bank is called *freeboard* (F.B.). Simon (1976) suggests a simple calculation of freeboard as follows:

$$F.B. = \sqrt{Cy} \qquad (8\text{-}16)$$

where $F.B.$ = freeboard (ft)

$\qquad y$ = normal depth of flow (ft)

$\qquad C$ = 1.5 for Q < 100 cfs; 2.0 for Q of 100–3,000 cfs; 2.5 for Q > 3,000 cfs

Example 8-6: Determining Channel Freeboard

A trapezoidal-shape channel has 2:1 side slopes and is to carry a flow of 50 cfs. The normal depth of flow in the channel, y, is 2 ft. Determine the freeboard needed and the total depth of the channel.

Solution

Coefficient: C = 1.5 for Q = 50 cfs

Freeboard: $F.B.$ = $\sqrt{Cy}$ = $\sqrt{1.5 \times 2.0}$ = 1.73 ft

Total channel depth: $Y = y + F.B.$ = 2 ft + 1.73 ft = 3.73 ft

Vegetation or other types of protection should be established on freeboard slopes to prevent erosion by wave effects or rain. If channels convey water periodically and for short intervals (e.g., drainage channels and emergency spillways), the entire channel should be planted with grass (U.S. Soil Conservation Service 1986). Many varieties of grass are suitable, such as Bermuda grass, Kentucky bluegrass, centipede grass, or crabgrass. It is generally recommended that native vegetation be used when available. The details of designing vegetated channels are discussed later in this chapter.

8.6.3 Channel Lining

Channels may be lined throughout or in vulnerable reaches such as bends, steep slopes, changes in width, reaches with unstable soil, and channel junctions to control erosion and scouring. Materials for channel lining include grass, rock riprap, fine clay, wood, concrete, asphalt, metal, and gabion. Today, one may even use geotextile materials, prefabricated and flexible lining systems made of plastic or rubber-based materials. However, channel linings other than grass are expensive. Biodegradable geotextile materials are often used to protect the freeboard slopes of newly built channels until vegetal cover is established.

For lined channels the height of the lining and the height of the bank

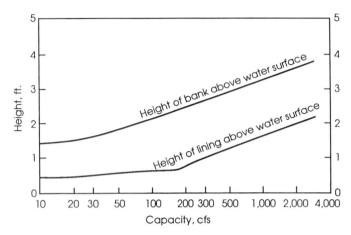

FIGURE 8-8. Recommended bank and lining heights above water surface. (*Source: U.S. Bureau of Reclamation 1952.*)

above the normal water surface depend on several factors. The U.S. Bureau of Reclamation (1952) developed a guide for channel lining for average bank height and height of lining above the water surface in relation to channel flow capacities (Fig. 8-8). Lining a channel with impervious materials prevents seepage losses and controls channel erosion. This is an important practice where water is pumped from a source and conveyed by an open channel built in pervious soils. When seepage loss from a channel is excessive, a part or all sections of the channel should be considered for lining. The seepage loss from an unlined channel may be calculated from the following equation, developed by the U.S. Bureau of Reclamation (1952):

$$\text{SEEP} = C \sqrt{\frac{Q}{v}} = C\sqrt{A} \tag{8-17}$$

where SEEP = seepage rate (cfs/mile)
$\quad\quad\quad$ C = seepage coefficient of channel material
$\quad\quad\quad\quad$ (from Table 8-6)
$\quad\quad\quad$ Q = flow rate (cfs)
$\quad\quad\quad$ v = mean velocity of channel flow (ft/sec)
$\quad\quad\quad$ A = cross-sectional area of flow (ft^2)

Example 8-7: Seepage Loss from an Unlined Channel

An unlined trapezoidal channel built in a sandy and gravelly soil receives 10 cfs of pumped water from a stream. The mean flow velocity of the

TABLE 8-6   Seepage Coefficients for Various Channel Materials

Soil Type	C
Clay and clayey loam	0.08
Sandy loam	0.13
Volcanic ash with sand	0.20
Sandy soil with rock	0.34
Sandy and gravelly soil	0.44

Source: U.S. Bureau of Reclamation (1952).

channel is 0.5 fps. The total channel length is 0.5 mile. Determine the seepage loss from this channel.

Solution: The seepage coefficient is $C = 0.44$ for a sandy and gravelly soil. The seepage loss per mile in the channel is

$$\text{SEEP} = 0.44 \times \sqrt{\frac{10}{0.5}} = 1.97 \text{ cfs/mile}$$

Then total seepage loss along the channel is

$$1.97 \text{ cfs/mile} \times 0.5 \text{ mile} = 1.0 \text{ cfs}$$

About 10% of the 10 cfs of pumped water will be lost by the seepage in the channel. Lining should be considered for this channel for more efficient conveyance of the pumped water. In addition to the control of erosion and seepage, other benefits of channel lining include lower maintenance cost, smaller channel cross section requirement, and steeper permissible bottom and side slopes. Economic factors that should be considered include costs of channel lining and pumping.

8.7 CHANNEL DESIGN

Three types of channel design are discussed: nonerodible channels, erodible channels, and vegetated channels. The objective of channel design is to determine the minimum dimensions of a channel that can carry the design flow without causing erosion or siltation in the channel. The most common shape of man-made channels is trapezoidal. Rectangular, triangular, parabolic, and semicircle shapes are also used.

8.7.1   Nonerodible Channels

Most nonerodible channels are lined with concrete, asphalt, clay, or plastic sheets that can resist erosion and prevent seepage. Concrete is the most common lining material because of its low cost and availability. Some unlined channels that are excavated in firm materials, such as rocks or clay bed, are considered nonerodible channels. The maximum permissible velocity is not used in designing nonerodible channels. However, the minimum permissible velocity should be considered where water carries silts. A mean velocity of 2–3 fps may be accepted as the minimum permissible velocity for nonerodible channels to prevent siltation (Chow 1959).

Design factors for nonerodible channels are flow capacity, roughness coefficient, channel shape, channel bottom and side slopes, minimum permissible velocity, hydraulically or economically efficient section, and freeboard. The design of a nonerodible channel for a given flow rate may consist of the following procedures.

1. Determine side-slope factor (z) from Table 8-4, roughness coefficient (n) from Table 8-1, and channel bottom slope (S).
2. Assume bottom width (b) and determine normal flow depth (y) using the best hydraulic section.
3. Determine hydraulic radius (R).
4. Calculate flow velocity and flow rate using the Manning equation.
5. Repeat steps 2 through 4 until calculated Q and design Q agree.
6. If the b/y ratio is satisfactory both economically and hydraulically, add freeboard to determine the final dimensions of the channel. Otherwise, adjust b and repeat steps 1–5.

The final dimensions should be decided on the basis of hydraulic efficiency and construction practicality. Figure 8-9 is a guide by the U.S. Bureau of Reclamation (1952) in selecting proper section dimensions of lined channels. The figure shows average bottom width and flow depth relative to flow capacity of a channel.

Example 8-8: Designing a Nonerodible Channel

A concrete-lined channel will be constructed to convey 30 cfs of water over a distance of 3,000 ft. The elevation difference between two ends of the channel is 4 ft, and the slope is relatively uniform. The water carries a moderate amount of silts. Determine the dimensions of a trapezoidal-shape channel.

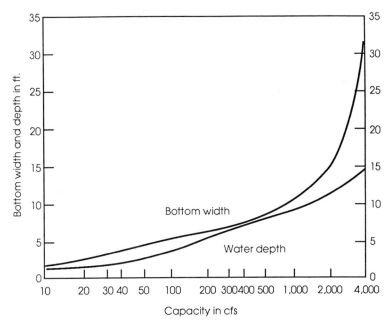

FIGURE 8-9.   Bottom width and flow depth of lined channels. (*Source: U.S. Bureau of Reclamation 1952.*)

Solution: Manning's roughness coefficient is $n = 0.015$. The channel bottom slope is

$$S = \frac{EL_u - EL_d}{L} = \frac{4\,\text{ft}}{3,000\,\text{ft}} = 0.0013 \quad \text{[from equation (8-10)]}$$

The best hydraulic section is calculated as follows, assuming an allowable side slope of $z = 0.5$ from Table 8-4. The side-slope angle is

$$\theta = \tan^{-1}\!\left(\frac{1}{z}\right) = \tan^{-1}\!\left(\frac{1}{0.5}\right) = 63.0°$$

Then, for the best hydraulic section

$$b = 2y \tan\!\left(\frac{63°}{2}\right) = 2y \tan 31.5° = 1.23y \quad \text{[from equation (8-11)]}$$

TABLE 8-7   Solution for Example 8-8: Design of a Nonerodible Channel

Trial No.	b (ft)	y (ft)	A (ft²)	P (ft)	R (ft)	v (fps)	Q (cfs)	Comment
1	4.0	3.23	18.3	11.26	1.62	4.9	89.7	High
2	2.0	1.61	4.51	5.6	0.81	3.11	14	Low
3	3.0	2.42	10.19	8.4	1.21	4.07	41.5	High
4	2.7	2.18	8.26	7.57	1.09	3.79	31.3	OK

Assume a bottom width $b = 4$ ft. Then the flow depth is $y = 4 \div 1.23 = 3.25$ ft and the cross-sectional area is

$$A = by + zy^2 = 4 \times 3.25 + 0.5 \times 3.25^2 = 18.3 \, \text{ft}^2$$

The wetted perimeter is

$$P = 4 + 2 \times 3.25 \sqrt{0.5^2 + 1} = 11.26 \, \text{ft}$$

The hydraulic radius is

$$R = \frac{A}{P} = \frac{18.3 \, \text{ft}^2}{11.26 \, \text{ft}} = 1.62 \, \text{ft}$$

From the Manning equation the flow velocity is

$$v = \frac{1.49}{0.015} \times 1.62^{2/3} \times 0.0013^{1/2} = 4.9 \, \text{fps}$$

$$Q = vA = 4.9 \, \text{fps} \times 18.3 \, \text{ft}^2 = 89.7 \, \text{cfs} \gg \text{design } Q$$

Repeat the above steps until the calculated Q agrees with the design Q. The problem is solved by trial and error, as shown in Table 8-7.

Trial 4 gives acceptable agreement between calculated Q and design Q. Freeboard for the final dimension is

$$F.B. = \sqrt{Cy} = \sqrt{1.5 \times 2.18} = 1.8 \, \text{ft} \quad [\text{from equation (8-16)}]$$

By incorporating the freeboard the final dimensions of the channel are determined as follows:

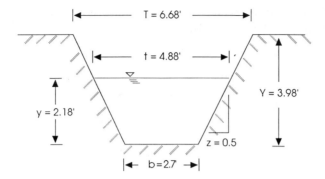

FIGURE 8-10.   Final dimensions of a nonerodible channel design in Example 8-8.

Total depth of the channel:

$$Y = y + F.B. = 2.18 + 1.8 = 3.98\,\text{ft}$$

Width of the water surface:

$$t = b + 2yz = 2.7 + 2 \times 2.18 \times 0.5 = 4.88\,\text{ft}$$

Top width of the channel:

$$T = b + 2Yz = 2.7 + 2 \times 3.98 \times 0.5 = 6.68\,\text{ft}$$

The final dimensions of the designed nonerodible channel are shown in Figure 8-10.

8.7.2 Erodible Channels

When channels are directly excavated in unstable materials, they are called *erodible* channels. Erodible channels are subject to scouring and erosion due to erosive forces of the flowing water. The design of erodible channels requires a stable channel condition that does not cause erosion of the channel bottom. The stability of erodible channels depends mainly on the properties of channel bed material and the flow velocity in the channel. The maximum permissible velocities calculated by equation (8-13) or given in Table 8-5 may be used as a criterion in designing erodible channels. The design procedure for an erodible channel for a given flow rate (Q) is given below. Trapezoidal-shape channels are considered in this discussion, since rectangular channels are not feasible for channels excavated in erodible materials.

1. Determine side-slope factor (z) from Table 8-4 and roughness coefficient (n) from Table 8-1.
2. Determine the maximum permissible velocity (v_{max}) from equation (8-13) or Table 8-5.
3. Calculate hydraulic radius (R) from the equation

$$R = \left(\frac{v_{max}n}{1.49S^{1/2}}\right)^{3/2} \tag{8-18}$$

4. Calculate cross-sectional area and wetted perimeter as follows:

$$A = \frac{Q}{v_{max}} \qquad P = \frac{A}{R}$$

5. Determine normal depth (y) and bottom width (b) by either of the following two methods, using the A and P values from step 4.
   a. Solve the following equations simultaneously for y and b:

   $$A = by + zy^2$$
   $$P = b + 2y\sqrt{z^2 + 1}$$

   b. Solve the following equations for y and b:

   $$y = \frac{-P \pm [P^2 + 4A(z - 2\sqrt{z^2 + 1})]^{1/2}}{2(z - 2\sqrt{z^2 + 1})} \tag{8-19}$$

   $$b = \frac{A - zy^2}{y} \tag{8-20}$$

6. Add a proper freeboard to calculate the final dimensions of the channel.

The final dimensions may be modified for practicality in terms of construction, land availability, and topography. The b/y ratio for erodible channels may be less than 1.0, or it may be 10 or higher (Hansen 1980). Some erodible-channel designs result in extremely large b/y ratios, which may not be feasible for the construction conditions. In such cases consideration should be given for a lining or a steep channel slope with drop structures to reduce the b/y ratio.

Example 8-9: Designing an Erodible Channel

Suppose that an 800-ft-long channel to convey 50 cfs of clear water is to be built in an excavated stiff clay. Depending on the route, available slopes of the channel are 0.3% and 1.2%. Assume no lining of the channel is planned. Find the required dimensions of a trapezoidal-shape channel.

Solution: The roughness coefficient is $n = 0.025$ for the excavated earth, assuming a clean winding channel (Table 8-1). The side-slope factor is $z = 1.5$ to prevent erosion (Table 8-4). The maximum permissible channel slope in erodible soils ranges from 0.4% to 0.9% for channels carrying 10–100 cfs flow and from 0.25 to 0.4% for over 100 cfs (Simon 1986). Hence, the route with 0.3% should be used. The maximum permissible velocity is $v_{max} = 3.75$ fps (Table 8-5). The required hydraulic radius of the given channel is then

$$R = \left(\frac{v_{max}n}{1.49S^{1/2}}\right)^{3/2} = \left[\frac{3.75 \times 0.025}{1.49 \times 0.003^{1/2}}\right]^{3/2} = \left(\frac{0.094}{0.082}\right)^{3/2} = 1.23\,\text{ft}$$

$$A = Q \div v_{max} = 50\,\text{cfs} \div 3.75\,\text{cfs} = 13.3\,\text{ft}^2$$

$$P = A \div R = 13.3\,\text{ft} \div 1.23\,\text{ft} = 10.8\,\text{ft}$$

To determine b and y, the equations for A and P will be solved simultaneously.

$$A = 13.3\,\text{ft}^2 = by + zy^2 = by + 1.5y^2$$

$$P = 10.8\,\text{ft} = b + 2y\sqrt{z^2 + 1} = b + 2y\sqrt{1.5^2 + 1} = b + 3.6y$$

Rearrange the above equation for b:

$$b = 10.8 - 3.6y$$

Substitute this into the equation for A. Then

$$A = 13.3\,\text{ft}^2 = (10.8 - 3.6y)y + 1.5y^2 = 10.8y - 3.6y^2 + 1.5y^2$$
$$= 10.8y - 2.1y^2$$

Rearrange to obtain the standard quadratic equation form:

$$2.1y^2 + (-10.8)y + 13.3 = 0$$

Solve the equation using the general solution of a quadratic equation,

$$y = \frac{-b \pm \sqrt{b^2 - 4ac}}{2a} \tag{8-21}$$

for $a = 2.1$, $b = -10.8$, and $c = 13.3$:

$$y = \frac{-(-10.8) \pm \sqrt{(-10.8)^2 - 4 \times 2.1 \times 13.3}}{2 \times 2.1}$$

$$= \frac{10.8 \pm 2.2}{4.2} = 2.0 \text{ ft or } 3.1 \text{ ft}$$

or use equation (8-19):

$$y = \frac{-10.8 \pm [10.8^2 + 4 \times 13.3(1.5 - 2\sqrt{1.5^2 + 1})]^{1/2}}{2(1.5 - 2\sqrt{1.5^2 + 1})}$$

$$= \frac{-10.8 \pm 2.2}{-4.2} = 2.0 \text{ ft or } 3.1 \text{ ft}$$

For a water depth of $y = 3.1$ ft,

$$A = 13.3 \text{ ft}^2 = by + zy^2 = 3.1b + 1.5 \times 3.1^2$$

Solving for b, we obtain

$$3.1b = 13.3 - 14.4 = -1.1$$

which means b is negative and thus not feasible. For $y = 2.0$ ft,

$$A = 13.3 \text{ ft}^2 = 2.0b + 1.5 \times 2.0^2$$

$$2.0b = 13.3 - 6.0 = 7.3$$

Then $b = 7.3 \div 2.0 = 3.65$ ft.
   The problem can also be solved by trial and error (see Table 8-8), using b and A as functions of y:

$$b = 10.8 - 3.6y \qquad A = 10.8y - 2.1y^2$$

Then $Q = v_{max}A = 3.75A$. This solution gives the same values of y and b as the analytical solution does. The final design dimensions are

TABLE 8-8   Solution for Example 8-9: Disigning an Erodible Channel

Trial No.	y (ft)	b (ft)	b/y	A (ft²)	Q (ft³)	Comments
1	1.0	7.2	7.2	8.7	32.6	Low
2	1.5	5.4	3.6	11.5	43.0	Low
3	2.0	3.6	1.8	13.2	49.5	OK

$$t = b + 2yz = 3.65 + 2 \times 2.0 \times 1.5 = 10.5\,\text{ft}$$

$$F.B. = \sqrt{Cy} = \sqrt{1.5 \times 2.0} = 1.7\,\text{ft}$$

$$Y = y + F.B. = 2.0 + 1.7 = 3.7\,\text{ft}$$

$$T = b + 2Yz = 4.5 + 2 \times 3.7 \times 1.5 = 15.6\,\text{ft}$$

Figure 8-11 shows the final dimensions of the designed trapezoidal-shape channel.

Example 8-10: Determining Maximum Permissible Channel Slope of an Emergency Spillway

A trapezoidal-shape emergency spillway for a pond is limited to a bottom width of 40 ft, is 0.5 ft deep (excluding freeboard), and has 2:1 side slopes. The design runoff is estimated at 75 cfs. The channel will be lined with coarse gravel. Find the maximum permissible slope to carry the design runoff.

Solution: The cross-sectional area of the channel at the design runoff is

$$A = by + zy^2 = 40 \times 0.5 + 2 \times 0.5^2 = 20.5\,\text{ft}^2$$

The wetted perimeter is

$$P = b + 2y\sqrt{z^2 + 1} = 40 + 2 \times 0.5\,\sqrt{2^2 + 1} = 42.2\,\text{ft}$$

The hydraulic radius is

$$R = A \div P = 20.5 \div 42.2 = 0.49\,\text{ft}$$

The required velocity for the design runoff in the channel is

$$v = Q \div A = 75 \div 20.5 = 3.66\,\text{fps} < v_{\text{max}}$$

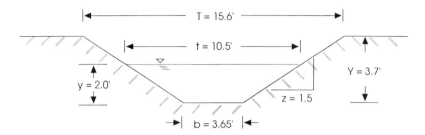

FIGURE 8-11.   Final dimensions of an erodible channel design in Example 8-9.

From Table 8-1, Manning's roughness coefficient $n = 0.025$. The maximum permissible slope of the spillway is calculated by solving the Manning equation with the value of R found above and $v_{max} = 4.00$ fps for coarse gravel taken from Table 8-5. The channel section is adequate to carry the design runoff at lower than v_{max}. Then, from Manning's equation,

$$S = \left(\frac{nv_{max}}{1.49R^{2/3}}\right)^2 = \left(\frac{0.025 \times 4.00}{1.49 \times 0.48^{2/3}}\right)^2 = 0.012 \text{ or } 1.2\%$$

Notice that the spillway problem was solved directly, so it was less time-consuming than problems requiring iterative solutions. Chow (1959) and Simon (1976) provide nomographs that may be used in lieu of iterative procedures for solving trapezoidal channel design problems.

8.7.3 Vegetated Channels

Vegetation or grass cover in an erodible channel provides erosion resistance mainly by increasing channel stability and reducing flow velocity. The design of vegetated channels is more complex than that of other channels because of the variation of channel roughness with flow depth and vegetation height. The roughness coefficient often is related to the retardance class of the vegetation, which is a function of the product of the mean velocity of flow (v) and the hydraulic radius (R), as shown in Figure 8-12. Gwinn and Ree (1979) developed the following approximate equation to calculate the roughness coefficient as a function of vR:

$$n = \frac{1}{2.1 + 2.3x + 6 \log_e(vR)} \tag{8-22}$$

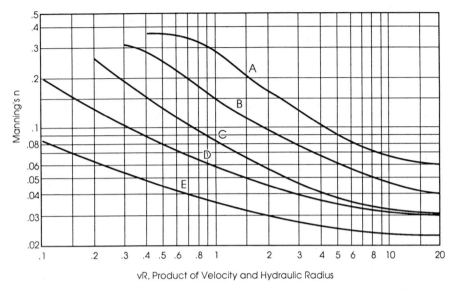

FIGURE 8-12. *n-vR* curves of retardance classes for vegetated channel design. (*Source: U.S. Soil Conservation Service 1986.*)

where n = Manning's roughness coefficient
       v = flow velocity (ft/sec)
       R = hydraulic radius (ft)
       x = −0.5 for retardance class A
            2 for retardance class B
            5 for retardance class C
            7 for retardance class D
            11 for retardance class E

The types of vegetation recommended for various areas in the United States are given in Table 8-9. Cost and availability of seed are important factors for selecting vegetation. Another consideration should be how well they will survive until a good coverage is obtained. In many cases native or locally well adapted vegetation is recommended. The classification of vegetation cover as to retardance class is given in Table 8-10, which is based on the kind, height, and condition of the vegetation.

Vegetated channels are used to carry natural concentrations of runoff or the discharge from diversion channels or emergency spillways. They are not recommended for carrying runoff from large storms, which may flow over long time periods. Continuous, long-duration flow in a vegetated channel may result in poor or even dead vegetation conditions. Vegetated

TABLE 8-9   Recommended Vegetation for Vegetated Channels in the United States

Geographic Area	Vegetation
Northeastern	Kentucky bluegrass, redtop, tall fescue, white clover
Southeastern	Kentucky bluegrass, tall fescue, Bermuda, brome, reed canary
Upper Mississippi	Brome, reed canary, tall fescue, Kentucky bluegrass
Western Gulf	Bermuda, King Ranch bluestem, native grass mixture, tall fescue
Southwestern	Intermediate wheatgrass, western wheatgrass, smooth brome, tall wheatgrass
Northern Great Plains	Smooth brome, western wheatgrass, redtop switch grass, native bluestem mixture

Source: U.S. Soil Conservation Service (1986).

TABLE 8-10   Classification of Vegetative Cover as to Flow Retardance

Retardance Class	Vegetative Cover	Condition
A	Weeping love grass	Excellent stand, tall (average 30 in.)
	Reed canary grass	Excellent stand, tall (average 36 in.)
B	Smooth brome grass	Good stand, mowed (average 12–15 in.)
	Bermuda grass	Good stand, tall (average 12 in.)
	Tall fescue	Good stand, unmowed (average 18 in.)
	Grass mixture	Good stand, uncut (average 20 in.)
	Reed canary grass	Good stand, uncut (average 12–15 in.)
	Blue grama	Good stand, uncut (average 13 in.)
C	Bahia grass	Good stand, uncut (6–8 in.)
	Bermuda grass	Good stand, mowed (average 6 in.)
	Redtop	Good stand, headed (15–20 in.)
	Grass mixture in summer	Good stand, uncut (6–8 in.)
	Centipede grass	Very dense cover (average 6 in.)
	Kentucky bluegrass	Good stand, headed (6–12 in.)
D	Bermuda grass	Good stand, cut to 2.5-in. height
	Red fescue	Good stand, headed (12–18 in.)
	Buffalo grass	Good stand, uncut (3–6 in.)
	Grass mixture in fall and spring	Good stand, uncut (4–5 in.)
	Kentucky bluegrass	Good stand, cut to 2-in. height
		Very good stand before cutting
E	Bermuda grass	Good stand, cut to 1.5-in. height or burned stubble

Source: U.S. Soil Conservation Service (1986).

channels should also not be used to carry continuous flows such as those from springs. The runoff for a 10-year-return-period storm is a sound basis for designing vegetated channels (Schwab et al. 1981).

The ability of vegetative cover to resist erosion is limited by the type and condition of the vegetation, soil type, and channel slope. A uniform cover of vegetation on a channel is important to protect the channel. Table 8-11 lists permissible velocities for vegetated channels. The permissible velocity is a velocity above which the channel may experience erosion. The Manning equation is solved to determine channel dimensions by using permissible velocity for a given channel.

The design of vegetated channels should consider two vegetation conditions: (i) channel stability when the grass is short, which is critical to erosion; and (ii) channel capacity when the grass is tall, which increases the depth of flow due to reduced flow velocity. Considering these two conditions, the following steps may be used to design a vegetated channel for a given discharge.

a. Design for Stability
>   Step 1. Determine design discharge, channel slope, vegetation type (Table 8-9), and soil type.
>
>   Step 2. Select a retardance class for a short-grass condition (Table 8-10) and the permissible velocity for the given design conditions (Table 8-11).
>
>   Step 3. Assume a value of Manning's roughness coefficient to estimate the corresponding vR value from the $n-vR$ curve (Fig. 8-12). A good starting value for n is 0.05 to 0.10. Calculate R from the vR value and the permissible velocity found in step 1.
>
>   Step 4. Calculate a new velocity (v) from the Manning equation, using n and the R found in step 2.
>
>   Step 5. Compare the permissible velocity from step 1 with the calculated v. Repeat steps 3 and 4 until the two velocities agree.
>
>   Step 6. Compute the cross-sectional flow area (A) and the wetted perimeter (P) with the values of Q, v, and R.
>
>   Step 7. Determine the channel bottom width (b), and flow depth (y) by solving the equations for A and P simultaneously or by solving equations (8-19) and (8-20).

b. Design for Capacity
>   Step 1. Select a retardance class for a tall-grass condition (Table 8-10).
>
>   Step 2. Assume a flow depth higher than the flow depth found in part a in order to increase channel flow capacity. Compute A and R, using the new flow depth and the bottom width found in part a.

TABLE 8-11   Permissible Velocity for Vegetated Channels

Type of Vegetation	Slope Range (%)	Permissible Velocity[a] (ft/sec)	
		Erosion-Resistant Soils[b]	Easily Eroded Soils[c]
Bermuda grass	<5	8	6
	5–10	7	4
	>10	6	3
Bahia grass Kentucky bluegrass Smooth brome Tall fescue	<5 5–10 >10	7 6 5	5 4 3
Grass mixture[d] Reed canary grass[d]	>5 5–10	5 4	4 3
Weeping love grass Redtop Alfalfa Red fescue	<5[e]	3.5	2.5
Common lespedeza[f] Sudan grass[f]	<5[g]	3.5	2.5

Source: U.S. Soil Conservation Service (1986).

[a] Use velocities exceeding 5 ft/sec only where good covers and proper maintenance can be obtained.
[b] Mixture of cohesive fine-grain soils and coarse-grain soils.
[c] Soils other than erosion-resistant soils.
[d] Not on slopes steeper than 10% except for side slopes in combination with a stone, concrete, or gabion center section.
[e] Not on slopes steeper than 5% except for side slopes in combination with a stone, concrete, or gabion center section.
[f] Annuals—use on mild slopes or as temporary protection until permanent covers are established.
[g] Use on slopes steeper than 5% is not recommended.

Step 3. Calculate the velocity for the dimensions found in step 2 ($v = Q \div A$), and determine n from the n–vR curve (Fig. 8-12).

Step 4. Compute the new velocity using the Manning equation and compare with the velocity found in step 3. Repeat steps 2 and 3 until the two velocities agree.

c. Determine the final dimensions of the vegetated channel by adding freeboard. A freeboard of 0.3 ft is generally recommended for a vegetated channel.

Graphical solutions for designing vegetated channels under various conditions were developed by the U.S. Soil Conservation Service (1986).

Example 8-11: Design of a Vegetated Channel

A vegetated emergency spillway for a fish pond is needed to carry a peak flow of 50 cfs. The slope is limited to 4.9% on easily erodible soil. The channel will be covered with Kentucky bluegrass. The final dimension of a trapezoidal-shaped, vegetated channel with side slopes of 4:1 will be determined.

Solution

Design for stability

The retardance class for mowed Kentucky bluegrass is D (Table 8-10). From Table 8-11, the permissible velocity is $v_p = 5$ ft/sec for Kentucky bluegrass and 4.9% slope on an easily eroded soil. Assume Manning's roughness coefficient is $n = 0.05$. From Figure 8-12, $vR = 1.5$ for retardance class D and Manning's $n = 0.05$. Then the hydraulic radius R is

$$R = vR \div v_p = 1.5 \, \text{ft}^2/\text{sec} \div 5 \, \text{ft/sec} = 0.3 \, \text{ft}$$

and the flow velocity is

$$v = \frac{1.49}{n} R^{2/3} S^{1/2} = \frac{1.49}{0.05} \times 0.3^{2/3} \times 0.049^{1/2}$$

$$= 3.0 \, \text{ft/sec} \ll v_p$$

Adjust n and recalculate the flow velocity until the calculated and permissible velocities agree. Remember that the roughness coefficient (n) is inversely related to flow velocity. The results of this iteration are shown in Table 8-12.

Section factors for the final velocity in Table 8-12 are

$$A = Q \div v = 50 \div 4.9 = 10.2 \, \text{ft}^2$$

$$P = A \div R = 10.2 \div 0.5 = 20.4 \, \text{ft}$$

Equations (8-19) and (8-20) will be used to find b and y.

$$y = \frac{-20.4 \pm [20.4^2 + 4 \times 10.2(4 - 2\sqrt{4^2 + 1})]^{1/2}}{2(4 - 2\sqrt{4^2 + 1})}$$

$$= \frac{-20.4 \pm 15.6}{-8.5} = 0.57 \, \text{ft or } 4.2 \, \text{ft}$$

TABLE 8-12   Solution for Example 8-11: Design of a Vegetated Channel for Stability

Trial No.	n	vR	R (ft)	v (ft/sec)	Comment
1	0.05	1.5	0.3	3.0	Low
2	0.04	3.0	0.6	5.9	High
3	0.045	2.0	0.4	4.0	Low
4	0.040	2.5	0.5	4.9	OK

TABLE 8-13   Solution for Example 8-11: Design of a Vegetated Channel for Capacity

Trial No.	Y (ft)	A (ft²)	R (ft)	Estimated Velocity (ft/sec)	vR	n	Calculated Velocity (ft/sec)	Comment
1	0.7	12.88	0.6	3.88	2.33	0.053	4.43	High
2	0.6	10.8	0.53	4.63	2.45	0.051	4.24	Low
3	0.65	11.83	0.56	4.23	2.37	0.052	4.31	High

For $y = 0.57\,\text{ft}$,

$$b = \frac{10.2 - 4 \times 0.57^2}{0.57} = 15.6\,\text{ft}$$

For $y = 4.2\,\text{ft}$, $b = -60.4$, which is not a feasible solution. The dimensions needed for stability are thus
$y = 0.57\,\text{ft}$ (depth of flow)   and $b = 15.6\,\text{ft}$ (bottom width)

Design for capacity
The retardance class of unmowed Kentucky bluegrass is C from Table 8-10. Assume a flow depth (y) of 0.7 ft to increase channel flow capacity and calculate channel parameters accordingly. The cross-sectional area is

$$A = by + zy^2 = 15.6 \times 0.7 + 4 \times 0.7^2 = 12.88\,\text{ft}^2$$

The wetted perimeter is

$$P = b + 2\,y\sqrt{z^2 + 1} = 15.6 + 2 \times 0.7\sqrt{4^2 + 1} = 21.4\,\text{ft}$$

The hydraulic radius is

$$R = A \div P = 12.88 \div 21.4 = 0.6\,\text{ft}$$

The estimated flow velocity is

$$v = Q \div A = 50 \div 12.88 = 3.88 \, \text{ft/sec}$$

Calculate the vR value to estimate Manning's roughness coefficient from Figure 8-12.

$$vR = 3.88 \times 0.6 = 2.33$$

From Figure 8-12, the estimated roughness coefficient (n) is 0.053 for the calculated vR. A new flow velocity will be calculated with this roughness coefficient and compared with the estimated flow velocity (3.88 ft/sec).

$$v = \frac{1.49}{0.053} \times 0.6^{2/3} \times 0.049^{1/2} = 4.43 \, \text{ft/sec} \gg 3.88 \, \text{fps}$$

This calculated flow velocity is higher than the estimated velocity for the assumed flow depth. Adjust the flow depth and recalculate the flow velocity until the two velocities agree. The results of this iteration are given in Table 8-13.

Final dimensions
Bottom width: $b = 15.6 \, \text{ft}$
Flow depth: $y = 0.65 \, \text{ft}$
Total channel depth: $Y = y + F.B. = 0.65 + 0.3 = 0.98 \, \text{ft}$
Top width of flow: $t = b + 2yz = 15.6 + 2 \times 0.65 \times 4 = 20.8 \, \text{ft}$
Total top width of channel: $T = b + 2Yz = 15.6 + 2 \times 0.98 \times 4 = 23.5 \, \text{ft}$

The final dimensions of the vegetated channel are shown in Figure 8-13.

The final selection of the channel shape and the dimensions depends on the specifics and circumstances at the project site.

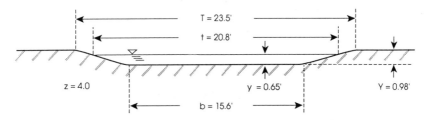

FIGURE 8-13.   Final dimensions of a vegetated channel design in Example 8-11.

9

Pipe Flows

9.1 INTRODUCTION

Pipe flows exist when a closed conduit of any kind is flowing full. In pipe flow the cross-sectional area of the flow is the same as the cross-sectional area of the conduit. The water surface is not exposed to the atmosphere, and the pressure in the pipe may be equal to, greater than, or lower than atmospheric pressure. Pressure differences in a pipeline cause water to flow through the pipe.

Pipelines are widely used to distribute water from wells or other sources to points of use in aquaculture. Pipes are manufactured from many materials, with concrete, plastic, steel, galvanized iron, and cast iron used most frequently in aquaculture applications. In pipeline design it is essential to know how much energy is necessary to force water through the pipe at desired discharge and pressure. The necessary energy is related to several factors, and several approaches have been taken to assess the energy loss in pipeline design problems. The scientific approach to pipeline design is discussed in this Chapter because it provides a way of explaining factors influencing pipe flow. However, empirical design methods are normally used in practice, and an empirical procedure for pipeline design is also discussed.

9.2 HYDRAULIC PRINCIPLES OF PIPE FLOWS

Pipelines must be under a pressure by either elevation difference or by pumping to force water flow in the pipe. The pressure (p) at any depth

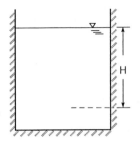

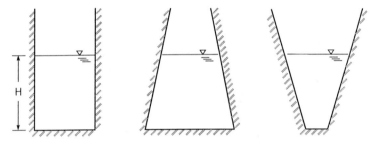

FIGURE 9-1.   Static hydraulic pressure of various shapes of water body.

below the water surface is called the *static pressure* and is equal to the product of the water depth (H) and the unit weight of water (ω):

$$p = \omega H \quad \text{or} \quad H = \frac{P}{\omega} \tag{9-1}$$

where p = static pressure
H = water depth or pressure head
ω = unit weight of water

Pressures are the same as long as the water depths are the same for any shape of water body, as illustrated in Figure 9-1. Water weighs approximately 62.4 pounds per cubic foot (lb/ft^3) at 50°F, and the static pressure of water at a depth of 1 ft is $62.4\,lb/ft^2$. Water pressure is often expressed in pounds per square inch (psi) or in feet of water. They are related as follows:

$$1 \text{ foot of water} = 62.4 \, \text{lb/ft}^2 \times \frac{1 \, \text{ft}^2}{144 \, \text{in.}^2} = 0.43 \, \text{psi}$$

or

$$1 \, \text{psi} = 2.31 \, \text{ft of water}$$

It is convenient to remember these conversion factors to study pipe flows.

Example 9-1: Determining Pressure Head of Water

Water is delivered by a pipeline from a pond located 50 ft above the outlet of the pipe. Find the water pressure at the end of the pipe, assuming no energy loss along the pipe.

Solution

Pressure head: $H = 50 \, \text{ft}$
Pressure (psi): $p = 50 \, \text{ft}/(2.31 \, \text{ft/psi}) = 21.6 \, \text{psi}$

The actual pressure at the end of the pipe is lower than 21.6 psi because energy loss occurs as water flows through the pipe. When water flows through a pipeline, energy is lost due to friction between the water and the pipe wall, as well as local disturbance created by changes in pipe diameter and by fittings (elbows, tees, valves, etc.). The principles used in pipe-flow problems are the conservation of energy and the continuity equation.

9.2.1 The Conservation of Energy in Pipe Flow

In understanding pipe flow, the principle of conservation of energy is important. It states that the sum of the pressure head (H_p), velocity head (H_v), and elevation head (H_e) above a reference plane at any point in a pipe flow is equal to the sum of the corresponding heads at any point downstream plus the hydraulic head loss (H_L) between the two points (Fig. 9-2). The total energy head at a point in a pipe flow is

$$H_t = H_p + H_v + H_e \tag{9-2}$$

Then, according to conservation of energy,

$$H_{t1} = H_{t2} + H_L \tag{9-3}$$

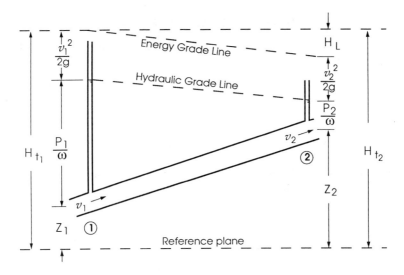

FIGURE 9-2.   Energy heads and the conservation of energy in a pipe flow.

or

$$\frac{P_1}{\omega} + \frac{v_1^2}{2g} + Z_1 = \frac{P_2}{\omega} + \frac{v_2^2}{2g} + Z_2 + H_L \qquad (9\text{-}4)$$

where v = flow velocity (ft/sec)
g = gravitational acceleration ($g = 32.2\,\text{ft/sec}^2$)
H_L = hydraulic head loss (ft)

and subscripts 1 and 2 denote designated points upstream and down-stream in the pipeline, respectively.

Equation (9-4) is a modified form of the well-known Bernoulli equation, where it is assumed that the head loss is so small that it may be neglected with small error.

The hydraulic head loss, or simply head loss, includes head losses due to pipe friction and disturbance created by fittings, changes in size and direction, and all other fixtures in the pipeline. The magnitude of the total head loss depends on the surface roughness of pipe walls, pipe size and length, and fittings and changes installed between the two designated points of the pipeline. For most pipe materials, aging also increases surface roughness, and consequently head loss, because of corrosion and abrasion.

9.2.2 The Continuity Equation

The flow rate of water in a pipe is the product of flow velocity and pipe cross-sectional area:

$$Q = Av \qquad (9\text{-}5)$$

where Q = flow rate (ft^3/sec or cfs) and A = cross-sectional area of pipe flow (ft^2).

When flow rate through a given pipeline stays constant, the flow is called *steady*. If steady flow occurs at all sections of a pipeline as shown in Figure 9-3, the following condition exists:

$$Q = A_1 v_1 = A_2 v_2 = \ldots = A_n v_n \qquad (9\text{-}6)$$

where subscripts $1, 2, \ldots, n$ represent different pipe sections. Equation (9-6) is called the continuity equation. In Figure 9-3, $v_1 < v_2 < v_3$ because $A_1 > A_2 > A_3$. This condition satisfies the continuity of flow in the pipe.

With a few exceptions, most pipes are circular. The cross-sectional area of a circular pipe is π times the square of the radius (r), but it is convenient to have an equation for the cross-sectional area of a pipe that is based on diameter (D) rather than radius. The equation is

$$A = \pi r^2 = \pi \left(\frac{D}{2}\right)^2 = \frac{\pi D^2}{4} = 0.785\, D^2 \qquad (9\text{-}7)$$

where r = inside radius of pipe (ft)
        D = inside diameter of pipe (ft)
        π = 3.14

A common unit of measure for pipe flow is gallons per minute (gpm), so it is necessary to convert the flow rate in gpm to cfs to calculate flow velocity in feet per second (fps):

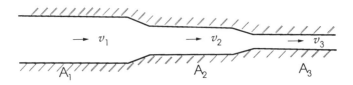

FIGURE 9-3.   Definition sketch of the continuity of flow in a pipeline.

$$1\,\text{ft}^3/\text{sec} \times 7.48\,\text{gal/ft}^3 \times 60\,\text{sec/min} \approx 450\,\text{gpm}$$

Thus, the flow rate in gpm divided by 450 gives the approximate flow rate in cfs.

9.2.3 Pressure Head (Hp)

Pressure in pipe flow may be thought of as the height of water in a vertical pipe (piezometer) inserted into the pipeline as shown in Figure 9-2. This is called the *pressure head* (H_p) and is expressed as

$$H_p = \frac{p}{\omega} \tag{9-8}$$

Standard atmospheric pressure, or simply atmospheric pressure, at sea level is 14.7 psi. Hence, atmospheric pressure may be expressed in feet of water as follows:

$$\text{Atmospheric pressure (ft)} = 14.7\,\text{psi} \times 2.31\,\text{ft/psi} = 34\,\text{ft}$$

When a vertical tube is standing in water, the water would rise 34 ft in the tube at the standard atmospheric pressure when total vacuum is exerted in the tube. It is usually necessary to have a certain amount of pressure available at the discharge point of a water distribution system unless the pipe discharges directly into the air.

9.2.4 Elevation Head (He)

Water located at an elevation above a reference level has a potential energy with respect to the reference plane. *Elevation head* (H_e) is the vertical distance (Z) to a point in a pipeline above the reference plane (Fig. 9-2):

$$H_e = Z \tag{9-9}$$

There may be elevation heads on the inlet and outlet sides of a pipeline. Unlike open-channel flow, a pipeline may have downhill or uphill slope as long as the sum of pressure and elevation heads at the outlet is higher than that of the inlet. For example, the inlet might be 5 ft above the reference plane, and the outlet might be 15 ft above the reference plane. The pressure and/or elevation difference between two points causes water flow in a pipeline; the upstream section has higher head than the downstream section. This difference is the hydraulic grade line shown in Figure 9-2.

9.2.5 Velocity Head (H_v)

The flow velocity is related to the energy head available to produce flow and the gravitational acceleration. This relationship is expressed as the velocity-head (H_v) equation. Flow velocity as a function of head is

$$v = \sqrt{2gH} \qquad (9\text{-}10)$$

$$H_v = \frac{v^2}{2g} = 0.015\, v^2 \qquad (9\text{-}11)$$

The hydraulic gradient in an open-channel flow is the slope of the water surface. In pipe flow, it is the line connecting the water levels in piezometers along the pipe, as shown in Figure 9-2. In both flows, the energy gradient is above the hydraulic gradient a distance equal to the velocity head.

Example 9-2: Determining Velocity Head

Calculate flow velocity in a 4-in.-diameter pipe carrying 225 gpm of water.

Solution: The pipe diameter is

$$D = \frac{4\,\text{in.}}{12\,\text{in./ft}} = 0.33\,\text{ft}$$

The cross-sectional area is:

$$A = 0.785\,D^2 = 0.785(0.33\,\text{ft})^2 = 0.085\,\text{ft}^2$$

The flow rate is

$$Q = \frac{225\,\text{gpm}}{450\,\text{gpm/cfs}} = 0.50\,\text{cfs}$$

The flow velocity is

$$v = \frac{Q}{A} = \frac{0.50\,\text{ft}^3/\text{sec}}{0.085\,\text{ft}^2} = 5.88\,\text{ft/sec}$$

The velocity head is

$$H_v = \frac{v^2}{2g} = \frac{(5.88\,\text{ft/sec})^2}{2 \times 32.2\,\text{ft/sec}^2} = 0.52\,\text{ft}$$

Flow velocity in pipelines is usually low, and velocity head frequently is negligible compared with other heads when determining the total head of a pipeline.

9.3 HEAD LOSSES IN PIPE FLOWS

In pipeline design, one of the main considerations is head loss. Two types of head losses are normally considered, namely, head loss due to friction (H_f) between the pipe wall and the flowing water, and head loss due to disturbance (H_m) created by fittings, changes in size and flow direction, and other fixtures in the pipe. These head losses influence the selection of pipe size and type and the economy of pipelines. The magnitude of the head losses depends on the surface roughness of pipe walls, flow rate, pipe size and length, and types of fittings and changes. Choosing a small pipe results in low initial cost, but operational cost may be excessive because of the high energy cost due to high head losses. Methods of calculating head losses and selecting optimal pipe sizes are discussed next.

9.3.1 Head Losses due to Pipe Friction

Head loss due to friction, or simply friction loss (H_f), is the energy loss resulting from turbulence created at the boundary between the pipe surface and the flowing water. In a straight, full-flowing pipe without fittings, the rate of friction loss is constant along the pipe. There are many methods to calculate friction loss, but the Darcy–Weisbach and Hazen–Williams equations are the most widely used.

The Darcy–Weisbach Equation
The Darcy–Weisbach equation, which permits the estimation of friction loss in a pipe, is

$$H_f = f\frac{Lv^2}{D2g} \tag{9-12}$$

where H_f = head loss due to friction or friction loss (ft)
     f = pipe friction factor (dimensionless)

L = pipe length (ft)
D = pipe diameter (ft)

Equation (9-12) shows that friction loss increases as pipe frictional forces opposing flow, pipe length, or flow velocity increase. However, friction loss decreases as pipe diameter increases. With known flow velocity, pipe diameter, and pipe friction factor, the Darcy–Weisbach equation is used to calculate the friction loss for any length of pipe that is straight and flowing full.

The pipe friction factor (f) depends on the Reynolds number (R_e) of the flow and the roughness of the inside of the pipe. The motion of water in a pipe does not depend on any single variable, but on a number of variables that are combined and called the Reynolds number:

$$R_e = \frac{Dv}{\nu} \qquad (9\text{-}13)$$

where R_e = Reynolds number (dimensionless) and ν = kinematic viscosity of water (ft²/sec). Values of kinematic viscosity (ν) and the unit weight of water (ω) at various temperatures are provided in Table 1-2. The kinematic viscosity decreases as temperature increases. It is fairly common practice in aquaculture to use a kinematic viscosity value of 1.0×10^{-5} ft²/sec for solving pipe-flow problems, because water flowing in pipes is often around 70–80°F during aquaculture operations.

Example 9-3: Determining Reynolds Number (R_e)

A 4-in.-diameter steel pipe has a flow of 200 gpm. The water temperature is 74°F. Calculate the Reynolds number.

Solution: The flow rate is

$$Q = \frac{200\,\text{gpm}}{450\,\text{gpm/cfs}} = 0.445\,\text{cfs}$$

The pipe diameter is

$$D = \frac{4\,\text{in.}}{12\,\text{in./ft}} = 0.33\,\text{ft}$$

The cross-sectional area of the pipe is

$$A = 0.785\,D^2 = 0.785(0.33\,\text{ft})^2 = 0.085\,\text{ft}^2$$

The flow velocity is

$$v = \frac{Q}{A} = \frac{0.445\,\text{cfs}}{0.085\,\text{ft}^2} = 5.23\,\text{ft/sec}$$

The Reynolds number, by equation (9-13), is thus

$$R_e = \frac{0.33 \times 5.23}{1.0 \times 10^{-5}} = 1.73 \times 10^5$$

The Reynolds number may be small for low velocities in small-diameter pipes; a Reynolds number less than 2,300 indicates layered or laminar flow (Simon 1976). In laminar flow, the water flows in horizontal layers without mixing of the layers. The velocities of the layers are not necessarily the same in laminar flow. Higher Reynolds numbers indicate turbulent flow. There are no layers in the flow pattern for turbulent flow, and the water molecules mix back and forth throughout the field of flow. The Reynolds number increases as turbulence increases, and more energy is required to sustain turbulent flow than laminar flow. Obviously, pipe friction losses increase as the Reynolds number increases. If flow is laminar, the frictional factor for pipe flow is

$$f = \frac{64}{R_e} \qquad (9\text{-}14)$$

Laminar flow seldom occurs in pipe flow under operational conditions. Usually, pipe flow is turbulent, and a factor called the relative roughness is required in addition to the Reynolds number for determining the pipe friction factor. The relative roughness (ε) is defined as

$$\varepsilon = \frac{e}{D} \qquad (9\text{-}15)$$

where ε = relative roughness (dimensionless) and e = absolute roughness (ft). Upon visual examination, the bore of a pipe may appear perfectly smooth. Microscopic examination, however, reveals that the surface actually has many tiny projections. The absolute roughness is the average distance that these projections extend into the pipe bore. Absolute roughness values given in Table 9-1 vary with the type of material from which the pipe is constructed.

For pipes made of the same material relative roughness decreases as pipe diameter increases. For example, relative roughness values for 4-in.- and 12-in.-diameter commercial steel pipes are

TABLE 9-1   Absolute Rhouhness (*e*) of Pipe Materials

Type of Pipe	*e*(ft)
Corrugated metal	0.1–0.2
Riveted steel	0.003–0.03
Concrete	0.001–0.01
Wood stave	0.0006–0.003
Cast iron	0.00085
Galvanized steel	0.0005
Asphalted cast iron	0.0004
Commercial steel	0.00015
Glass, copper, plastic	0.000005

Source: Robert L. Daugherty, Joseph B. Franzini, and E. John Finnemore, *Fluid Mechanics with Engineering Applications*, Copyright © 1985 by McGraw-Hill, Inc., New York, p. 225. Reprinted by permission of McGraw-Hill, Inc.

$$\varepsilon = \frac{0.00015\,\text{ft}}{0.33\,\text{ft}} = 0.00045 \quad \text{and} \quad \varepsilon = \frac{0.00015\,\text{ft}}{1.0\,\text{ft}} = 0.00015$$

respectively. Once the Reynolds number and relative roughness are known, the pipe friction factor may be determined from the *Moody diagram* (Fig. 9-4). Use of the Moody diagram is discussed as follows in Example 9-4.

Example 9-4: Application of Moody Diagram

A 4-in.-diameter steel pipe carries 200 gpm. The pipe flow has $R_e = 1.73 \times 10^5$ and $\varepsilon = 0.00045$ (from Example 9-3 and preceding calculations). Enter the Moody diagram at $R_e = 1.73 \times 10^5$ on the abscissa, and project a line upward until it intersects a relative roughness line equal to 0.00045. Relative roughness is given on the right-hand ordinate. Notice that lines labeled 0.0004 and 0.0006 are drawn, but 0.00045 must be determined by interpolating between those two lines. Mark the intersection of the Reynolds number and relative roughness line. Now, project a line from this point horizontally to the abscissa until it intersects the friction factor scale along the left-hand ordinate. A friction factor of approximately 0.019 is read from the friction factor scale.

Example 9-5: Application of the Darcy–Weisbach Equation

Consider the 4-in. steel pipe carrying 200 gpm described in Examples 9-3 and 9-4. The total pipe length is 200 ft. Find the head loss in the pipe due to friction.

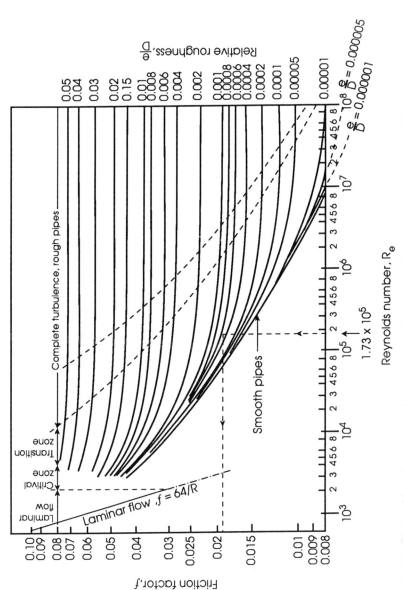

FIGURE 9-4. Pipe friction factor (f) versus Reynolds number (R_e) and relative roughness (ε) in pipe flows (Moody diagram).

Solution: From Examples 9-3 and 9-4 the following values are obtained:

Pipe diameter: $D = 0.33\,\text{ft}$
Flow velocity: $v = 5.23\,\text{ft/sec}$
Pipe friction factor: $f = 0.019$

Then by equation (9-12) friction loss is

$$H_f = 0.019 \times \frac{200\,\text{ft}}{0.33\,\text{ft}} \times \frac{(5.23\,\text{ft})^2}{2 \times 32.2\,\text{ft/sec}^2} = 4.89\,\text{ft}$$

This procedure is known as the scientific method for estimating friction loss in pipe-flow problems. The method requires several calculations, and the Moody diagram is not easy to use. Empirical methods for estimating friction loss in pipes are more commonly used in aquaculture problems.

The Hazen–Williams Equation

The Hazen–Williams technique for estimating head loss due to pipe friction (H_f) is based on an equation developed from empirical data on pipe flow. The equation for circular pipes is

$$Q = 0.285\,Cd^{2.63}S^{0.54} \tag{9-16}$$

Solving for S gives

$$S = \left(\frac{Q}{0.285\,Cd^{2.63}}\right)^{1.85} \tag{9-17}$$

where Q = flow rate (gpm)
   C = Hazen–Williams coefficient (dimensionless)
   d = inside diameter of pipe (in.)
   S = slope of energy grade line (ft/ft)

The slope of the energy grade line due to friction loss between two points in a pipeline as shown in Figure 9-2 is

$$S = \frac{H_f}{L} \tag{9-18a}$$

where L is the pipe length in feet between the two points considered. Then the head loss due to pipe friction for a pipe length, L, is

$$H_f = S \times L \tag{9-18b}$$

TABLE 9-2   Hazen–Williams Coefficients for Pipe Friction Loss

Type of Pipe	Hazen–Williams Coefficient
Extremely smooth	140
Very smooth	130
Asbestos–cement	140
Brass or copper	130–140
Galvanized iron	120
Plastic	150
Steel	110
Tarcoated enamel-lined	150
New unlined	140–150
Riveted, new	110
Riveted, old	95
Cast iron	110
New	130
5-year old	120
10-year old	110
20-year old	100
40-year old	65–80
Cement-lined	130–150
Tar-coated	115–135
Rubber-lined hose	135
Vitrified clay	110
Ordinary brick	100

Values of the Hazen–Williams coefficient for different pipe materials are given in Table 9-2. Examination of equation (9-17) reveals that head loss due to friction for a particular discharge decreases as pipe diameter increases. Notice also that for a given pipe diameter, friction loss increases with increasing discharge. It is obvious that pipe diameter is a critical factor in determining the friction loss that must be overcome to provide a required discharge and pressure at the pipeline outlet. Friction loss may be reduced by selecting a larger-diameter pipe. However, the cost of pipe increases with increasing diameter, so the energy saving resulting from using larger pipe must be compared with the corresponding cost increase.

Example 9-6: Application of Hazen–Williams Equation

For the pipeline in Example 9-5, determine the friction loss in the pipeline by the Hazen–Williams equation.

Solution: From Table 9-2 the Hazen–Williams coefficient is C = 140 for new unlined steel pipe. The slope of the energy grade line, from equation (9-17), is

$$S = \left(\frac{200}{0.285 \times 140 \times 4^{2.63}}\right)^{1.85} = 0.023 \, \text{ft/ft}$$

Total friction loss for the 200 ft pipe is

$$H_f = S \times L = 0.023 \, \text{ft/ft} \times 200 \, \text{ft} = 4.6 \, \text{ft}$$

This solution shows that the Hazen–Williams equation provides a friction loss slightly different from that obtained by the Darcy–Weisbach equation in Example 9-5. The difference could be due to selection of the Hazen–Williams coefficient and the pipe roughness value for the Darcy–Weisbach equation. Graphical solution of the Hazen–Williams equation is shown in Figure B-1 (see Appendix B).

Pipe manufacturers have solved the Hazen–Williams equation for different pipe types, diameters, and flow rates, and have tabulated the resulting friction loss values and flow velocities for easy use. The results for steel or cast-iron pipe are presented in Table B-1 (Berkeley Pump Co. 1987). Aged pipe conditions should be used in the calculation of a design coefficeint C = 100 for 10-year-old steel or 18-year-old cast iron), since old pipelines usually have higher friction loss than new pipelines. Thus, the use of data for old pipe in design calculation is a good idea because head loss increases with time. The pipe diameters listed in the appendix are nominal diameters that are not necessarily the same as the inside diameters. Data for larger diameters are found from the references. Tables B-2 and B-3 show similar calculations for Schedule 40 and Schedule 80 plastic pipes (coefficient C = 150). Schedule 80 plastic pipe has a thicker wall than Schedule 40 pipe. To use these appendixes, first find the friction head loss for the flow rate and pipe diameter. This value is in feet of friction loss per 100 ft of pipe length, so the friction loss for a pipe length is determined as follows:

$$H_f = H_f \text{ per } 100 \, \text{ft} \times \frac{\text{pipe length (ft)}}{100 \, \text{ft}} \tag{9-19}$$

Example 9-7: Application of Manufacturer's Friction Loss Data

A 4-in.-diameter, 200-ft-long, steel pipe (C = 100) is carrying 200 gpm. Find the friction loss for the total pipe length.

Solution: The friction loss H_f per 100 ft of 4-in.-diameter straight pipe is 4.3 ft from Table B-1. The friction loss for 200 ft is

$$H_f = 4.3\,\text{ft} \times \frac{200\,\text{ft}}{100\,\text{ft}} = 8.6\,\text{ft}$$

This answer is greater than the value determined in both Examples 9-5 and 9-6. The difference is because absolute roughness value and the Hazen–Williams coefficient used in the examples are for new pipes while the friction loss values in Table B-1 are for aged pipes with a coefficient of $C = 100$.

9.3.2   Head Losses due to Fittings and Other Fixtures

In addition to the head loss due to pipe friction the total head loss (H_L) in pipe flow includes various head losses due to disturbance created by changes in size and direction, and numerous fittings installed in the pipeline. The head loss due to these conditions is often called *minor loss* (H_m) because it is often negligible for long pipelines compared with the pipe friction loss. However, minor loss is an important part of the total head loss for short pipelines with various fittings and changes. Safe design of pipelines requires an estimation of the minor loss. In a pipe-flow analysis, minor-loss coefficients (k_m) are determined for various changes, fittings, and other fixtures, such as valves, bends, tees, elbows, size changes, diffusers, and strainers. The coefficient for a particular item is multiplied by the kinetic energy of the pipe flow to estimate the minor loss (Bouthillier 1981):

$$H_m = k_m \frac{v^2}{2g} \tag{9-20}$$

where H_m = minor loss (ft) and k_m = minor-loss coefficient (dimensionless). The sum of all minor losses is then added to the pipe friction loss (H_f) to find the total head loss (H_L) in the pipeline.

The minor-loss coefficients for various pipe fittings and changes commonly used in pipelines are listed in Table B-4. For convenience, pipe manufacturers list minor losses in terms of equivalent lengths of the same-sized, straight pipes, as shown in Figure B-2. To use Figure B-2 first find a dot that represents the fittings and changes of interest. Connect the dot to the pipe diameter given on the right-hand-side line. The value intersected by this line on the centerline provides the minor loss of the fittings in terms of equivalent length of the same size pipe.

Example 9-8: Determining Minor Losses

A 2-in., 50-ft, steel pipeline ($C = 100$) is carrying 40 gpm and contains two standard elbows, one 45° elbow, and two fully open gate valves. Find the total minor loss of the pipeline.

Solution: The minor losses in terms of equivalent length of 2-in. straight pipe for the fittings are taken from Figure B-2:

Fittings	No. of Fittings	Equivalent Length of Straight Pipe (ft)
Standard elbow	2	5.5 ft × 2 = 11.0
45° elbow	1	2.5 ft × 1 = 2.5
Gate valves (fully open)	2	1.2 ft × 2 = 2.4
Total		15.9

The friction loss per 100 ft for 2-in. steel pipe carrying 40 gpm is 5.6 ft (from Table B-1). The total minor loss in the pipe is

$$H_m = 5.6 \, \text{ft} \times \frac{15.9 \, \text{ft}}{100 \, \text{ft}} = 0.89 \, \text{ft}$$

The total head loss in the 50-ft steel pipe is

$$H_L = H_f + H_m = 5.6 \, \text{ft} \times \frac{50 \, \text{ft}}{100 \, \text{ft}} + 0.89 \, \text{ft} = 3.69 \, \text{ft}$$

The minor loss is thus too large to be neglected in the short pipe.

9.3.3   Total Hydraulic Head of Pipe Flows

The total hydraulic head in a pipeline includes pressure required at the outlet, total head loss, elevation head, and velocity head. The total head available at the inlet side of a pipeline must be higher than the total head required in the pipeline. A pump or elevation head is provided at the pipe inlet to pressurize and deliver water through pipelines. The hydraulic head required for a pipeline on the discharge side of a pumping system is called the *discharge head*. The head required to lift water from source to the pump impeller is called the *suction head*. The discharge head plus the suction head is the total hydraulic head or total dynamic head required for the pump system. Details of the total dynamic head of pump systems are discussed in Chapter 11.

9.3.4 Peak Water Demand for Aquaculture Ponds

The demand for water from an aquaculture system is not uniformly distributed over time, and the greatest demand is used in determining system capacity. There is no set procedure for estimating peak demand; it depends on the way water will be used. For example, if water is to be used in filling and maintaining a pond, the peak demand is based on the flow rate necessary to fill the pond in a specific length of time. The discharge needed to keep the pond full to meet the water losses (e.g., evaporation and seepage) would be much less than the demand for initially filling the pond.

As an example, suppose a fish-holding facility which has ten 10,000 gal earthen ponds for holding fish. A fairly small flow will replace the lost water by evaporation and seepage in the ponds, but at times it may be necessary to flush water through the ponds to reduce concentrations of waste products. A flushing rate of one pond volume per hour is considered adequate for maintaining fish at high density, and it is anticipated that all ponds may be in use at once. The peak demand is (10 ponds × 10,000 gal/pond) ÷ 1 hr = 100,000 gph, or 100,000 gph ÷ 60 min/hr = 1,670 gpm. The main supply line must be designed to deliver a minimum of 1,670-gpm flow to meet the peak demand for the flushing operation. As the water discharges directly into the pond, no pressure is needed at the outlet. However, energy must be provided to the pipeline to overcome all head losses.

9.4 DESIGN TECHNIQUES OF PIPELINES

The solutions to the following problems illustrate pipeline designs for aquaculture applications. The energy required to force a design discharge through the pipe from a point at the upstream end to a downstream point is the head difference between the two points, which is the total energy head. In pipeline design the amount of energy required to force the necessary discharge through the pipeline and deliver the water at the desired pressure must be determined. This amount of energy is usually provided by elevation head or energy from pumps. Basically three types of problems are involved in pipeline design:

1. Determine the total head loss for the given pipe type and size carrying a known flow rate and pressure at the outlet.

2. Determine the flow rate for the given head and pipe type and size.
3. Determine the size of pipe required for the given type of pipe, head, and flow rate.

9.4.1 Determining a Pipe Size

A discharge of 400 gpm at 15 psi is needed at a fish hatchery. A storage reservoir will provide the water. The pipeline must cover a distance of 2,800 ft, and Schedule 40 plastic pipe will be used. The water surface in the reservoir at minimum level is 65 ft above the pipe outlet. The pipeline will have one standard elbow, an ordinary entrance, and two gate valves. The pipe size will be determined to deliver the design flow.

Solution: The total head required in the system must be smaller than the total available head. The total head available for the system is $H_t =$ 65 ft. The pressure head required at the hatchery is $H_p = 15$ psi $\times$ 2.31 ft/psi $= 34.6$ ft. Then the energy head available for the head loss in the pipe is 65 ft $-$ 34.6 ft $= 30.4$ ft. First, neglecting minor losses, the maximum permissible head loss per 100 ft of pipe is

$$\frac{30.4 \text{ ft}}{2,800 \text{ ft}} \times 100 \text{ ft} = 1.09 \text{ ft}$$

Reference to Table B-2 shows that 6-in. diameter Schedule 40 plastic pipe carrying 400 gpm has a friction loss of 1.03 ft/100 ft and a flow velocity of 4.43 ft/sec. This is less than the maximum permissible head loss. The friction loss for the straight pipe length of 2,800 ft is

$$H_f = 1.03 \times \frac{2,800}{100} = 28.84 \text{ ft}$$

The minor losses in the 6-in. pipe from Figure B-2 are calculated and added to the friction loss.

Fittings	No. of Fittings	Equivalent Length of Straight Pipe (ft)
Ordinary entrance	1	9.0
Standard elbow	1	16.0
Gate valves (fully open)	2	3.5 ft $\times$ 2 = 7.0
Total		32.0

The total minor loss is equivalent to 32 ft of the 6-in. pipe or

$$H_m = 1.03\,\text{ft} \times \frac{32\,\text{ft}}{100\,\text{ft}} = 0.33\,\text{ft}$$

The total head loss in the pipe is $H_L = 28.84 + 0.33 = 29.14\,\text{ft}$. The velocity head of the system is $0.015 \times 4.43^2 = 0.29\,\text{ft}$ [from equation (9-11)]. The total hydraulic head required in the system is thus $H_t = 34.6 + 29.14 + 0.29\,\text{ft} = 64.3\,\text{ft}$. This value is lower than the total available head of 65 ft; therefore, the desired discharge can be delivered with Schedule 40, 6-in. plastic pipe. However, because of the small difference between available and required heads, it should be recommended for this system to use the next larger pipe, perhaps 8-in. pipe, considering additional head loss in the future. In this example the minor loss and velocity head are less than 1% of the total head and can be considered negligible. However, unless one has enough experience to know when minor loss and velocity head are negligible, it is best to calculate them.

9.4.2 Optimizing a Water Supply System

A fish farm will consist of eight, 5-acre ponds of 3-ft average depth that will be filled and maintained by water pumped from a stream. The layout for the ponds and pipeline is shown in Figure 9-5. There will be 3,070 ft of straight pipe, seven standard tees, one standard elbow, eight gate valves, and eight 20-ft-long inlet pipes. The gate valve will be installed at the inlet pipe of each pond. The pumping level when the stream is at its lowest level is expected to be no more than 20 ft below the elevation of the inlet point for the most distant and highest pond. To allow for future expansion, a pressure head of 30 ft will be provided at the last discharge point. During operation one pond at a time will be filled, and a filling time of five days is required.

Solution: The discharge required to fill a pond in five days is

$$\frac{5\ \text{acres} \times 325{,}872\ \text{gal/acre-ft} \times 3\,\text{ft}}{5\ \text{days} \times 1{,}440\ \text{min/day}} = 679\,\text{gpm}$$

A flow rate of 750 gpm is selected to provide a small safety factor (about 10%). The maximum head that the pump must operate against to supply the furthermost pond results from the 20-ft elevation at the pumping site, 3,070-ft straight pipe, one standard elbow, one open gate valve, the straight run of seven standard tees, and the 30 ft of pressure head. Assume that demand for filling a pond in five days will be much

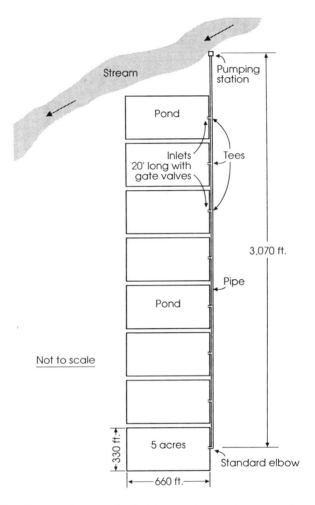

FIGURE 9-5. Concept diagram of optimizing a water supply system of an aquaculture project in Section 9.4.2.

greater than the demand for maintaining water levels in all eight ponds. Therefore, only the demand for filling the furthermost pond will be considered to determine the maximum head requirement.

The pipeline will be constructed of steel pipe. From Table B-1 the friction losses per 100 ft for different nominal pipe diameters carrying 750 gpm are shown in Table 9-3. Minor losses will be determined for

TABLE 9-3   Solution for Determining Pipe Friction Losses in Section 9.4.2

Pipe Diameter (in.)	Velocity (ft/sec)	Velocity Head (ft)	Friction Loss H_f (ft)	
			Per 100 ft	For 3,070 ft
4	18.9	5.6	49.50	1,519.6
6	8.34	1.08	6.74	206.9
8	4.81	0.36	1.77	54.3
10	3.05	0.15	0.61[a]	18.7

[a] Interpolated.

TABLE 9-4   Solution for Determining Pipe Minor Losses in Section 9.4.2

Pipe Diameter (in.)	Equivalent Length of Straight Pipe (ft)				Friction Loss per 100 ft (from previous table)	Minor Loss H_m[b] (ft)
	Standard Elbow (1)[a]	Run of Standard Tee (7)	Run of Gate Valve	Total		
4	11	7 × 5 ft = 35	2.3	48.3	49.50	23.9
6	16	7 × 7 ft = 49	3.5	68.5	6.74	4.6
8	21	7 × 10 ft = 70	4.5	95.5	1.77	1.7
10	26	7 × 13 ft = 91	5.7	122.7	0.61	0.75

[a] Number in parentheses is number of fittings.
[b] H_m = (total equivalent length ÷ 100 ft) friction loss per 100 ft.

different pipe diameters as shown in Table 9-4. Adding the friction loss and the total minor loss to the elevation head (20 ft) and pressure head (30 ft) required at the end of the pipeline for different pipe sizes gives the following values for the total head requirement in the pipeline:

Pipe diameter (in.)	Total Head Requirement (ft)
4	1593.0
6	261.5
8	106.0
10	69.5

In Chapter 11, methods for determining power requirements of a pump system are discussed; for now, assume that the pump will be 70% efficient and use the following methods. Water power requirement for a pump system is

$$WP = \frac{Q \times H}{3,960}$$

Brake power requirement for a power unit is

$$BP = \frac{WP}{E_p}$$

where BP = brake power of power unit (hp)
$\quad\quad WP$ = water power (hp)
$\quad\quad Q$ = flow rate (gpm)
$\quad\quad H$ = total head (ft)
$\quad\quad E_p$ = pump efficiency (fraction)

The power requirement of a pump to pump 750 gpm through the different pipe sizes is

Pipe Diameter (in.)	Water Power (hp)	Brake Power (hp)
4	303.0	433.0
6	50.0	71.0
8	20.0	29.0
10	13.0	19.0

Selection of a pipe diameter must consider the cost differential between using a larger pipe size with a smaller pump and power unit and a smaller pipe size with a larger pump and power unit. These computations can only be made after consultation with suppliers of water supply equipment. In this problem, the 10-in. pipeline requires the least power to operate. But initial cost for this pipeline may be higher than others due to higher pipe cost.

9.4.3   Determining a Drain Pipe Size

A 2-acre pond will contain 12 acre-ft of water. The pond will have an average water depth of 6 ft above the drain pipe installed on the pond bottom. A Schedule 80 plastic pipe will be 60 ft long, and a gate valve will be attached at the inlet end of the pipe as shown in Figure 9-6. The exit of the drain pipe will be 1 ft lower than the intake. A pipe diameter is to be selected that will permit the pond to be drained within four days. The

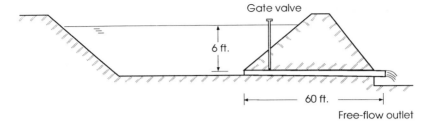

FIGURE 9-6.   Concept diagram of determining drain pipe size of a fish pond in Section 9.4.3.

pipe will consist of an ordinary entrance, a gate valve, and free-flow outlet.

Solution: The water surface is 6 ft above the entrance at the beginning, but the water surface declines as water drains. For a simple calculation it is assumed that the average head above the entrance to the drain is 3 ft, and there is an additional 1-ft elevation difference between the entrance and exit of the drain pipe. For simplicity the sum of the two values will be used as the average available head to drain water in the pond. Therefore, an average of 4 ft of head is available to force water through the drain. Figuring only the pipe friction loss, the maximum allowed head loss per 100 ft is

$$\frac{4\,\text{ft}}{60\,\text{ft}} \times 100\,\text{ft} = 6.67\,\text{ft}$$

The necessary discharge rate is

$$\frac{12\,\text{acre-ft} \times 325{,}872\,\text{gal/acre-ft}}{4\,\text{days} \times 1{,}440\,\text{min/day}} = 679\,\text{gpm}$$

From Table B-3 (Schedule 80 plastic pipe), a 6-in.-diameter pipe carries 700 gpm with a friction loss of 3.58 ft per 100 ft. This is lower than the maximum available head loss in the pipe. Notice also that 5-in.-diameter pipe is too small because friction loss exceeds the available head. The friction loss for the 60-ft-long straight pipe is

$$H_f = 3.58\,\text{ft} \times \frac{60\,\text{ft}}{100\,\text{ft}} = 2.15\,\text{ft}$$

The minor losses for pipe fittings for the 6-in.-diameter pipe from Figure B-2 are

Pipe Fitting	Number of Fittings	Equivalent Length of Straight Pipe (ft)
Ordinary entrance	1	9.0
Open gate valve	1	3.5
Total		12.5

The total minor loss is

$$H_m = 3.58 \, \text{ft} \times \frac{12.5 \, \text{ft}}{100 \, \text{ft}} = 0.45 \, \text{ft}$$

Thus, the total head loss through the 6-in. drain pipe is

$$H_L = H_f + H_m = 2.15 + 0.45 = 2.60 \, \text{ft}$$

This value is less than the 4 ft of average head available to force water through the pipe. A Schedule 80, 6-in.-diameter plastic pipe will allow the pond to drain in less than four days. The available head used in this solution (4 ft) might not be an accurate assumption, since the flow rate is not linearly related to the head in a pipe flow. However, this solution would not be too far off from the exact solution for the mostly shallow ponds used in aquaculture.

9.4.4 Determining Discharge Head Requirement

A pump is required to supply 5,000 gpm of water from a canal to a pond. The maximum length of 16-in.-diameter steel pipe will be 150 ft, and the greatest lift between the water surface in the canal and the pipe outlet is 10 ft, as shown in Figure 9-7. The pipe has three 45° elbows, and water will be freely discharged at the pipe outlet. The total head required to deliver the water will be calculated.

Solution: Friction loss per 100 ft for the 16-in. steel pipe is $H_f/100 \, \text{ft} =$ 2.54 ft (from Table B-1). The total friction loss through the pipe is then $H_f = 2.54 \, \text{ft} \times (150 \, \text{ft}/100 \, \text{ft}) = 3.81 \, \text{ft}$. The minor loss for three 45° elbows of 16-in. pipe is a 57-ft (3×19 ft) equivalent length of straight pipe (from Figure B-3). Hence, the total minor loss is $H_m = 2.54 \, \text{ft} \times$ ($57 \, \text{ft}/100 \, \text{ft}) = 1.45 \, \text{ft}$. The total head loss in the pipe is

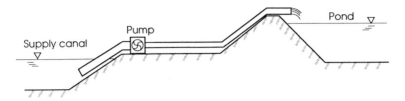

FIGURE 9-7. Concept diagram of determining pumping head requirement of a pipeline in Section 9.4.4.

$$H_L = H_f + H_m = 3.81 \, \text{ft} + 1.45 \, \text{ft} = 5.26 \, \text{ft}$$

Adding the elevation head (10 ft) and the velocity head to the pipe head loss gives the total head requirement in the pipeline. No pressure is required at the pipe outlet. The flow rate is

$$Q = 5,000 \, \text{gpm} \div 450 \, \text{gpm/cfs} = 11.1 \, \text{cfs}$$

The cross-sectional area is

$$A = 0.785D^2 = 0.785(16 \, \text{in.} \div 12 \, \text{in./ft})^2 = 1.4 \, \text{ft}^2$$

The flow velocity is

$$v = Q \div A = 11.1 \, \text{cfs} \div 1.4 \, \text{ft}^2 = 8.0 \, \text{ft/sec}$$

The velocity head is

$$H_v = \frac{v^2}{2g} = \frac{(8 \, \text{ft/sec})^2}{2 \times 32.2 \, \text{ft/sec}^2} = 0.96 \, \text{ft}$$

The total discharge head that should be supplied by the pump is

$$
\begin{aligned}
H_d &= H_L + H_e + H_v + H_p \\
&= 5.26 \, \text{ft} + 10 \, \text{ft} + 0.96 \, \text{ft} + 0 \, \text{ft} \\
&= 16.22 \, \text{ft}
\end{aligned}
$$

9.5 WATER HAMMER

A sudden closure of valve in a pipeline may cause a drastic increase of pressure in the pipeline. This phenomenon, called *water hammer*, is a relatively short and instantaneous pressure increase that may be great enough to cause pipe failure. The increased pressure due to water hammer at a sudden valve closure is calculated as

$$\Delta p = \frac{\rho v c}{144} \tag{9-21}$$

where Δp = increased pressure by water-hammer effect (psi)
ρ = density of water (1.94 slug/ft^3 at 70°F)
v = initial velocity of pipe flow (ft/sec)
c = speed of the pressure wave in a water pipe, 2,000–4,000 ft/sec for normal pipe dimensions (Linsley and Franzini 1979).

The total pressure due to the water-hammer effect is $p + \Delta p$, where p is the static pressure in the pipe. Delaying the valve closing time reduces water-hammer effect. The minimum closure time (t_m) to reduce the water-hammer pressure is calculated as

$$t_m = \frac{2L}{c} \tag{9-22}$$

where t_m = minimum closure time to reduce water-hammer pressure (sec) and L = pipe length (ft). The time of valve closure must be longer than t_m to reduce water-hammer pressure. The water-hammer pressure developed by a gradual closure of the valve when the closure time (t_c) is longer than the minimum time (t_m) is approximately

$$P_h \approx \frac{t_m}{t_c} \Delta p \tag{9-23}$$

where P_h = water-hammer pressure at $t_c > t_m$ (psi) and t_c = valve closure time (sec).

   Pipelines must be designed and operated to sustain the pressure due to the water-hammer effect. This can be done by installing pressure-relief valves to eliminate excessive pressure, installing thick-walled pipe, or by operating the valves so that any rapid closure is avoided. The water-hammer phenomenon is discussed by Parmakian (1955).

Example 9-9: Determining Water-Hammer Pressure

A 6-in. PVC pipe (Schedule 80) carries 500 gpm at 50 psi at the pipe outlet. The pipe delivers water from a reservoir to a fish pond and is 2,500 ft long. The maximum pressure developed in the pipe due to a sudden closure of a valve at the pipe outlet will be determined.

Solution

Flow rate: $Q = 500\,\text{gpm} \div 450\,\text{gpm/cfs} = 1.11\,\text{cfs}$
Cross-sectional area: $A = 0.785 D^2 = 0.785(6/12)^2 = 0.196\,\text{ft}^2$
Flow velocity: $v = Q \div A = 1.11\,\text{cfs} \div 0.196\,\text{ft}^2 = 5.66\,\text{ft/sec}$

Assume $c = 3{,}000\,\text{ft/sec}$. Increased pressure in the pipe due to the water-hammer effect is

$$\Delta p = \frac{1.94\,\text{slug/ft}^3 \times 5.66\,\text{ft/sec} \times 3{,}000\,\text{ft/sec}}{144} = 229\,\text{psi}$$

Thus, the total pressure in the pipeline after the sudden closure is

$$p + \Delta p = 50\,\text{psi} + 229\,\text{psi} = 279\,\text{psi}$$

The minimum closure time to reduce water-hammer pressure is

$$t_m = \frac{2L}{c} = \frac{2 \times 2{,}500\,\text{ft}}{3{,}000\,\text{ft/sec}} = 1.67\,\text{sec}$$

If the valve is closed in less than 1.67 sec, the pipe will be affected by the full water-hammer effect. Therefore, the pressure in the pipe is 229 psi higher than the normal pressure. When the closure time is 10 sec, the pressure developed in the pipe is

$$P_h = \frac{t_m}{t_c}\,\Delta p = \frac{1.67\,\text{sec}}{10\,\text{sec}} \times 229\,\text{psi} = 38\,\text{psi}$$

As shown in the example, the pressure in the pipeline due to water hammer is excessive at a sudden valve closure and may seriously damage the pipeline unless it is designed to sustain the pressure. In the example, it must take longer than 1.67 sec of closure time to reduce the water-hammer effect. In many practices, the sudden closure of valves takes place by mistake or through negligence of the valve operator.

9.6 CULVERTS

Culverts are pipe spillways installed to carry water under roads, railways, canals, or embankments. They are designed to pass the design flow without building an excessive water depth at the inlet side. The water-surface elevation at the inlet must be low enough so that there is no

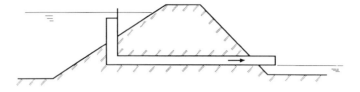

a) Drop inlet pipe spillway of embankment

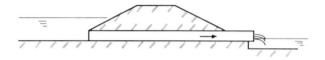

b) Sharp entrance highway cross culvert

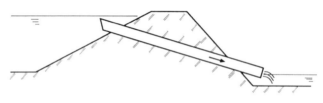

c) Hooded-inlet pipe spillway of embankment

FIGURE 9-8.   Uses of culverts to drain water across barriers.

damage to the embankment due to overflow. The cross section of culverts may be rectangular, square, or circular, although a circular shape is the most common in aquaculture. Culverts are commonly made of corrugated metal, corrugated plastic, and concrete pipes. Corrugation improves the structural strength of the pipe and helps to prevent pipe slippage. However, corrugation increases pipe roughness, which reduces flow capacity. Corrugated plastic pipes with smooth-surface inner lining are commercially available. The smooth-surface inner lining improves flow capacity by reducing the roughness coefficient to a value similar to that of a plastic pipe. Figure 9-8 shows typical uses of culverts to drain water across embankments and other barriers.

It is important to understand the neutral slope (S_n) to analyze culvert flows. The *neutral slope* is a hydraulic slope in which the head loss due to friction is equal to the head loss due to an elevation change of the culvert (Schwab et al. 1981). The neutral slope for a small angle θ is calculated as

$$S_n = \tan\theta = \sin\theta = \frac{H_f}{L} = k_c\frac{v^2}{2g} \qquad (9\text{-}24)$$

where S_n = neutral slope (ft/ft)
$\quad H_f$ = friction loss (ft)
$\quad L$ = culvert length (ft)
$\quad k_c$ = friction loss coefficient for culvert flowing full
$\quad \theta$ = slope angle of culvert

For a circular culvert

$$k_c = \frac{5,087n^2}{d^{4/3}} \qquad (9\text{-}25)$$

and for a square culvert

$$k_c = \frac{29.16n^2}{R^{4/3}} \qquad (9\text{-}26)$$

where d
$\quad R$ = inside diameter of circular culvert (in.)
$\quad n$ = hydraulic radius of square culvert (ft)
$\quad\quad$ = Manning's roughness coefficient of culvert

Solutions for k_c are given in Tables B-5 and B-6 for common circular and square culverts for various Manning's roughness coefficients. Manning's roughness coefficients for pipe materials are given in Table 9-5.

The characteristics of culvert flow are complicated because of the variables that influence the flow regimes. The flow regimes of culverts are determined by conditions at the inlet and outlet and by the culvert slope. The headwater may be above or below the top of the inlet, and the tailwater may be above or below the pipe outlet. Culvert flows are classified as pipe flow, orifice flow, open-channel flow, and weir flow. The possible flow regimes through culverts are illustrated in Figure 9-9. The solution of a culvert problem is primarily a determination of the flow regime that will occur under the given culvert conditions.

1. Pipe flow dominates when the inlet and outlet are submerged (Fig. 9-9a).
2. When the inlet is submerged but the outlet is not, two types of flow occur. If the culvert slope (S) is milder than the neutral slope of the

TABLE 9-5   Manning's Roughness Coefficient (n) for Pipe Materials

Pipe Material	n
Brass, copper or glass	0.009–0.013
Wood stave	0.010–0.013
Concrete	
Smooth	0.011–0.013
Rough	0.013–0.017
Vitrified or common clay	0.011–0.017
Cast iron	
Clean, coated	0.011–0.013
Clean, uncoated	0.012–0.015
Dirty	0.015–0.035
Brick (cement mortar)	0.012–0.017
Commercial steel	0.012–0.017
Riveted steel	0.013–0.017
Corrugated metal	
Ring	0.020–0.026
Helical	0.013–0.015
Corrugated plastic tubing	0.014–0.018
Asbestos cement	0.009

Source: V.T. Chow, *Open-Channel Hydraulics*, Copyright © 1959 by McGraw-Hill, Inc., New York, selected from p. 110. Reprinted by permission of McGraw-Hill, Inc.

culvert (S_n), pipe flow dominates (Fig. 9-9b). If the culvert slope is steeper than the neutral slope, orifice flow occurs (Fig. 9-9c). The inlet is considered submerged when the inlet water depth is greater than 1.2 times the culvert diameter (D) above the inlet invert.

3. When the inlet and outlet are not submerged, the flow is open-channel (Fig. 9-9d) or weir (Fig. 9-9e). If the culvert slope is too mild to carry the maximum inlet flow at the required depth, open-channel flow occurs. Weir flow prevails when culvert slope is greater than that required to move the flow through the inlet.

We will discuss pipe and orifice flows because they are the most common types of culvert flow in aquaculture projects. The equation used to calculate pipe flow of a culvert is derived from the conservation of energy [equation (9-4)] and the continuity equation [equation (9-6)]. For a pipe-flow culvert with a free-flow outlet the equation is

$$Q = A \sqrt{\frac{2gH}{1 + k_m + k_c L}} \tag{9-27a}$$

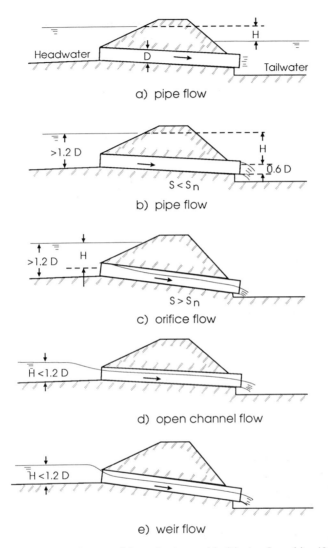

FIGURE 9-9. Possible flow conditions of culverts: (a), (b) pipe flow, (c) orifice flow, (d) open-channel flow, and (e) weir flow.

where
Q = flow rate (cfs)
H = effective head causing flow in culvert (ft)
A = cross-sectional area of culvert flow (ft^2)

k_m = minor-loss coefficient, which may include losses due to entrance, bend, and other fittings installed in the culvert

k_c = head loss coefficient due to friction for culvert flowing full (Tables B-5 and B-6)

L = culvert length (ft)

Most culverts used in aquaculture ponds for storm-water spillway are straight and no fittings are installed. For culverts with submerged outlets, the equation is

$$Q = A \sqrt{\frac{2gH}{k_x + k_m + k_c L}} \tag{9-27b}$$

where k_x is the exit coefficient. Since k_x is 1.0 according to Table B-4, equation (9-27b) is the same as equation (9-27a). The effective head of a culvert must be correctly determined as shown in Figure 9-9. The effective head for a free-flow outlet is the vertical distance between the headwater surface and 0.6 times the culvert diameter above the outlet invert. For a submerged outlet, the effective head is the vertical distance between the headwater and tailwater surfaces.

If a culvert has a slope steeper than the neutral slope and the outlet is not submerged, the culvert maintains orifice flow. The flow rate is calculated from the orifice-flow equation

$$Q = CA\sqrt{2gH} \tag{9-28}$$

where

C = orifice coefficient, 0.45 to 0.75, depending on the entrance shape ($C = 0.65$ for a square-edged entrance)

H = effective head causing flow (ft); the vertical distance between the headwater surface and the center of the culvert inlet

Flow regimes of a culvert with free-flow outlet may also be determined by calculating flow rates for pipe and orifice flows and comparing them. The lower flow rate of the two governs the culvert. When a culvert is short, orifice flow governs even if the culvert slope is milder than the neutral slope. In this case, orifice flow should be calculated and compared with the conduit capacity as pipe flow.

Under normal conditions of a free-flow culvert, pipe flow should be maintained because it provides greater effective head than does orifice flow. A culvert that requires a steep slope (up to 30%) may need some devices or modification at the inlet to develop pipe flow. The hood inlet

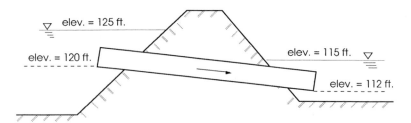

FIGURE 9-10.   Illustration of Example 9-10; an outlet-submerged culvert.

with a suitable antivortex device, shown in Chapter 13, makes the culvert prime and flow full. Hood inlets are simple and easy to install with common, commercially available corrugated metal, corrugated plastic, or concrete pipe. Beasley et al. (1960) reported that the hood inlet for a typical culvert requires the water level above the top of the culvert invert to be 40% of the culvert diameter to eliminate vortex and maintain pipe flow.

Example 9-10: Determining the Flow Rate of a Submerged Culvert

A circular corrugated metal culvert of 18-in. diameter will be installed across a small impoundment as shown in Figure 9-10. The straight culvert is 50 ft long and has a inward projecting entrance. The maximum headwater surface elevation allowed during a storm is 125 ft, and the elevation of the inlet invert is 120 ft. The outlet is submerged in a stilling basin, which has a water-surface elevation of 115 ft. Find the flow capacity that the culvert can carry while maintaining a 125 ft maximum water surface.

Solution: Since the inlet and outlet are submerged, the culvert has pipe flow during the storm. The cross-sectional area of the culvert is

$$A = 0.785D^2 = 0.785(18 \text{ in.} \div 12 \text{ in./ft})^2 = 1.77 \text{ ft}^2$$

Effective head causing the flow: $H = 125 \text{ ft} - 115 \text{ ft} = 10 \text{ ft}$
Entrance coefficient: $k_e = 0.8$ (from Table B-4)
Coefficient for bend: $k_b = 0.0$ for a straight culvert
Minor-loss coefficient: $k_m = k_e + k_b = 0.8 + 0.0 = 0.8$
Manning's coefficient: $n = 0.025$ for a corrugated metal pipe (from Table 9-5)

From equation (9-25) the conduit head loss coefficient of the circular culvert is

$$k_c = \frac{5{,}087 \times 0.025^2}{18^{4/3}} = \frac{3.18}{47.13} = 0.067$$

A similar value is found for k_c from Table B-5. Then

$$Q = 1.77 \sqrt{\frac{2 \times 32.2 \times 10}{1 + 0.8 + 0.067 \times 50}}$$

$$= 1.77 \sqrt{\frac{644}{5.15}} = 19.8 \, \text{cfs}$$

The proposed culvert can handle a storm that generates 19.8 cfs or less while maintaining the water-surface elevation lower than 125 ft at the inlet.

Example 9-11: Determining the Flow Rate of a Free-Flow Culvert

The culvert of Example 9-10 will have no backwater at the outlet, but will be a free-flow outlet. If the elevation at the outlet invert is 110 ft, find the flow rate.

Solution: Since the outlet has free-flow, the culvert maintains either pipe or orifice flow. First, assume pipe flow. The effective head is

$$H = 125 \, \text{ft} - 110 \, \text{ft} - 0.6(18 \, \text{in.} \div 12 \, \text{in./ft}) = 14.1 \, \text{ft}$$

From Example 9-10, the conduit head loss coefficient is $k_c = 0.067$, and the minor-loss coefficient is $k_m = k_e + k_b = 0.8$. The culvert capacity as pipe flow is

$$Q = 1.77 \sqrt{\frac{2 \times 32.2 \times 14.1}{1 + 0.8 + 0.067 \times 50}}$$

$$= 1.77 \sqrt{\frac{908}{5.15}} = 23.5 \, \text{cfs}$$

The flow velocity is

$$v = 23.5 \, \text{cfs} \div 1.77 \, \text{ft}^2 = 13.3 \, \text{fps}$$

From equation (9-24) the neutral slope is

$$S_n = 0.067 \times \frac{(13.3\,\text{fps})^2}{2 \times 32.2\,\text{ft/sec}^2} = 0.18$$

The culvert slope (S) for a small slope may be determined as

$$S = \frac{\text{elev. at inlet invert} - \text{elev. at outlet invert}}{\text{total length of culvert}}$$

$$= \frac{120\,\text{ft} - 110\,\text{ft}}{50\,\text{ft}} = 0.2 > S_n$$

Therefore, pipe flow may not prevail in the culvert.
    Now assume orifice flow. The effective head is

$$H = 125\,\text{ft} - 120\,\text{ft} - 0.5(18\,\text{in.} \div 12\,\text{in./ft}) = 4.25\,\text{ft}$$

Assume that the entrance coefficient is 0.65 for the square-edged inlet.
The orifice flow, from equation (9-28), is

$$Q = 0.65 \times 1.77 \sqrt{2 \times 32.2 \times 4.25} = 19.0\,\text{cfs}$$

The orifice-flow culvert can carry up to 19.0 cfs while maintaining the maximum water-surface elevation of 125 ft at the inlet when the outlet is not submerged. Examples 9-10 and 9-11 illustrate that the submerged outlet of the culvert may increase its flow capacity. The reason is the increased effective head of the pipe-flow culvert even with the higher head loss in the pipe flow. Often, a stilling basin is installed at the culvert outlet to keep it submerged.

Example 9-12: Selecting a Culvert Size to Carry a Design Flow

For the culvert layout in Example 9-10 a corrugated metal pipe will be selected to carry 30 cfs flow. Determine the available minimum size of this culvert.

Solution: The culvert has pipe flow because the inlet and outlet are submerged. From Example 9-10 the following data are obtained:

Effective head: $H = 10\,\text{ft}$
Minor-loss coefficient: $k_m = 0.8$

TABLE 9-6   Solution for Example 9-12: Selecting a Culvert Size to Carry a Design Flow

| Trial No. | Culvert Size | | | k_c | Q (cfs) | Comment |
	d (in.)	D (ft)	A (ft²)			
1	18	1.5	1.77	0.067	19.8	Low
2	24	2.0	3.14	0.046	39.4	High
3	22	1.8	2.64	0.052	31.9	OK

The culvert flow is calculated from equation (9-27a) as follows:

$$Q = A \sqrt{\frac{2gH}{1 + k_m + k_c \times L}} = 0.785D^2 \sqrt{\frac{2 \times 32.2 \times 10}{1 + 0.8 + k_c \times 50}}$$

The friction loss coefficient of the culvert, from equation (9-25), is

$$k_c = \frac{5,087n^2}{d^{4/3}}$$

In these equations, Q is a function of culvert size and k_c, and k_c is a function of culvert size only. Since Q and k_c are both functions of culvert size, there is no simple way to determine it. The trial-and-error method shown in Table 9-6 may be used to solve this problem. In this solution, n = 0.025 will be used for the corrugated metal culvert. Therefore, the required minimum size of culvert is 22 in. in diameter to carry 30 cfs flow from a storm while maintaining the allowed water-surface elevation at the inlet.

9.7   SIPHONS

A siphon, which frequently is used in aquaculture, is a pipe spillway. A siphon can be used to transfer water from one level over a small increase in elevation to a lower level (Fig. 9-11). When the elevation of the siphon outlet or water surface at the submerged outlet is lower than the water-surface elevation at the inlet, water flows through the siphon and over the barrier to discharge. The pressure at points A and B in Figure 9-11 is atmospheric pressure. Atmospheric pressure pushes water through the siphon when the pipe between A and B is below atmospheric pressure. For this purpose a siphon must be properly primed (full of water before positioned for operation). The effective head that causes flow through the

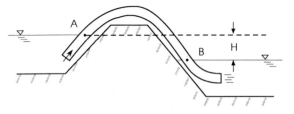

a) Submerged siphon

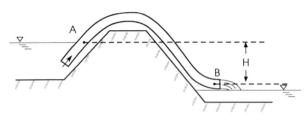

b) Free flow siphon

FIGURE 9-11.   Definition sketch of siphon flows: (a) outlet-submerged siphon; (b) free-flow siphon.

siphon is the elevation difference between the water surfaces at the inlet and outlet of the siphon when the outlet is submerged (Fig. 9-11a). At a free-flow outlet, the effective head is the vertical distance measured between the water surface at the inlet and the center of the outlet end (Fig. 9-11b).

The resistance opposing flow in a siphon is head loss through the pipe due to friction, inlet, and bend. The flow rate of a siphon tube depends on its size, length, type of pipe material, and the available head. Siphon problems may be solved by the same techniques used to solve a pipe-flow culvert for a long siphon or an orifice-flow problem for a short siphon. Small aluminum or plastic siphon tubes are often used to deliver water over the bank of ditches or canals. Figure 9-12 may be used to estimate the flow rates for small aluminum siphon tubes installed as shown in Figure 9-11. For an example, suppose a small aluminum siphon tube is used to supply water to a fish pond from a canal. The siphon is 4 in. in diameter and 8 ft long and installed as in Figure 9-11b. Water-surface elevation in the canal is maintained at 150 ft, and the elevation of the center of the free-flow outlet is 149.5 ft. The initial flow rate through the siphon tube when the outlet stays above the pond water level is calculated

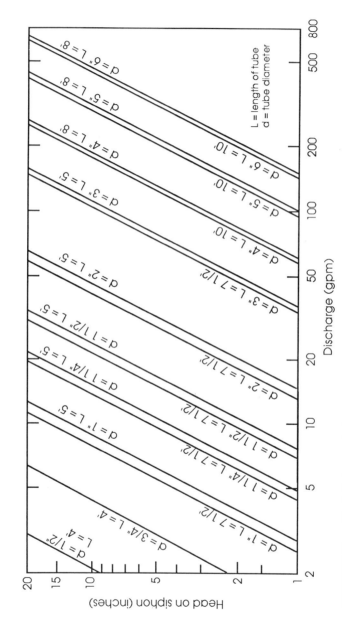

FIGURE 9-12. Head-discharge relationships of small aluminum siphon tubes similar to those shown in Figure 9-11. (*Source: U.S. Soil Conservation Service 1983.*)

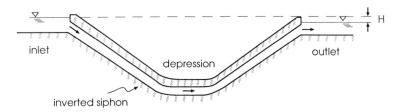

FIGURE 9-13.   Illustration of Example 9-14; an inverted siphon.

as follows. The effective head is $H = 150\,ft - 149.5\,ft = 0.5\,ft$ or 6 in. The flow rate (Q) from Figure 9-12 is 150 gpm. However, when the water level in the pond rises above the top of the siphon outlet, the flow rate decreases as the effective head decreases. Determining the total time to fill a pond under this situation is more complicated due to dynamic changes of the flow rate. One simple method is to determine the flowing time at various water levels above the top of the outlet up to the final water level in the pond. The total filling time is the sum of the calculated flowing times above the top of the outlet until the water reaches the final water level.

Inverted siphons are used to convey water under a road where minimum road clearance is available or to carry water under depressions or drainage ditches. Inverted siphons consist of an inlet, an outlet, and a conveyance section (Fig. 9-13). The pipe-flow equation for a culvert [equation (9-27a)] may be used to calculate the flow capacity of inverted siphons. Consideration should be given to head losses due to friction, angled turns, inlet, and outlet. The hydraulic head that causes water flow through the siphon is the elevation difference between the water surfaces at the inlet and outlet. Inverted siphons must be constructed of materials that can sustain the external load over the pipe as well as the hydraulic pressure. The design of inverted siphons in a water conveyance system is presented by Aisenbrey et al. (1978). The pipe-flow equation for a culvert may be used to determine the discharge through an inverted siphon. Where siphon length is short the orifice equation may be used for a simple solution.

10

Water Measurement

10.1 INTRODUCTION

Rapidly expanding aquacultural industries require more water from the limited resource. Knowing the amount of water delivered to a water use system will help reduce waste. Water measurement is important in effectively using water in an aquaculture system. This will also help aquaculturists share the resource fairly with other water users.

Many methods of measuring water are available. Some are complicated and require expensive measurement devices, and others are simple and inexpensive. In general, simple inexpensive methods are less accurate than expensive methods. Selection of a method should be determined by the availability of measuring devices, adequacy of the site, and accuracy required. Most water-control structures in open channels can be used to measure flow with a minimum effort if they have the hydraulic characteristics discussed in this chapter, which include drop structures, sluice gates, siphons, inverted siphons, culverts, inlet structures, diversion gates, and other structures.

10.2 UNITS

Water is measured in two ways: in motion and at rest. Water measured in motion is the discharge or flow rate (cfs or gpm). One cfs is equivalent to 448.8 gpm, but for simplicity 450 gpm may be used. In general, cfs is used to measure large flows, such as in canals and streams, and gpm is used for small flows, such as in pipes or from pumps or springs. Water measured at rest is a volume (in gal, ft^3, acre-in., or acre-ft). One acre-ft of water represents a volume of water that will uniformly cover 1 acre to a depth

TABLE 10-1   Common Units Used in Measuring Rate and Volume of Water Flow, and Their Conversion Factors

Volume
$1\,ft^3 = 7.48\,gal = 1{,}728\,in.^3 = 62.4\,lb$ at 50°F
$1\,gal = 231\,in.^3$
1 acre-in. = 1 in. of water which uniformly covers 1 acre of land
        $= 3{,}630\,ft^3 = 27{,}154\,gal$
1 acre-ft = 1 foot of water which uniformly covers 1 acre of land
        $= 43{,}560\,ft^3 = 325{,}850\,gal$

Flow Rate
$1\,ft^3/sec$ (cfs) $= 7.48\,gal/sec = 449\,gal/min$ (gpm)
$1\,gpm = 0.0023\,cfs = 1{,}440\,gal/day$ (gpd)
1 acre-in./hr $= 452\,gpm =$ approximately 1 cfs
1 cfs flowing for
    1 hr $= 0.992$ acre-in. or approx. 1 acre-in.
    12 hr $= 0.992$ acre-ft or approx. 1 acre-ft
    24 hr $= 1.984$ acre-ft or approx. 2 acre-ft

Note: General unit conversion factors are given in Appendix A.

of 1 ft. It is equivalent to $43{,}560\,ft^3$. Acre-feet or acre-inches are used to measure a large volume of water, whereas small volumes are measured in cubic feet or gallons; $1\,ft^3$ is approximately equivalent to 7.5 gal. Table 10-1 shows several common units for volume and rate of flow, along with useful conversion factors.

10.3 PRINCIPLES OF WATER MEASUREMENT

Measurement of a flow rate is based on application of the formula

$$Q = Av \tag{10-1}$$

where $Q =$ flow rate (ft^3/sec or cfs)
     $A =$ flow cross-sectional area (ft^2)
     $v =$ mean flow velocity (ft/sec or fps)

Therefore, measurement of a flow rate involves determination of the mean velocity and cross-sectional area of the flow. Measurement of a flow volume requires integration of the flow rate for the time period involved.

$$V = Qt \tag{10-2}$$

where $V =$ volume of flow (ft^3) and $t =$ time of flow (sec).

**Example 10-1: Determining the Discharge and Volume
of a Supply Canal Flow**

The mean flow velocity in a trapezoidal-shape canal is 1.5 fps. The canal
is 1.5 ft deep at the center and has a bottom width of 5 ft. Its side slope is
2:1. From Table 8-2 the cross-sectional area of the flow is

$$A = by + zy^2 = 5\,\text{ft} \times 1.5\,\text{ft} + 2(1.5\,\text{ft})^2 = 12\,\text{ft}^2$$

Then the flow rate is

$$Q = Av = 12\,\text{ft}^2 \times 1.5\,\text{fps} = 18\,\text{cfs}$$

Assuming that the flow rate stays constant, the total volume of water
flowing in the canal in one day is

$$V = Qt = 18\,\text{cfs} \times 24\,\text{hr} \times \frac{3,600\,\text{sec}}{1\,\text{hr}} = 1{,}555{,}200\,\text{ft}^3$$

The volume in acre-feet is

$$1{,}555{,}200\,\text{ft}^3 \times \frac{1\,\text{acre}}{43{,}560\,\text{ft}^2} = 35.7\,\text{acre-ft}$$

This water can uniformly cover 35.7 acres to a depth of 1 ft.

10.4   SMALL FLOWS

A simple but accurate flow measurement method is the volumetric-flow
method shown in Figure 10-1. This method involves measuring the time
required for a flow to fill a container of known volume. The volume
collected in the container divided by the flow time is then equal to the
flow rate:

$$Q\,(\text{gpm}) = \frac{\text{volume (gal)} \times 60\,\text{sec/min}}{\text{time of flow (sec)}} \qquad (10\text{-}3)$$

This method should be used for small flows, such as from springs and
small pumps. A larger container requires more time to fill, but provides a
more accurate measurement; though more time-consuming.

A short pipe can be installed at the outlet of a spring to use this
method for a spring-flow measurement (Fig. 10-1). In this case the capacity

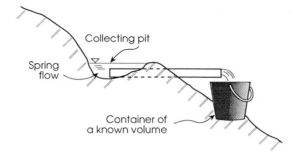

FIGURE 10-1.   Schematic of a volumetric-flow measurement.

of the pipe must be slightly larger than the flow rate from the spring, to avoid an overflow. More than one pipe may be needed to measure a large flow. The measurement should be conducted at least three times to obtain the average flow.

Example 10-2: Volumetric-Flow Measurement of a Spring Flow

A spring flow was measured with a 5-gal container. The flow was measured three times, and the measurements took 28, 23, and 25 sec, respectively, to fill the container. The average time to fill the container is

$$\text{Average filling time} = \frac{28 + 23 + 25}{3} = 25.3 \, \text{sec}$$

The flow rate is then

$$Q = \frac{5 \, \text{gal} \times 60 \, \text{sec/min}}{25.3 \, \text{sec}} = 11.9 \, \text{gpm}$$

10.5 WATER MEASUREMENT OF OPEN-CHANNEL FLOWS

Many methods have been developed for measuring open-channel flows either directly or indirectly with measuring devices or hydraulic structures. Flow volume is determined from equation (10-2).

10.5.1 Velocity–Area Methods

Velocity–area methods involve an actual measurement or an estimation of the cross-sectional area and the mean flow velocity. The flow rate is determined from equation (10-1). For an accurate measurement the cross section of a channel may be divided into several segments to determine the flow rate in each segment. The total flow rate of the channel is obtained by summing the flow rates of all segments.

Floating Methods

A simple but crude method of determining open-channel flow rates is made by measuring the velocity of a floating material moving with the water current. A known distance (normally 50–100 ft) of a relatively straight and uniform section of a channel is selected and marked on both ends. The time for a floating material (a colored wooden block) to travel the marked distance is measured. The velocity is determined as

$$v = C \times \frac{\text{travel distance (ft)}}{\text{travel time (sec)}} \tag{10-4}$$

where v = flow velocity (fps) and C = velocity correction factor given by Table 10-2. The velocity correction factor is necessary because the velocity measured by a floating method is that of the water at the channel surface. Under normal flow conditions the velocity at the channel surface is higher than the mean velocity in the channel. The velocity distribution in a channel section is discussed later.

The velocity should be measured three to five times to obtain the average velocity. The cross-sectional area of a channel is estimated by multiplying the average depth by the width of flow. The depth and width of flow may be estimated or actually measured to find a more accurate

TABLE 10-2   Velocity Correction Factors for Floating Methods

Average Flow Depth (ft)	C
1.0	0.66
2.0	0.68
3.0	0.70
4.0	0.72
5.0	0.74
6.0	0.76
9.0	0.77
12.0	0.78
15.0	0.79
>20.0	0.80

Source: U.S. Bureau of Reclamation (1975).

flow rate. The travel distance may be measured by pacing when the pace length of the operator is known. The advantage of this method is that it does not require a great deal of equipment other than a measuring tape, ground markers, and a floating material. A disadvantage is that experience is often required for a reliable measurement. In general, this method provides a quick but rough estimation of the flow rate in an open channel.

Example 10-3: Floating Method for Measuring an Open-Channel Flow

A 100-ft-long relatively straight section of a stream was marked to measure the flow rate by the floating method. The estimated width and mean depth of the flow are 13 ft and 1.5 ft, respectively. A red wooden block was dropped at the upstream end of the 100-ft section. The travel time of the block was recorded five times: 48, 43, 51, 55, and 49 sec. The average traveling time is thus $(48 + 43 + 51 + 55 + 49) \div 5 = 49.2$ sec. The velocity correction factor at 1.5-ft flow depth is 0.67, by interpolation from Table 10-2. The mean velocity of the flow is

$$v = 0.67 \times \frac{100\,\text{ft}}{49.2\,\text{sec}} = 1.36\,\text{fps}$$

The cross-sectional area of the stream is $A = 13\,\text{ft} \times 1.5\,\text{ft} = 19.5\,\text{ft}^2$. Hence, the flow rate is $Q = Av = 19.5 \times 1.36 = 26.5$ cfs.

Current Meter Methods
A device called a current meter can be used to reliably measure flow velocity and flow depth in an open channel. A wading-type current meter is composed of a rotating impeller, a revolution counter, and a wading rod. Two types of impellers are commonly used: propeller and cup. The impeller may be attached to a weight and cable instead of a rod to measure flows in channels that cannot be waded. Figure 10-2 shows the components of a cup-type current meter.

An impeller rotates when it is suspended in flowing water. The rate of revolution of the impeller is proportional to the flow velocity. An electric circuit is used to count the number of revolutions of the impeller in a given time period. Usually, the revolving impeller actuates a set of breaker points in the circuit, and clicks are detected by an earphone or counter. The number of revolutions is converted to flow velocity with the aid of a calibration chart provided by the manufacturer. Some current meters are equipped with an electronic gauge that directly converts the revolution rate of the impeller to flow velocity.

To measure an open-channel flow with a current meter, select a

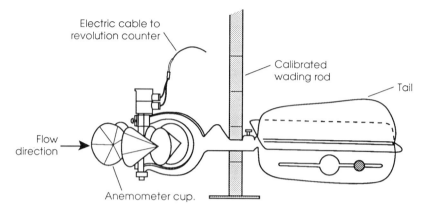

FIGURE 10-2.   Components of a cup-type current meter.

relatively straight and uniform channel section. A zero station or reference initial point is set on one bank of the channel and a measuring tape
is used to measure horizontal distance perpendicular to the direction
of flow. Current meter readings and flow depths are taken at regular
intervals from the channel edge. The intervals may be 2–10 ft, depending
on channel width. Close intervals should be used where there are abrupt
changes of velocity or depth of flow. The channel edges are counted as
the first and last measurement stations, respectively. The area between
two adjacent stations is a channel segment. Figure 10-3 shows the
cross section of a channel divided into segments for current meter
measurements.

The current meter should be held in front of the operator so that the
impeller faces into the flow. The operator should stand facing the stream
bank at arm's length from the meter, to minimize interruption to the flow
velocity to be measured. The flow depth is read on the wading rod. The
mean velocity is often measured at 0.6 of the flow depth measured from
the water surface. For more accurate measurement, a two-point method
is used, which employs the mean of velocities measured at 0.2 and 0.8 of
the depth from the water surface. Figure 10-4 shows the vertical distribution of velocity in a channel section and the depths at which velocities are
measured. As shown, the flow velocity increases from the channel bottom
to the surface. The maximum velocity occurs near the surface. The
surface velocity is not the maximum velocity but is higher than the mean
velocity of flow in the channel. The vertical variation of flow velocity in a
channel is due to friction caused by channel bottom materials and the air.

An electromagnetic probe can also be used to measure flow velocity.

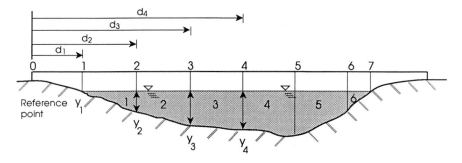

FIGURE 10-3.   Cross section of an open channel that has been divided for current meter measurements.

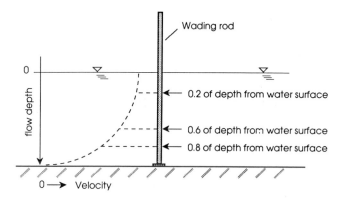

FIGURE 10-4.   Typical velocity distribution in an open-channel flow and setup positions for current meter measurements.

The probe is installed on a wading rod and connected to a display meter, which also supplies electric energy to the probe. A small voltage is created around the probe by flowing water. The voltage is transmitted to the meter, which directly shows the flow velocity since it is proportional to the voltage. This probe is used just like impeller-type meters.

The discharge of a channel segment is the product of the cross-sectional area of the segment and the mean velocity of flow in the segment. The total flow rate is the sum of the flow rates in all channel segments. As shown in Figure 10-3, y_2, y_3, and y_4 are the flow depths at three adjacent measurement stations; d_2, d_3, and d_4 are the respective distances of these stations from the reference point; and v_2, v_3, and v_4 are the respective mean velocities at these stations. The flow rates q_2 and q_3 of segments 2 and 3, respectively, are

$$q_2 = \left(\frac{y_2 + y_3}{2}\right)(d_3 - d_2)\left(\frac{v_2 + v_3}{2}\right) \qquad \text{(10-5a)}$$

$$q_3 = \left(\frac{y_3 + y_4}{2}\right)(d_4 - d_3)\left(\frac{v_3 + v_4}{2}\right) \qquad \text{(10-5b)}$$

In general, the flow rate of channel segment i is

$$q_i = \left(\frac{y_j + y_{j+1}}{2}\right)(d_{j+1} - d_j)\left(\frac{v_j + v_{j+1}}{2}\right) \qquad \text{(10-5c)}$$

where q = segment flow rate
  y = flow depth
  d = distance from reference point
  v = flow velocity
  i = segment number, $i = 1, \ldots, n$
  j = point of measurement, $j = 1, \ldots, n + 1$
  n = total number of segments

Then the total flow rate of the channel, Q, is

$$Q = \sum_{i=1}^{n} q_i \qquad \text{(10-6)}$$

Example 10-4: Flow Rate Measurement Using a Current Meter

A channel flow is measured with a current meter as shown in Figure 10-3. Flow depth, velocity, and distance from the reference point at each measuring point along with results of the calculation are given in Table 10-3. There are seven measuring points and six segments. Detailed calculation procedures are given in the table.

Velocity-Head Rod Methods
A velocity-head rod is a simple calibrated stick for measuring velocity and depth of flow in an open channel (Fig. 10-5). These rods are not recommended for velocities below 1.0 fps and higher than the critical velocity (U.S. Soil Conservation Service 1973a). The sharp edge of the rod is set facing into the flow. The flow depth is read on the stick, and the rod is then turned 180° to place the flat edge of the rod against the flow. A hydraulic jump is developed by the obstruction of the flat edge. Figure 10-5 illustrates the use of a velocity-head rod in a channel. The average water depth at the jump minus the flow depth is the velocity head of the flow, h_v. The flow velocity is calculated from the kinetic energy equation

TABLE 10-3   Data Obtained by a Current Meter Method Illustrated in Example 10-4

(a) Measuring Point Number	(b) Distance from Reference (ft)	(c) Flow Depth (ft)	(d) Depth at measurement[a] (ft)	(e) Time of Measurement (sec)	(f) No. of Revolutions	(g) Flow Velocity[b] (fps)	(h) Mean Velocity[c] (fps)	(i) Mean Flow Depth[d] (ft)	(j) Segment Width[e] (ft)	(k) Cross-sectional Area[f] (ft)	(l) q[g] (cfs)	(m) Segment Number
1	4	0				0					0	
2	10	1.5	0.9	54.2	40	1.65	0.83	0.75	6.0	4.5	3.74	1
3	15	3.0	1.8	59.8	60	1.98	1.82	2.25	5.0	11.25	20.48	2
4	20	3.9	2.34	48.4	60	2.4	2.19	3.45	5.0	17.25	37.78	3
5	25	4.4	2.64	46.0	40	1.95	2.18	4.15	5.0	20.75	45.24	4
6	30	4.2	2.52	43.2	40	2.07	2.01	4.3	5.0	21.50	43.22	5
7	32	0			0		1.04	2.1	2.0	4.20	4.37	6
										0		

Total discharge   154.83

[a] Single-depth method. Impeller is set at 0.6 of the total flow depth from the water surface.
[b] Velocity read from the manufacturer's calibration table.
[c] Column h: Mean of two adjacent velocities in column g.
[d] Column i: Mean of two adjacent water depths in column c.
[e] Column j: Width of segments. The value is the difference of two adjacent distances from the reference point shown in column b.
[f] Column k: Product of values in columns i and j.
[g] Column l: Product of values in columns h and k.

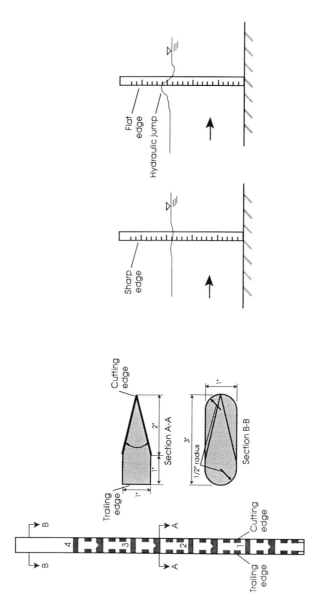

FIGURE 10-5. Components and field setup of a velocity-head rod. (*Source: U.S. Soil Conservation Service 1973a.*)

$$v = \sqrt{2gh_v} \qquad (10\text{-}7)$$

where v = velocity (fps)
$\quad\;\; g$ = gravitational acceleration (g = 32.2 ft/sec^2)
$\quad\;\; h_v$ = velocity head (ft)

One advantage of this method is the availability and low cost of the device. A velocity rod may be built locally with a wooden stick. Procedures for using a velocity-head rod are similar to those for using current meters.

10.5.2 Indirect Measurement of Open-Channel Flows

The hydraulic characteristics of an open channel and flow conditions in the channel can be used to indirectly determine flow rates in the channel. When flowing water in an open channel is structurally constricted, the flow rate at the constricted point is related to the flow depth at that point. This relationship is used to determine flow rates in open channels by measuring flow depths over specially designed structures installed in the channels.

Many devices have been developed and calibrated to determine relationships between flow depth and flow rate. The flow depth in an open channel can be measured with a staff gauge (Fig. 5-1), automatic stage recorder, potentiometer, or pressure transducer (Bos 1989). A staff gauge is a vertical scale marked in feet, tenths of a foot, and hundredths of a foot. The scale is often painted on a wooden or metal staff or directly on the vertical wall of a measuring structure. Figure 10-6 shows a flow-measuring device equipped with a staff gauge, a stilling well, and a float-type recording device. An automatic drum-type recorder uses a spring winding or battery-operated drums designed with interchangeable gear ratios for various recording-time intervals and gauge scales. The drum is connected to a pulley that rotates relative to the vertical movement of a float in a stilling well. The float moves in response to the water-level changes in the stilling well. The position changes of the float are recorded on a chart attached to the drum. Manual reading or digitizing of the chart is conducted to determine flow rates.

Potentiometers or pressure transducers are sometimes used instead of drum-type stage recorders. These devices are more advanced and require microprocessor based data loggers to record and analyze the data. A potentiometer connected to the float pulley in a system such as that shown in Figure 10-6 can be used to detect the rotational position of the

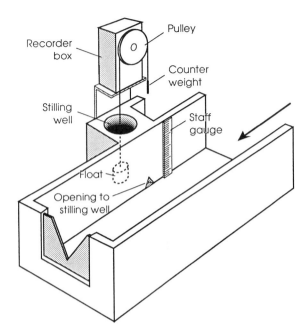

FIGURE 10-6.   An open-channel flow measuring station equipped with a V-notch weir.

pulley as a voltage. Voltage changes are recorded by a data logger. A pressure transducer can be used to detect water-depth changes in a water-measurement device. The transducer output is a voltage that is detected and recorded by a data logger. The data can be easily converted to effective head and analyzed to determine the flow rate and volume. Except for the staff gauge, these instruments require a stilling well to maintain a stable water-surface condition.

Structural devices for which the relationship between effective head and flow rate can be used to determine the flow rate in an open channel flow will be discussed. Proper construction and installation of these devices are critical for accurate measurement of flow rates. The devices must be equipped with side walls and cutoffs driven into the channel bottom and banks to prevent seepage losses. The details of using these and other devices to measure open-channel flows are given in several references, including Bos (1984) and Walker and Skogerboe (1973).

10.5.3 Orifice Plates

Orifice plates offer a simple method for measuring relatively small open-channel flows. An orifice is an opening (commonly circular or

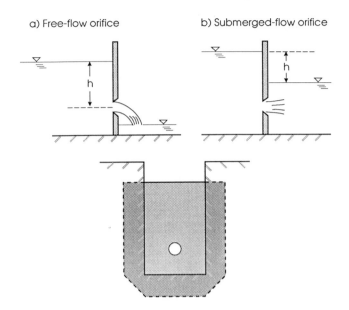

FIGURE 10-7.   A circular orifice plate under free- and submerged-flow conditions.

rectangular) with a closed perimeter through which water flows. An orifice plate is simple and inexpensive to build and easy to install. The plate is installed across the channel, and effective hydraulic heads are measured to determine the flow rates (Fig. 10-7). Flow velocities through the orifice are related to the effective hydraulic heads. The relationship between flow velocity and the effective head of an orifice is

$$v = C\sqrt{2gh} \tag{10-8}$$

where v = flow velocity (fps)
h = effective hydraulic head (ft)
C = orifice coefficient

Then the flow rate is

$$Q = Av = CA\sqrt{2gh} \tag{10-9}$$

where Q = flow rate (cfs) and A = cross-sectional area of orifice opening (ft^2).

An orifice may have a free-flow or a submerged flow on the downstream side of the orifice (Fig. 10-7). Effective hydraulic heads must be

TABLE 10-4   Coefficients for Sharp-Edged Orifices

Orifice Diameter (in.)	Coefficient	
	Free Flow	Submerged Flow
0.75	0.61	0.57
1.0	0.62	0.58
1.75	0.63	0.61
2.0	0.62	0.61
2.5	0.61	0.60
3.0	0.60	0.60
3.5	0.60	0.60
4.0	0.60	0.60
>4.0	0.60	0.60

Source: U.S. Soil Conservation Service (1973a).

measured according to the orifice conditions. The effective head of a free-flow orifice is the vertical distance between the water surface behind the orifice plate and the centerline of the orifice opening. For a submerged-flow orifice, the effective head is the vertical distance between the water surfaces at the upstream and downstream sides of the orifice. The orifice coefficient for a sharp-edged orifice varies from 0.57 to 0.64 (Bos 1989), depending on the size and flow condition of the orifice. The heads should be measured close to the orifice, but where the water surface is not turbulent. In general, free-flow orifices are not recommended in open channels with a mild slope because of the large head differences required for the device to operate properly. Table 10-4 shows orifice coefficients for various sizes of sharp-edged circular orifices. For a submerged-flow orifice the submergence should be complete to avoid difficulties in measuring effective heads.

Example 10-5: Flow Measurement Using an Orifice Plate

A submerged circular orifice is used to measure flow in the small ditch shown in Figure 10-8. The orifice has a 2-in.-diameter sharp-edged opening. The water surfaces at the upstream and downstream sides are 1.8 and 1.4 ft above channel bottom, respectively. The effective hydraulic head is $h = h_u - h_d = 1.8\,\text{ft} - 1.4\,\text{ft} = 0.4\,\text{ft}$. The cross-sectional area of the orifice is

$$A = \frac{\pi D^2}{4} = 0.785 D^2 = 0.785 \left(\frac{2\,\text{in.}}{12\,\text{in.}/\text{ft}} \right)^2 = 0.022\,\text{ft}^2$$

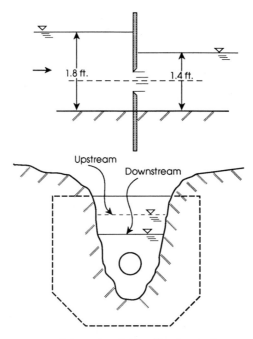

FIGURE 10-8.   Schematics of the orifice flow in Example 10-5.

The orifice coefficient is 0.61 for a 2-in. submerged orifice (Table 10-4). The flow rate is

$$Q = CA\sqrt{2gh} = 0.6 \times 0.022\sqrt{2 \times 32.2\,\text{ft/sec}^2 \times 0.4\,\text{ft}}$$
$$= 0.068\,\text{cfs or } 30\,\text{gpm}$$

10.5.4   Weirs

A *weir* consists of a barrier plate that constricts the flow in an open channel through a fixed-size opening. Common shapes of weirs are rectangular, trapezoidal, and triangular (Fig. 10-9), rectangular being the simplest. The profile of a sharp-crested rectangular weir is shown in Figure 10-10. The bottom edge of the barrier plate is called the *weir crest*, and flow depth over the crest is measured upstream from the crest and is called the *effective head*. The overflowing sheet of water is called the *nappe*. The effective head and the shape and size of a weir crest are used to determine the flow rate. The flow through a weir must maintain a free discharge for an accurate measurement. The water level on the down-

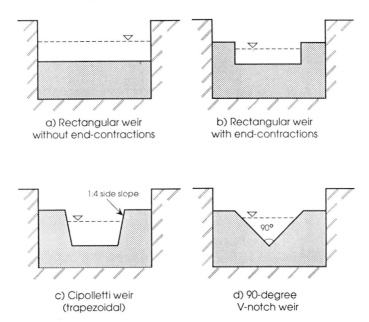

a) Rectangular weir
without end-contractions

b) Rectangular weir
with end-contractions

c) Cipolletti weir
(trapezoidal)

d) 90-degree
V-notch weir

FIGURE 10-9.   Various types of weirs for open-channel flow measurement: (a) suppressed rectangular weir; (b) end-contracted rectangular weir; (c) Cipolletti (trapezoidal) weir; and (d) 90° V-notch weir.

stream side should be low enough to maintain a free-flow. This requires some amount of head loss, which limits the use of a weir in mild-slope channels.

A rectangular weir may be installed and operated with or without end contractions, as shown in Figure 10-9. The flow rate of a completely end-contracted rectangular weir is computed as

$$Q = 3.33(L_w - 0.2h)h^{3/2} \qquad (10\text{-}10)$$

where Q = flow rate (cfs)
  L_w = length of weir notch at the crest (ft)
  h = effective hydraulic head (ft)

If a rectangular weir is not end-contracted, the width of the weir crest is approximately the same as that of the approaching channel. In this case, the flow rate is determined from the relation

$$Q = 3.33L_w h^{3/2} \qquad (10\text{-}11)$$

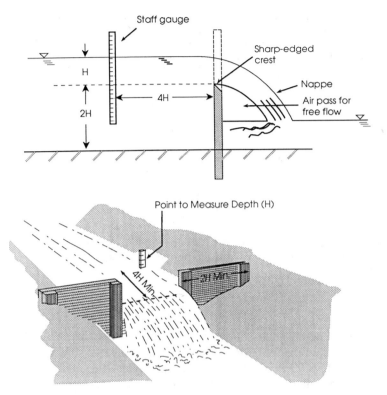

FIGURE 10-10.   Guidelines for weir installation and measurement; an example with an end-contracted rectangular weir. (*Source: U.S. Soil Conservation Service 1973a.*)

A rectangular weir without end contractions is called a *suppressed weir*. The difference between equations (10-10) and (10-11) is the calculation of the effective weir length. When a weir is end-contracted, the width of the flow narrows after water passes over the weir crest. The reduction ratio is approximately 10% of the effective head for each side of the weir (Hansen et al. 1980).

An end-contracted trapezoidal weir with 1:4 side slopes (Fig. 10-9) is called a *Cipolletti weir* after the developer's name. The flow rate over a Cipolletti weir is directly proportional to the crest length, and no correction for end contraction is needed; mathematically,

$$Q = 3.37L_w h^{3/2} \tag{10-12}$$

The Cipolletti weir can measure slightly larger flow rates than an end-contracted rectangular weir with the same crest length.

TABLE 10-5   Coefficients and Effective Head Correction
Factors for V-Notch Weirs as Function of Notch Angles

V-Notch Angle (deg)	Coefficient	ε^a (ft)
30	0.67	0.007
40	0.91	0.006
50	1.15	0.005
60	1.43	0.004
70	1.73	0.004
80	2.07	0.003
90	2.48	0.003

Source: Shen (1981).

^{a} Effective head correction factors.

Triangular weirs are used to measure small open-channel flows. The formula for the flow rate over a triangular V-notch weir is

$$Q = C(h - \varepsilon)^{5/2} \qquad (10\text{-}13a)$$

where C = triangular weir coefficient and ε = effective head correction factor for V-notch angle. Shen (1981) presents coefficients and effective head correction factors for various notch angles of the triangle weir (Table 10-5). A 90° V-notch weir is commonly used to measure flows less than 1 cfs. Smaller-angle V-notch weirs may be used to measure smaller flows. The common equation for the 90° V-notch weir shown in Figure 10-9 is

$$Q = 2.48h^{5/2} \qquad (10\text{-}13b)$$

The effective head correction factor is ignored in this equation.

All weir plates must be installed vertically, and the height of the crest above the channel bottom should be at least twice the effective head to be measured over the crest. The effective head should be measured at a distance four times the effective head upstream from the crest. The distances between the ends of the notch and the banks of the channel should not be less than twice the effective head. In general, the dimensions are determined by the designed maximum effective head to be measured by a weir. Guidelines for installation and measurement with a weir are shown in Figure 10-10.

Even though weirs are easy to build and install, they are not well adapted for water carrying silt, which deposits behind the weir plate. The rise in elevation of the channel bottom behind the weir plate in response to siltation may alter the conditions needed for accurate measurements. Debris carried in the water may also obstruct the weir-opening area. Hansen et al. (1980) provide increase in discharge due to silting behind weirs. They report that discharge increases up to 16% of the measured discharge, depending on the height and length of the deposit. In addition, weirs are not well adapted for use in channels with small grades because they do not provide adequate elevation differences between the water surfaces on the upstream and downstream sides of the weirs.

A weir may be built of wood, sheet metal, or concrete if the crest has a sharp edge. Because of potential damage from debris and rust pitting, extremely sharp-edged crests are not recommended (Ackers et al. 1978). The recommended crest shape for a weir is given in Figure 10-11. Some hydraulic structures in a channel may be directly used as weirs. In this case the condition of the structures, especially the shape of the crest, should be carefully studied to determine whether it meets the requirements for accurate flow measurements. In general, the flow rate over a non-sharp-edged notch is different from the flow rate determined with a weir equation. A hydraulic structure that has a broad notch should be evaluated before it is used for measurements. Often, a thin metal plate may be attached on the upstream side of the structure to provide weir-flow conditions (Fig. 10-12). In this case the height of the extension plate above the crest must be adjusted to allow air to enter below the nappe. The extension height must be limited so that it will not reduce the flow capacity of the structure.

Example 10-6: Flow Measurement with a Cipolletti Weir

A Cipolletti weir with a 1.0-ft crest width is installed at the head end of a supply canal as shown in Figure 10-13. The water stage read on a staff gauge installed 4 ft upstream from the crest is 2.3 ft. The gauge reading at the top of the weir crest is 1.4 ft. The effective head of the weir flow is $h = 2.3\,\text{ft} - 1.4\,\text{ft} = 0.9\,\text{ft}$. Then,

$$Q = 3.37L_w h^{3/2} = 3.37 \times 1.0\,\text{ft} \times (0.9\,\text{ft})^{3/2}$$
$$= 2.88\,\text{cfs}$$

10.5.5 Flumes

A *flume* is a specially designed hydraulic section installed in a channel to measure flow. Flumes are composed of an inlet (converging section), a throat (constricting section), and an outlet (diverging section). The throat section acts as a control, that is, a point in the flume where a unique

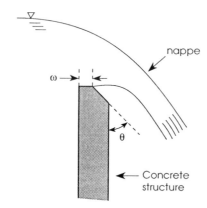

θ = maximum 45°
ω = maximum 1/8 inch

FIGURE 10-11.   Recommended geometry of the sharp edge of the weir crest. (*Source: Larry G. James*, Principles of Farm Irrigation System Design, *copyright © 1988 by John Wiley & Sons, Inc., New York, p. 152. Reprinted by permission of John Wiley & Sons, Inc.*)

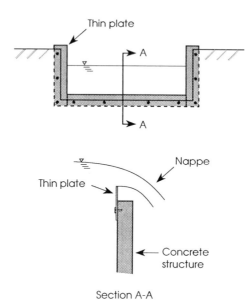

Section A-A

FIGURE 10-12.   A thin-plate extension installed on a hydraulic structure (a drop structure) for a weir-flow measurement.

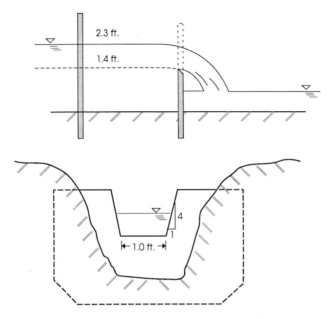

FIGURE 10-13.   Schematics of the Cipolletti weir in Example 10-6.

relationship exists between the water depth in the converging section and the flow rate. Cutoff walls should be attached to the flumes and driven into the channel bottom and bank to prevent seepage.

One of the most important advantages of flumes over weirs is that flumes are effectively used to measure flow in relatively shallow channels with flat bottoms. This can be done because of its low head loss requirement. Although flumes are normally operated under free-flow conditions, they can also accurately measure flow under submerged conditions. For submerged-flow conditions an additional head measurement at the downstream side of the throat section is required.

Several types of flumes have been developed. Each has its own unique configuration and is accurate when constructed and used properly. Flumes may be built of wood, concrete, sheet metal, plastic, or fiberglass. Most large flumes are constructed on site, whereas small flumes may be built in a shop and installed later.

Parshall Flume
The Parshall flume is a common type used in open channels (Fig. 10-14). Its most desirable feature is that it requires much less head loss than a weir does. It measures flow over a wide range (5 gpm to 3,000 cfs),

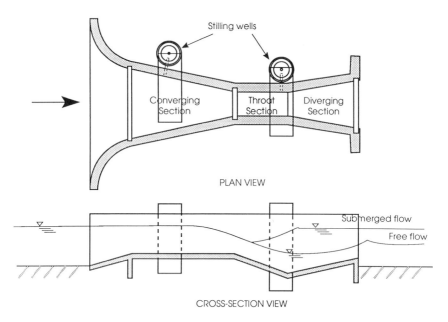

FIGURE 10-14.   Components of a Parshall flume.

depending on the size of the flume. Normally the Parshall flume is used to measure flow under free-flow conditions. Under these conditions, only measurement of the water depth at the inlet section is needed. Chow (1959) developed empirical equations for calculating flow rates through free-flow Parshall flumes with various throat widths. To use a Parshall flume it is desirable to maintain the free-flow conditions. When the water depth at the outlet section is greater than 60–70% of that at the inlet section above the bottom of the converging section (Fig. 10-14), the flume is considered submerged. This case requires a correction of the flow rates calculated by these equations since the submergence reduces the flow rate through the flume (Chow 1959).

The main disadvantage of the Parshall flume is the accurate design and installation required for satisfactory measurements. The Parshall flume has a contoured bottom, as shown in Figure 10-14, that must be built precisely. Details of the design and the calibrated flow rates of various sizes of Parshall flumes are given in the references (U.S. Soil Conservation Service 1973a; U.S. Bureau of Reclamation 1975).

Trapezoidal Flumes

A trapezoidal flume comprises approach, converging, throat, diverging, and exit sections with a flat bottom (Fig. 10-15). The design dimensions

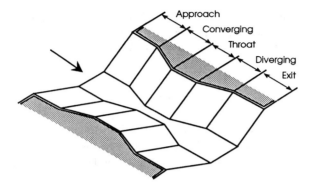

FIGURE 10-15.   Components of a trapezoidal flume. (*Source: American Society of Agricultural Engineers 1983.*)

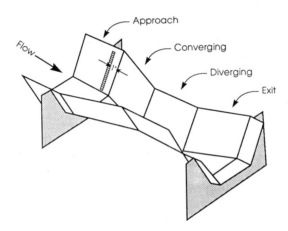

FIGURE 10-16.   Components of a 60° V-notch flume. (*Source: American Society of Agricultural Engineers 1983.*)

and calibration values for several standard trapezoidal flumes have been published by the American Society of Agricultural Engineers (1982). An advantage of the trapezoidal flume over the Parshall flume is the flat bottom, which makes it easy to design and install the flume on a channel bottom. Trapezoidal flumes can measure the flow rate with an accuracy of ±5% under free-flow conditions. A trapezoidal flume with a submergence up to 70% may be considered to be under a free-flow. Submergence is the ratio of water depth at the converging section to that at the diverging section above the flume bottom.

A trapezoidal flume with a 60° V-notch throat (Fig. 10-16) was developed for small flow measurements (Robinson and Chamberlain 1960). It was designed to measure flows from 1 to 26 gpm. The flow rate of this flume may be determined by the following equation (Walker and Skogerboe 1987):

$$Q = 0.75(h_u - 0.06)^{2.61} \qquad (10\text{-}14)$$

where Q = flow rate (gpm) and h_u = head reading along the sloping side of the flume upstream (in.). Various-sized, prefabricated trapezoidal and 60° V-notch flumes are commercially available (Plasti-Fab 1980).

Example 10-7: Flow Rate Measurement Using a 60° V-Notch Flume

A 60° V-notch flume is used to measure flow in a drainage ditch. The gauge reading on the sloping side of the inlet section is 1.35 in. The head at the downstream side is relatively lower than that at the upstream side. The flow through the flume is considered a free-flow. The flow rate Q is

$$Q = 0.75(h_u - 0.06)^{2.61} = 0.75(1.35 - 0.06)^{2.61} = 1.46\,\text{gpm}$$

Cutthroat Flumes

A *cutthroat* flume has a rectangular cross section, a flat bottom, and uniform inlet and outlet sections (Fig. 10-17). A constriction occurs at the intersection of the converging and diverging sections. The name was given by the developer (Skogerboe et al. 1967) because the flume has no throat section. A cutthroat flume is easier to design and build than Parshall or trapezoidal flumes because of its simple configuration. The flat bottom of the cutthroat flume allows it to be placed directly on a channel bed. Flow through a cutthroat flume is considered to be a free-flow when the water depth at the outlet is less than 60–80% of the inlet depth. The construction details of a cutthroat flume are also given in Figure 10-17. The flow rate under a free-flow condition is calculated from the equation

$$Q = KW^{1.025}h^n \qquad (10\text{-}15)$$

where Q = flow rate (cfs)
   K = free-flow coefficient
   n = free-flow exponent
   h = effective head measured at the converging section (ft)
   W = throat width (ft)

The coefficient K and exponent n are determined from known flume lengths. Table 10-6 shows the coefficients and exponents for cutthroat

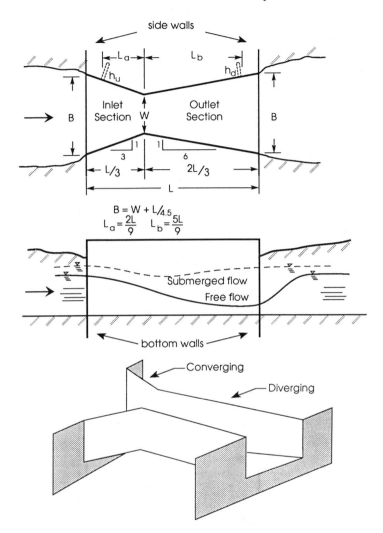

FIGURE 10-17.   Components of a cutthroat flume. (*Source: Skogerboe et al. 1973.*)

flumes. Measurement of the water depth at the outlet section is required to determine the flow rate under a submerged condition. The equation and related coefficients for determining the flow rates under submerged conditions are given by Walker and Skogerboe (1987).

Example 10-8: Application of a Cutthroat Flume

A cutthroat flume with 0.5-ft throat is installed in a small conveyance canal as shown in Figure 10-18. The total length of the flume is 3.0 ft. The

TABLE 10-6   Coefficients and Exponents for Cutthroat Flumes under Free-Flow Conditions

Length of Flume (ft)	K	n
9.0	3.5	1.56
8.0	3.6	1.58
6.0	3.65	1.64
4.5	3.98	1.72
4.0	4.05	1.74
3.0	4.5	1.84
2.0	5.0	1.97
1.5	6.1	2.15

Source: Skogerboe et al. (1973).

effective heads measured at the diverging and converging sections are 0.32 ft and 0.2 ft, respectively. Since the submergence level is 0.2 ft ÷ 0.32 ft = 62.5%, the flow is considered a free-flow. The flume coefficient K is 4.5, and the exponent n is 1.84 (Table 10-6). The flow rate is

$$Q = KW^{1.025}h^n = 4.5 \times 0.5^{1.025} \times 0.32^{1.84}$$
$$= 4.5 \times 0.49 \times 0.123 = 0.271 \, \text{cfs}$$

H-flumes

There are three types of H-flumes: HS, H, and HL. The S and L refer to small and large flow capacities. These flumes can accurately measure very small to relatively large flows, provided they are built and installed according to the developer's guidelines (U.S. Department of Agriculture 1979). One of the advantages of H-flumes over other flumes is the concentrated discharge that can be collected for water sampling. However, these flumes require a head loss to maintain a free-flow condition.

A combination of an H-flume and a Coshocton-type runoff sampler is used to measure discharge and collect aliquot samples of the discharge (Fig. 10-19). The samples may be used for water-quality studies or other uses. Design details of H-flumes and runoff samplers along with calibration values of the flow rates are given by the U.S. Department of Agriculture (1979).

10.5.6   Dilution Methods

By determining the dilution level of a tracing agent injected at an upstream location in a channel, we can determine the discharge without knowing

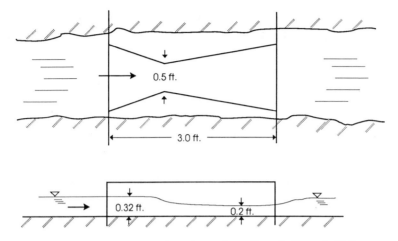

FIGURE 10-18.   Illustration of a cutthroat flume in Example 10-8.

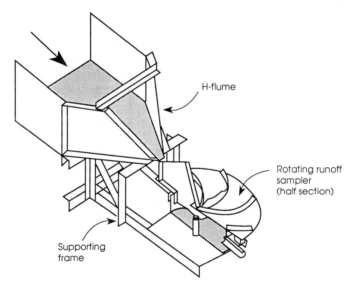

FIGURE 10-19.   An H-flume and a Coshocton-type runoff sampler. (*Source: U.S. Department of Agriculture 1979.*)

the cross-sectional area of the channel (U.S. Bureau of Reclamation 1975). Tracing agents include chemicals, dyes, and fluorescent materials whose concentrations in water can be determined. This method should be used only where the tracer material is completely mixed before samples are collected. This produces a uniform concentration of the tracer at the sampling point. The discharge is determined from the equation

$$Q = q\left(\frac{c_1 - c_2}{c_2 - c_0}\right)$$

(10-16)

where Q =flow rate (gpm)
q =injection rate of water mixed with tracer (gpm)
c_1 =concentration of tracer in water injected at an upstream point (ppm)
c_2 =concentration of tracer in water sampled at a downstream point (ppm)
c_0 =concentration of tracer in water before injection (ppm)

Example 10-9: Application of the Dilution Method

A tracer material is injected into a canal at a concentration of 100,000 ppm. The water mixed with the tracer is injected at a rate of 1.5 gpm. The water in the canal does not contain the tracer material at the time of injection. Water samples were collected from a bridge located 1.2 miles downstream from the injection site. No inflow or outflow was found along the 1.2-mile reach. A drop structure and a flume were found along this reach. They would aid in mixing the tracer material with the canal water. Five water samples were collected, and the average concentration of the tracer was found to be 130.0 ppm. The flow rate is

$$Q = q\left(\frac{c_1 - c_2}{c_2 - c_0}\right) = 1.5\,\text{gpm}\left(\frac{100,000\,\text{ppm} - 130.0\,\text{ppm}}{130.0\,\text{ppm} - 0.0}\right)$$

$$= 1,152\,\text{gpm or } 2.57\,\text{cfs}$$

10.6   WATER MEASUREMENT OF PIPE FLOWS

Many devices are used to measure pipe flows. They primarily determine flow velocity through pipelines, since the cross-sectional area of a pipe is readily determined. Most devices can directly read the flow rate or total volume of the flow, or both. An accurate measurement of the flow through a pipeline is important for managing water and maintaining the pipeline. Any unexpected changes in flow rate in a pipeline may signal a system failure. Most flowmeters cause moderate amounts of head losses in pipelines.

10.6.1 Propeller Flowmeters

A propeller flowmeter is a rotating mechanical flowmeter used in pipe-lines (Fig. 10-20a). It has a propeller that has a rotating velocity proportional to the flow velocity through the confined area of a known pipe size. The meter is accurate and easy to install in any closed conduit, and covers a wide range of flows. However, it is expensive and requires frequent maintenance when the water carries large amounts of sand or debris, which can easily wear the propeller, thus reducing accuracy and increasing maintenance costs.

10.6.2 Pitot Tubes

A pitot tube is used to determine flow velocity in a pipeline (Fig. 10-20b). It is easy to install, accurate over a wide range of flows, and causes very small head losses. A pitot tube consists of a tube with a right-angle bend directly pointing upstream. The tube is divided into two sections connected to manometers. The inner tube measures the total energy head (velocity and pressure heads) of the flow, and the outer tube measures the pressure head in the pipe. This difference in head readings equals the velocity head of the flow. The flow velocity through an orifice is determined from the velocity head as follows:

$$v = C\sqrt{2g(h_i - h_o)} = C\sqrt{2g\ \Delta h} \qquad \text{(10-17a)}$$

where v = flow velocity (fps)
C = flow coefficient
h_i = head reading of inner tube (ft)
h_o = head reading of outer tube (ft)
Δh = head difference (ft)

Then the flow rate is

$$Q \text{ (cfs)} = CAv = CA\sqrt{2g(h_i - h_o)} = CA\sqrt{2g\ \Delta h} \qquad \text{(10-17b)}$$

The flow coefficient C = 1.0 for a well-designed pitot tube shown in Figure 10-20b (James 1988). Then the flow rate is

$$Q = 7.22A\sqrt{\Delta h} \qquad \text{(10-17c)}$$

where Q = flow rate (gpm)
A = cross-sectional area of pipe (in.^{2})
Δh = head difference (in.)

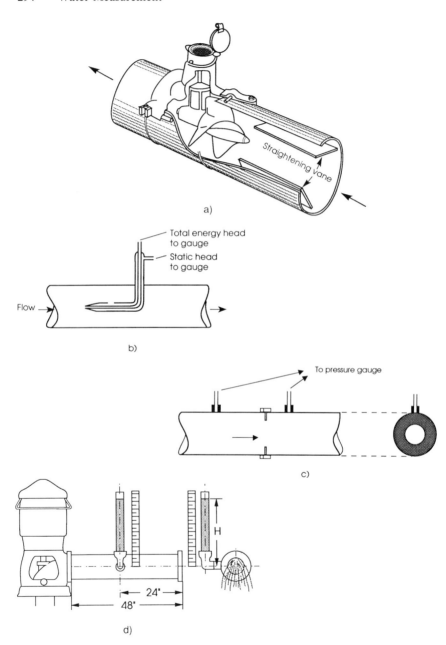

a)

b)

c)

d)

FIGURE 10-20. Flow-measuring devices for pipe flows: (a) propeller-type flowmeter; (b) pitot tube; (c) circular pipe orifice; and (d) circular pipe-end orifice installed at the end of a pump outlet.

Flow velocity in a pipe varies across the pipe; velocity increases toward the center, and the maximum velocity exists at the center of the pipe where the influence of the pipe friction is a minimum. When a low level of accuracy is acceptable, the head reading may be obtained from a single measurement at a point located three quarters of the pipe radius from the pipe center. For obtaining high accuracy, the head readings should be measured at several locations across the pipe, since the pitot tube measures velocity at only one point and velocity varies across the pipe diameter. Pitot tubes are commercially available. They typically have several strategically located tubes with holes facing upstream to measure various velocity heads across the pipe and a tube to measure the pressure head. The readings are used to determine, the average velocity head across the pipe diameter, and a device displays the mean velocity on an attached calibrated gauge.

Pitot tubes are not suitable for measuring low-velocity flows because head differences are small and reading errors may affect the results. Clean water in a pipeline is important, because suspended foreign materials may plug the holes.

Example 10-10: Application of a Pitot Tube

A well-designed pitot tube ($C = 1.0$) is installed in an 8-in.-inside-diameter PVC pipe. The differences in head readings between the inner and outer tubes at two equidistant locations across the half-section of the pipe are 2.3, 4.0, and 6.8 in. The average difference of the head readings is

$$\frac{2.3\,\text{in.} + 4.0\,\text{in.} + 6.8\,\text{in.}}{3} = 4.37\,\text{in.}$$

The cross-sectional area of the pipe is

$$A = 0.785D^2 = 0.785(8.0\,\text{in.})^2 = 50.0\,\text{in.}^2$$

The flow rate is

$$Q = 7.22A\sqrt{\Delta h} = 7.22(50.0\,\text{in.}^2)\sqrt{4.37\,\text{in.}} = 755\,\text{gpm}$$

10.6.3   Pipe Orifices

A pipe orifice is a thin plate with a square-edged opening installed in a pipeline to measure the flow. Most pipe orifices have a circular hole (Fig. 10-20c). The operational principle of a pipe orifice is that the pressure in the pipe drops between the immediate upstream and downstream sides of the orifice due to increased velocity through the orifice. The pressure drop across the orifice is related to the flow velocity through the orifice. The pressure heads at the upstream and downstream sides of the orifice are measured, and the flow rate is determined from the equation

$$Q = CA\sqrt{2g(h_u - h_d)} = CA\sqrt{2g\,\Delta h} \qquad (10\text{-}18a)$$

where Q = flow rate (cfs)
A = cross-sectional area of orifice opening (ft²)
C = coefficient of pipe orifice
h_u = upstream head (ft)
h_d = downstream head (ft)
Δh = head difference (ft)

or

$$Q = 7.22CA\sqrt{\Delta h} \qquad (10\text{-}18b)$$

where Q = flow rate (gpm)
A = cross-sectional area of orifice opening (in.²)
Δh = head difference (in.)

The pressure heads should be measured at distances of 1.0 and 0.5 pipe diameters from the orifice at the upstream and downstream sides, respectively. The coefficient of pipe orifice varies as a function of the ratio of the orifice diameter to the pipe diameter. The coefficient typically varies from 0.6 to 0.7, depending on the ratio.

A pipe orifice is often used to measure flow rates from the open end of a pipe such as a pump outlet (Fig. 10-20d). The pipe outlet must be level and a free-flow must be maintained. A manometer for measuring the head should be placed about 2 ft upstream from the orifice. No fittings should be allowed closer than 2 ft from the manometer. The discharge may be determined by the pipe orifice equation [equation (10-18)], where the head at the downstream side is zero. Table 10-7 shows the coefficients of circular pipe-end orifices for various ratios of orifice diameter to pipe diameter.

Example 10-11: Flow Measurement Using a Pipe-End Orifice

A 6-in.-diameter pipe-end orifice is installed at the end of a 10-in.-diameter free-flow pump outlet. The head reading of the orifice is 12.5 in.

TABLE 10-7   Coefficients for Pipe-End Orifices

Ratio of Orifice Diameter to Pipe Diameter	Pipe-End Orifice Coefficient C
0.5	0.58
0.55	0.60
0.6	0.62
0.65	0.64
0.7	0.67
0.75	0.71
0.8	0.76

Source: U.S. Soil Conservation Service (1973a).

The orifice coefficient is 0.62 from Table 10-7 for an orifice-to-pipe-diameter ratio of 0.6. The cross-sectional area of the orifice opening is

$$A = 0.785D^2 = 0.785(6.0\,\text{in.})^2 = 28.0\,\text{in}^2$$

The flow rate is then

$$Q = 7.22 \times 0.62 \times (28.0\,\text{in.}^2)\sqrt{12.5\,\text{in.}} = 443\,\text{gpm}$$

10.6.4   Coordinate Methods

Coordinate methods are often used to measure the flow rate from a pipe or a pump outlet under a free-flow condition (Greeve 1928). The methods are not very accurate, but they are simple and require no measuring devices. The shape of a jet of flow from the end of a pipe represents the kinetic energy of the flow as the water freely departs the pipe. As shown in Figure 10-21, coordinate methods may be used whether the pipe is flowing full vertically upward, horizontal, or has some angle from the horizontal. Also the methods can be used for a partial flow under these flow conditions.

The coordinates of the pipe flow are measured as distances from the end of the pipe. The coordinates of a flow from pipes flowing vertically upward are measured by the height of the jet above the top of the pipe outlet (Fig. 10-21a). For a flow from pipes flowing full horizontally, the coordinates are measured as horizontal and vertical distances from the end of the pipe (Fig. 10-21b). For convenience, the horizontal distance is measured from the end of the pipe to where the vertical distance measures 12 in. from the top of the jet. Figures 10-21a and b show

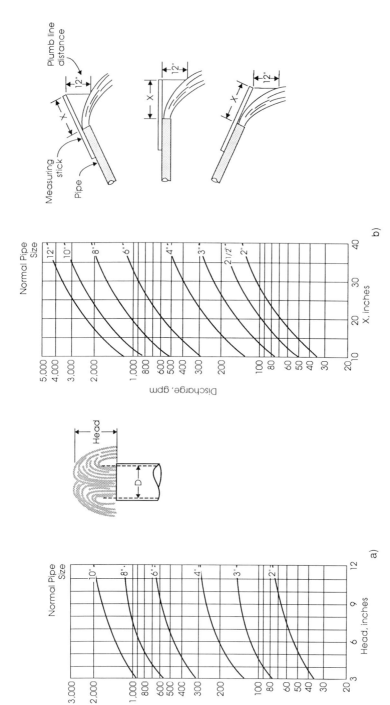

FIGURE 10-21. Discharge relationships of pipe flows using coordinate methods (a) for a vertical pipe and (b) horizontal and inclined pipes.

TABLE 10-8   Correction Factors for the Coordinate Method of Measuring Partial Flows from Horizontal Pipes

Ratio of Flow Depth to Pipe Diameter	Correction Factor
0.0	0.000
0.5	0.019
0.10	0.052
0.2	0.142
0.3	0.253
0.4	0.375
0.5	0.500
0.6	0.627
0.7	0.747
0.8	0.858
1.0	1.000

Source: U.S. Soil Conservation Service (1973a).

graphical relationships between the coordinates of flow and flow rates of various pipe sizes for full-flow conditions. For a partial-flow pipe, the flow rate is corrected by multiplying the flow rate by a correction factor. The correction factors vary by the ratio between the depth of a partial flow to the pipe diameter. Table 10-8 gives correction factors for a range of values of this ratio.

10.7 WATER MEASUREMENT FOR CULVERTS AND SIPHONS

When water is conveyed through a culvert or a siphon in an open channel, the flow rate may be determined from the culvert equation [equation (9-27)]. For short siphon tubes, the orifice equation [equation (10-18)] may be applied with a slightly modified orifice coefficient. Because of its length, shape, and inlet and outlet conditions, flow through a siphon tube causes more energy loss than flow through orifices. Figure 9-12 gives effective head-discharge relationships for small siphon tubes. When the outlet is submerged, the effective head is the vertical distance between the water surfaces at the upstream and downstream ends of the tube. For a free-flow condition, the effective head is the vertical distance from the upstream water surface to the centerline of the outlet end. See Chapter 9 for more details of calculating flows through these structures.

11

Pumps and Pumping

11.1 INTRODUCTION

Pumps are used widely in aquaculture, and their initial purchase price and operation and maintenance costs are major expenses in many aquaculture enterprises. Pumps are mechanical devices that impart mechanical energy to water to lift and pressurize while overcoming energy losses during conveyance of water. The mechanical energy usually comes from electric motors or internal combustion engines. Where these sources are not available, solar, wind, animal, and human power are also used. Each pump is designed for a particular application, and for efficient operation a pump must be matched to the type of application.

Pumps can be classified into two types, which generally describe how energy is applied to fluid: dynamic and displacement. Displacement pumps include rotary and reciprocating pumps. Rotary pumps use gears, vanes, lobes, or screws to convey fluid from the inlet to outlet of the pump. Pumps that use the back-and-forth motion of mechanical parts, such as pistons or diaphragms, to move and pressurize the fluid are called reciprocating pumps. These pumps are mainly used to move low discharge of clean fluid. Since these pumps are seldom used in water conveyance systems as a principal pump, they will not be discussed in this chapter. Readers interested in such pumps should consult Hicks and Edwards (1971) or Karassik et al. (1986). The most common type of dynamic pump is the centrifugal pump. As its name implies, a centrifugal pump depends on centrifugal force to impart energy to water. This chapter reviews the basic features of centrifugal pump design and performance, discusses power units, and demonstrates how centrifugal pumps may be selected for a specific application.

11.2 CENTRIFUGAL PUMPS

A centrifugal pump consists primarily of a specially formed casing with inlet and outlet and a rotating impeller attached at its center of rotation to a drive shaft (Fig. 11-1). Its carefully shaped impeller rotates rapidly inside a close-fitting casing. Centrifugal force spins water outward while drawing replacement water through the inlet. Water spun outward is directed by the casing into the outlet. The casing is shaped so that it gradually converts velocity head to pressure head. The shafts may be connected to a power unit directly or indirectly by pulley and belt systems, power takeoff (PTO) shafts, or other means. Power units and connection methods are discussed later. Centrifugal pumps are designed with respect to type of intake, number of stages, shape of casing, type of impeller, and position of shaft.

Impellers may be single or double suction; that is, water may enter from one or both sides of the impeller (Fig. 11-2). A pump with one impeller is a single-stage pump; one with two or more impellers in series is a multistage pump. Each stage of a multistage pump behaves like a separate pump. Water from the first stage is pumped to the second stage, and so on, and the final discharge can be no greater than the discharge of the first stage. However, each stage adds pressure, so the total head developed is directly proportional to the number of stages. Obviously, power requirements increase in direct proportion to the number of stages.

Impellers have three or more curved vanes that sling water from their surfaces at high velocity. A large number of vanes favors development of high head, whereas fewer vanes are used for a large discharge at low head. Vanes are always curved backward from the direction of flow. Impellers may be open, semiopen, or enclosed (Fig. 11-3). Enclosed

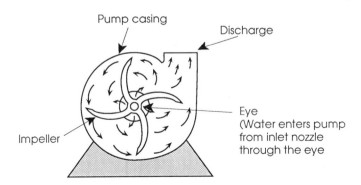

FIGURE 11-1.   Operating principle of centrifugal pump.

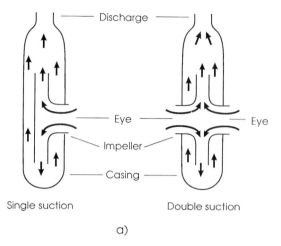

Single suction   Double suction

a)

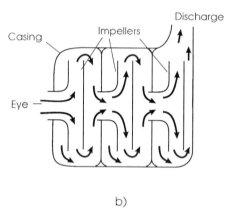

b)

FIGURE 11-2. Suction types of pump impeller: (a) single-stage pump, and (b) multistage pump.

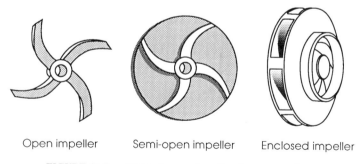

Open impeller   Semi-open impeller   Enclosed impeller

FIGURE 11-3. Type of pump impellers by construction.

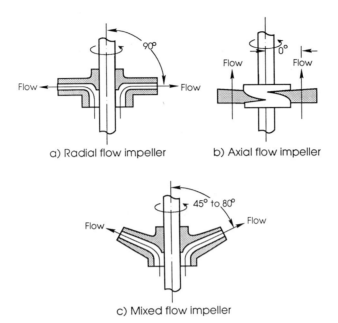

a) Radial flow impeller   b) Axial flow impeller

c) Mixed flow impeller

FIGURE 11-4.   Classification of pump impellers by direction of flow: (a) radial-flow, (b) axial-flow, and (c) mixed-flow. (*Source: Larry G. James*, Principles of Farm Irrigation System Design, *copyright © 1988 by John Wiley & Sons, Inc., New York, p. 152. Reprinted by permission of John Wiley & Sons, Inc.*)

impellers are useful only for pumping clean water, for they clog easily. Semiopen and open impellers are best suited for pumping water containing large amounts of fibrous, suspended solids.

Pumps are also classified as radial-, axial-, and mixed-flow pumps, depending on the direction of flow relative to the axis of rotation (Fig. 11-4). The true centrifugal pump forces water out at right angles to the axis of rotation of its impeller (Fig. 11-4*a*). This type of pump is called a radial-flow pump, commonly referred to as a centrifugal pump. Axial-flow pumps (commonly referred to as propeller pumps) use a propeller to move water parallel to the axis of propeller rotation (Fig. 11-4*b*). Mixed-flow pumps have impellers that use both centrifugal force and the lifting action to move water with an angle between radial-flow and axial-flow pumps (Fig. 11-4*c*). It is most convenient to treat radial-flow, mixed-flow, and axial-flow pumps under the broad category of centrifugal pumps. Radial-flow pumps are often used for pumping against high head requirements. Mixed-flow and axial-flow pumps are suitable for applications requiring large discharges at low head requirements. Axial-flow pumps

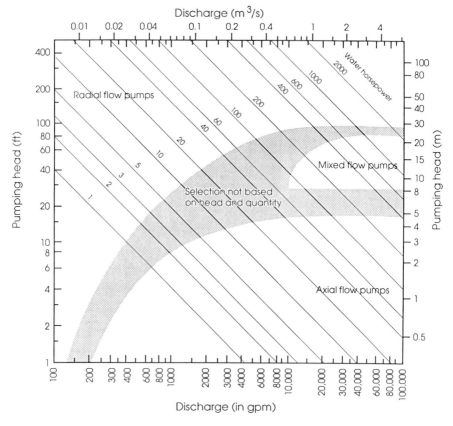

FIGURE 11-5.   Graphical guideline for pump selection by discharge-head relationships.
(*Source: U.S. Soil Conservation Service 1973b.*)

provide lower head than do mixed-flow pumps. The types of impellers
and their head-discharge relationships are depicted in Figure 11-5, which
may be used to select pump types for specific discharges and head
requirements.

Pump casings are single-volute, double-volute, or diffuser types (Fig.
11-6). The pump casing contains an intake nozzle that directs water into
the eye of the impeller, and the casing converts the velocity energy into
pressure energy as the water flows toward the discharge nozzle. Energy
conversion is accomplished with either a volute or a diffuser casing. In
volute casing the passage for flow expands gradually. The progressively
enlarging cross-sectional area of the passage causes reduction in velocity,
and the reduced velocity head is changed to pressure head. Both single-

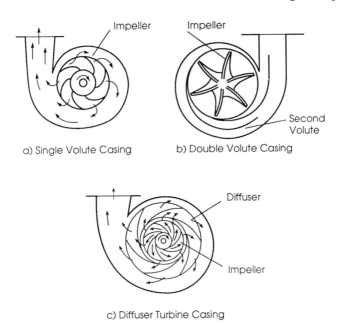

a) Single Volute Casing   b) Double Volute Casing

c) Diffuser Turbine Casing

FIGURE 11-6.   Type of pump casings: (a) single volute, (b) double-volute, and (c) diffuser. (*Source: Larry G. James*, Principles of Farm Irrigation System Design, *copyright © 1988 by John Wiley & Sons, Inc., New York, p. 152. Reprinted by permission of John Wiley & Sons, Inc.*)

and double-volute casings are in common use. In diffuser casing, stationary diffusion vanes are installed around the impeller. Velocity is reduced by the diffusion vanes, so conversion of velocity energy to pressure occurs within the diffusion vanes. Pumps with diffuser casing are often called turbine pumps. Most single-stage horizontal pumps have a volute casing, but multistage pumps may have either volute or diffuser casing.

Centrifugal pumps are also classified as vertical or horizontal, depending on whether the pump shaft is vertical or horizontal to the ground surface or the mounting surface (Fig. 11-7). The horizontal type has a vertical impeller connected to a horizontal shaft. The vertical pump has a horizontal impeller connected to a vertical shaft. The horizontal pumps lift water into their impeller position, so they must be installed close to the supply waterline (Fig. 11-7a). The need to prime is a disadvantage of horizontal pumps. They are best suited for surface-water pumping unless water-level fluctuation is excessive (greater than 20 ft). Suction head and priming are discussed later. In case of excessive water-level fluctuation, a vertical or turbine pump may be recommended. The

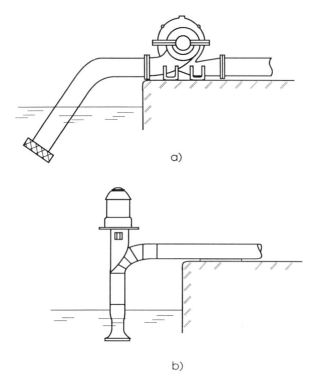

a)

b)

FIGURE 11-7.   Installation position of centrifugal pumps: (a) horizontal pump, and (b) vertical pump.

impellers of vertical-type pumps are usually lowered below the supply water level, and the vertical shaft is extended to the surface where power is applied (Fig. 11-7b). Any centrifugal pump can be built with either a vertical or horizontal shaft. However, the vertical turbine pump is built specifically for vertical operation. Vertical turbine pumps are multistage pumps of comparatively small diameter. They are designed for use in wells and will give long and dependable service if properly installed and maintained. However, they are usually more expensive than horizontal pumps and are more difficult to inspect and repair. Figure 11-8 shows a multistage turbine pump. A deep-well turbine pump close-coupled to a submersible electric motor is called a submersible pump. It is more efficient than other types and is used in very deep wells where long shafts are not practical.

Due to their nonpositive action, centrifugal pumps with impellers located above the supply water level must be primed before the pump will

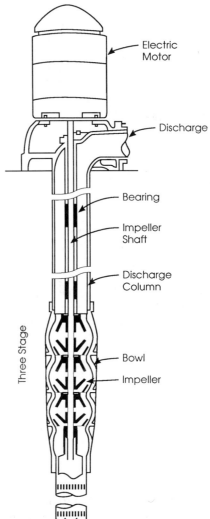

FIGURE 11-8.   A multistage vertical turbine pump.

operate. They will not lift water from a source unless the casing and suction pipe are both full of water before starting the pump. Priming can be done by adding water to the suction side in the pump or by removing the air from the suction side. The lowered pressure in the suction area due to removed air will cause the water to rise from the source by atmospheric pressure. Priming methods include (a) outside water supply, (b) hand-priming pump, (c) dry vacuum pump, and (d) exhaust or

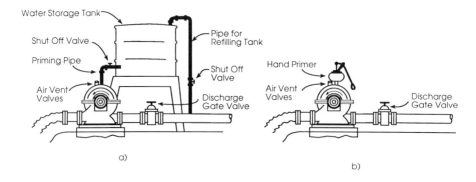

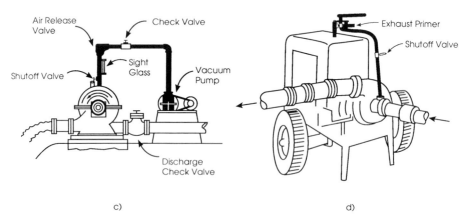

FIGURE 11-9.   Priming methods of centrifugal pumps: (a) outside supply, (b) hand pump, (c) dry vacuum pump, and (d) exhaust primer. (*Source: U.S. Soil Conservation Service 1970.*)

manifold primer of internal combustion engines. These methods are shown in Figure 11-9. Since deep-well turbine pumps are installed below the supply water line, priming is not necessary.

In summary, a centrifugal pump has two basic parts: the impeller and the casing. Design of the impeller and casing depends on the head and discharge requirements for the intended application. The impeller is attached to a shaft supported by bearings. The bearings are supported by the pump casing, and the casing must be packed around the shaft to prevent water from leaking to the outside. Casing wearing rings fitted between the impeller and casing prevent water that has passed through the impeller from leaking back into the suction area at the impeller eye. A cross-sectional view of a radial-flow pump is provided in Figure 11-10.

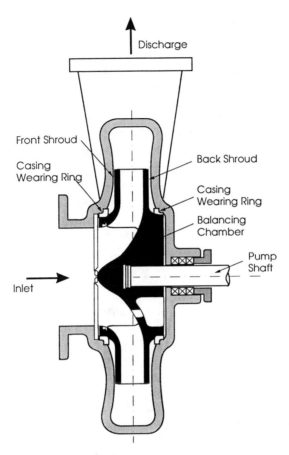

FIGURE 11-10.   Cross-sectional view of a radial-flow pump.

11.3 PERFORMANCE OF CENTRIFUGAL PUMPS

Several parameters describe the performance of centrifugal pumps: discharge (Q), total dynamic head (TDH) or pumping head, power requirement, pump efficiency (E_p), net positive suction head (NPSH), and specific speed (N_s). They should be well defined to select the best pump for an application.

11.3.1 Pumping Heads

The energy that a pump imparts to the water is referred to as *pumping head*. The pumping head is usually expressed in feet. The pumping head that a pump must provide includes the velocity, operating pressure, elevation, and energy losses on the suction and discharge sides of the pump.

The total pumping head is the sum of the total suction head and the total discharge head. The total suction head represents the energy that the pump expends in moving water from the pumping water level to the centerline of the impeller. The energy that the pump must supply to move water from the centerline of the impeller to the discharge point at the necessary discharge and pressure is the total discharge head. Pumps are designed to operate within a certain range of head-discharge combination. A change in head influences discharge, or vice versa. Pump manufacturers provide data on relationships of head, power, efficiency, and required net positive suction head ($NPSH_r$) against discharge of individual pumps. The data are plotted on a graph and called the pump performance curves, or simply pump curves. The pump performance curves are described later.

Net Positive Suction Head

Rotation of the impeller reduces pressure at the pump intake below atmospheric pressure, and the atmospheric pressure acting on the water surface forces water into the pump. Negative pressure or vacuum (pressure below atmospheric pressure) in the pump intake is called *suction*. The maximum suction lift is limited to the height of a column of water that can be supported by the atmospheric pressure. Atmospheric pressure at sea level is 14.7 psi, which is equal to a column of water 34 ft high at 70°F. It is useful to remember that head in feet of water is equivalent to 2.31 times pressure in pounds per square inch, and pressure in pounds per square inch is equivalent to head in feet of water times 0.43.

The maximum suction lift cannot be realized in centrifugal pump applications (Simon 1976). If pressure in the intake nozzle is lowered below the vapor pressure of water, the water, in effect, boils. Vapor bubbles form in the intake stream and pass through the pump impeller. When water reaches areas in the pump where pressure is higher, the bubbles collapse. This phenomenon causes cavitation of pump impellers. Cavitation often is recognized by loud popping noises in the pump. The discharge head may be reduced by impeller cavitation, and the impulsion of vapor bubbles may cause pitting and metal fatigue on the impeller and casing. Continued cavitation and pitting on a pump impeller can severely damage the pump, and must be avoided. Cavitation is avoided when the head at the pump inlet is great enough to prevent the pressure within the pump impeller from dropping below vapor pressure.

The amount of head required to prevent the formation of vapor-filled cavities of water within the impeller center is called the *required net positive suction head* (NPSH_r). NPSH_r is a function of pump design and experimentally determined and published for each pump by the manufacturers. The available net positive suction head (NPSH_a) for a pump application must be higher than the NPSH_r provided for the pump. To compute NPSH_a, it is necessary to measure the vertical distance from the water surface to the impeller eye or static suction head (H_e), and to determine total head loss (H_L) for pipe, valves, fittings, foot valves, and other items, and the velocity head (H_v) on the suction side. Thus,

$$\text{NPSH}_a = H_b - H_{vp} - (H_e + H_L + H_v)_s = H_b - H_{vp} - H_s \quad (11\text{-}1)$$

where H_b = expected absolute barometric pressure (ft)
   H_{vp} = vapor pressure of water (ft)
   H_s = total suction head, where $H_s = H_e + H_L + H_v$ on suction side (ft)

The vapor pressure of water is a function of water temperature and the expected absolute barometric pressure varies as a function of altitude. These values are found in Tables 1-3 and 2-1. If NPSH_a is lower than the NPSH_r for a pump, it must be increased by decreasing the vertical distance that water is lifted. Some increase in NPSH_a may also be obtained by increasing the suction pipe diameter to reduce friction loss and velocity head. As shown in Figure 11-11, NPSH_r increases and

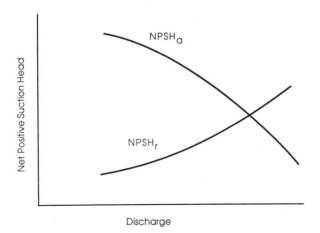

FIGURE 11-11.   Typical relationship between discharge and net positive suction head (NPSH). Discharge must be limited to maintain higher NPSH_a than NPSH_r for a pump to prevent cavitation.

$NPSH_a$ decreases as flow rate increases for a pump. For $NPSH_a$ to be higher than $NPSH_r$, the discharge by the pump must be lower than the discharge at the intersection of $NPSH_a$ and $NPSH_r$.

To illustrate NPSH, suppose that the manufacturer's $NPSH_r$ for a pump is 6 ft and the calculated total suction head for the pumping system is 21 ft. The pump is used at 6,000 ft above the mean sea level (m.s.l.) to pump water at 60°F. From Tables 1-3 and 2-1, $H_{vp} = 0.517$ in. Hg × 1.13 = 0.58 ft of water and $H_b = 23.98$ in. Hg or 27.10 ft of water. Then, $NPSH_a = 27.10$ ft − 0.58 ft − 21.0 ft = 5.52 ft. Next, compare $NPSH_a$ with $NPSH_r$ for the pump. In this example $NPSH_a$ is lower than the $NPSH_r$ by 0.48 ft. To prevent cavitation of the impeller, one must reduce the suction lift by lowering the pump location or by using larger suction pipe.

Often pump owners are interested in the location of a pump station that does not cause cavitation problems. From equation (11-1), the static suction head (the vertical distance from the supply water level to the impeller center) for a pump may be calculated as

$$(H_e)_s = H_b - H_{vp} - (H_L + H_v)_s - NPSH_r - Fs \qquad (11\text{-}2)$$

where Fs is a safety factor for additional head losses in the future, which may be taken as 2.0 ft (Schwab et al. 1981).

TABLE 11-1   Practical Static Suction Lift[a] (vapor pressures of water at the selected temperature are given in parentheses, all units in feet)

		Temperature (°F)					
Altitude	Atmospheric Pressure	50 (0.41)	60 (0.59)	70 (0.84)	80 (1.17)	90 (1.61)	100 (2.19)
0	33.9	23.4	23.3	23.1	22.9	22.6	22.2
500	33.3	23.0	23.0	22.8	22.5	22.2	21.8
1,000	32.7	22.6	22.5	22.4	22.0	21.8	21.4
1,500	32.1	22.2	22.1	21.9	21.6	21.4	20.9
2,000	31.5	21.8	21.6	21.5	21.2	20.9	20.5
3,000	30.4	21.0	20.9	20.7	20.5	20.2	19.8
4,000	29.3	20.5	20.1	20.0	19.7	19.4	19.0
5,000	28.2	19.5	19.3	19.2	18.9	18.7	18.2
6,000	27.2	18.8	18.6	18.4	18.2	17.9	17.5
7,000	26.2	18.0	17.9	17.8	17.5	17.2	16.8
8,000	25.2	17.4	17.2	17.0	16.8	16.5	16.1
9,000	24.2	16.7	16.6	16.4	16.1	15.8	15.4

[a] 70% of theoretical maximum suction lift. 0.7 × (atmospheric pressure − vapor pressure).

Pump manufacturers usually recommend that suction lift be limited to 70% of the theoretical maximum suction lift (Pair et al. 1983). Table 11-1 was developed with this recommendation at various altitudes and water temperatures. The table may be used to find the practical suction head under normal operational conditions of centrifugal pumps. For the pump's safety, the maximum distance from the surface of supply water to the impeller center during pumping must be used as the static suction head in the calculation. This is especially important for a pumping system that pumps water from varying elevations of the water surface such as wells, creeks, or small surface ponds. From Table 11-1, the pump in the example is safe if located no higher than 18.3 ft above the lowest level of the supply water.

Total Discharge Head

To determine total discharge head (H_d) of a pumping system, it is necessary to measure the vertical distance from the centerline of the pump to the end of the discharge pipe (H_e), to estimate head losses resulting from pipe, valves, elbows, exits, and other fittings (H_L), to calculate the velocity head (H_v) in the discharge line, and to establish the desired pressure head at the point where water is released (H_p). The total discharge head (H_d) is

$$H_d = (H_e + H_L + H_v + H_p)_d \tag{11-3}$$

where all terms are in feet.

Total Dynamic Head (TDH)

The total pumping head that a pump must provide is called the *total dynamic head* (TDH), or simply the pumping head. It is sum of the total suction head (H_s) and the total discharge head (H_d):

$$\text{TDH} = H_s + H_d = (H_e + H_L + H_v)_s + (H_e + H_L + H_v + H_p)_d \tag{11-4a}$$

The sum of the elevation heads (H_e) on the suction and discharge sides is called the *total static head* (H_t):

$$H_t = (H_e)_s + (H_e)_d \tag{11-4b}$$

The total static head represents the elevation difference between the supply water level and the point of discharge. Compared with other energy heads, H_L and H_v on the suction side may be neglected in calculating TDH. Thus,

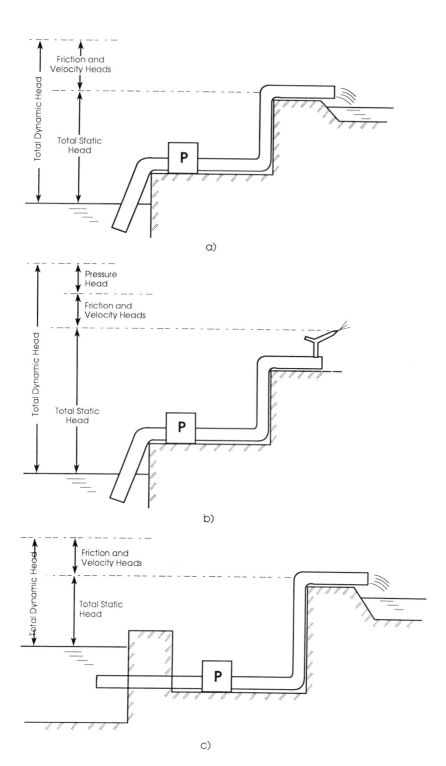

a)

b)

c)

$$TDH = H_t + (H_L + H_v + H_p)_d \qquad \text{(11-4c)}$$

There may be some variation in the way that the TDH is computed. For example, if water enters the pump under pressure (i.e., pump located below the supply water line), the static suction head, $(H_e)_s$, is negative and must be subtracted in the equation. If the discharge outlet is located lower than the supply waterline the static discharge head, $(H_e)_d$, is negative. Velocity head is quite often omitted in determining TDH because of its small value compared with other heads. It is usually recommended to limit flow velocity in a pipeline to lower than 5.0 ft/sec, which has a velocity head of 0.4 ft. Pressure head is omitted if no pressure is needed at the discharge point (i.e., free discharge at the pump outlet). These are illustrated in Figure 11-12.

Example 11-1: Determining NPSH$_a$ and TDH of a Pumping System

A pumping system is planned to deliver 1,200 gpm of water from the supply canal to a pond. No pressure is required because the outlet has a free flow. The minimum elevation of the canal water surface is 1,125 ft above m.s.l., and the elevation of the free discharge point is 1,152 ft above m.s.l. The elevation at the pump impeller is 1,131 ft above m.s.l. The suction pipe is 20 ft long, 12-in.-diameter steel pipe, and the discharge pipe is 120 ft long and 10-in.-diameter Schedule 40 PVC pipe. The suction pipe has a 45° elbow and a strainer bucket entrance with a foot valve. The discharge pipe has two 45° elbows and an ordinary exit. NPSH$_a$ and TDH will be determined as follows.

Solution: In this solution all parameters shown in equation (11-4a) will be calculated to demonstrate their significance.

1. Total suction head (H_s)
   a. Static suction head $H_e = 1,131 \text{ ft} - 1,125 \text{ ft} = 6 \text{ ft}$
   b. Head losses
      The friction loss of 12-in. steel pipe to carry 1,200 gpm is 0.58 ft per 100 ft (Table B-1). Friction loss of the 20-ft pipe is

$$H_f = 0.58 \text{ ft}/100 \text{ ft} \times 20 \text{ ft} = 0.12 \text{ ft}$$

◄FIGURE 11-12.   Methods to determine total dynamic head (TDH) for various pump setups.

The minor losses due to fittings from Figure B-2 are

Fitting	Number	Equivalent Length of 12-in.-Diameter Pipe (ft)
45° elbow	1	15
Strainer bucket entrance with a foot valve[a]	1	80
Total		95

[a] The head loss in a strainer bucket entrance with a foot valve is approximately the same as for a full open check valve.

$$H_m = 0.58\,\text{ft}/100\,\text{ft} \times 95\,\text{ft} = 0.55\,\text{ft}$$

The total head loss is

$$H_L = H_f + H_m = 0.12\,\text{ft} + 0.55\,\text{ft} = 0.67\,\text{ft}$$

c. Velocity head
The velocity head through the suction pipe is calculated as follows:

$$A = \frac{\pi}{4}D^2 = 0.785\left(\frac{12\,\text{in.}}{12\,\text{in.}/\text{ft}}\right)^2 = 0.79\,\text{ft}^2$$

$$Q = \frac{1,200\,\text{gpm}}{450\,\text{gpm/cfs}} = 2.67\,\text{cfs}$$

$$v = \frac{Q}{A} = \frac{2.67\,\text{cfs}}{0.79\,\text{ft}^2} = 3.38\,\text{ft/sec}$$

$$H_v = \frac{v^2}{2g} = 0.0155v^2 = 0.0155 \times 3.38^2 = 0.18\,\text{ft}$$

d. Total suction head

$$H_s = 6\,\text{ft} + 0.67\,\text{ft} + 0.18\,\text{ft} = 6.85\,\text{ft}$$

As shown in this solution, H_v and H_L are small compared with the static suction head.

2. Total discharge head (H_d)
   a. Static discharge head

$$H_e = 1,152\,\text{ft} - 1,131\,\text{ft} = 21\,\text{ft}$$

b. Head losses
   Friction loss of 10-in. Schedule 40 PVC pipe to carry 1,200 gpm is
   0.26 ft per 100 ft by interpolation (Table B-2). The friction loss of
   120-ft pipe is

$$H_f = 0.26\,\text{ft}/100\,\text{ft} \times 120\,\text{ft} = 0.31\,\text{ft}$$

Minor losses due to fittings from Figure B-2 are

Fittings	Number	Equivalent Length of 10-in.-Diameter Pipe (ft)
45° elbow	2	2 × 12 = 24
Ordinary exit[a]	1	0
Total		24

[a] Free-flow pipe outlet has no head loss.

$$H_m = 0.26\,\text{ft}/100\,\text{ft} \times 24\,\text{ft} = 0.06\,\text{ft}$$

Total head loss is

$$H_L = 0.31\,\text{ft} + 0.06\,\text{ft} = 0.37\,\text{ft}$$

c. The velocity head is calculated as follows:

$$A = 0.785(10\,\text{in.} \div 12\,\text{in./ft})^2 = 0.55\,\text{ft}^2$$
$$v = Q \div A = 2.67\,\text{cfs} \div 0.55\,\text{ft}^2 = 4.85\,\text{ft/sec}$$
$$H_v = 0.0155 \times 4.85^2 = 0.36\,\text{ft}$$

d. No pressure is required at the pipe outlet; $H_p = 0.0$.
e. The total discharge head is

$$H_d = 21\,\text{ft} + 0.37\,\text{ft} + 0.36\,\text{ft} + 0.0 = 21.7\,\text{ft}$$

3. The total dynamic head is

$$\text{TDH} = H_s + H_d = 6.85\,\text{ft} + 21.7\,\text{ft} = 28.55\,\text{ft}$$

As shown in this solution, the pumping system has very small heads
required for friction and velocity on the suction and discharge sides
compared with TDH. However, for a long-discharge pipe the friction loss
may become significantly large.

The $NPSH_a$ for the pumping system is calculated as follows. From the solution in step 1, the total suction head (H_s) required is 6.85 ft. The absolute barometric pressure at the pump station is 32.5 ft by interpolation (Table 1-3). The vapor pressure of water at 70°F is 0.84 ft (Table 2-1). Then

$$NPSH_a = 32.5\,ft - 0.84\,ft - 6.85\,ft = 24.81\,ft$$

The $NPSH_r$ of the pump at 1,200 gpm flow must be lower than 24.81 ft − 2.0 ft = 22.81 ft, considering the safety factor in equation (11-14c) to avoid cavitation of the impeller.

11.3.2   Power Requirements and Pump Efficiency

The efficiency of a pump (E_p) is a ratio of energy output of the pump to energy input to the pump. The energy output of a pump is calculated from the discharge and the total dynamic head, and the energy input is the power applied to the pump shaft by a power unit. The energy output of a pump is less than the energy input because of mechanical inefficiency of the pump. The amount of energy applied to the water may be termed *water power* (*WP*). The power of a pump system is expressed in horsepower (hp) or kilowatts (kW). Mathematically,

$$WP = \frac{Q \times TDH}{3,960} \tag{11-5}$$

where WP = water power (hp)
Q = pump discharge rate (gpm)
TDH = total dynamic head (ft)

To convert horsepower to kilowatts, multiply the result by 0.746, for 1.0 hp = 0.746 kW, or 1.0 kW = 1.34 hp.

The term *brake power* (*BP*) is used to refer to power applied to a pump shaft. Torque and shaft speed of a pump are used to calculate the brake power of a power unit. The torque applied to a pump shaft can be measured with a torque transducer or prony brake. The brake power in horsepower is

$$BP = \frac{T \times RPM}{5,252} \tag{11-6}$$

where BP = break power (hp)
      T = torque (ft-lb)
   RPM = rotational speed of pump shaft (revolutions per minute, rpm)

The pump efficiency (E_p) in decimal is

$$E_p = \frac{WP}{BP} \qquad (11\text{-}7)$$

Combining equations (11-5) and (11-7) gives

$$BP = \frac{WP}{E_p} = \frac{\text{TDH} \times Q}{3,960 \times E_p} \qquad (11\text{-}8)$$

Manufacturers determine pump efficiencies and brake powers over a range of operating conditions and provide these data for individual pumps.

Example 11-2: Determining Power Requirements and Pump Efficiency

To illustrate the calculation of pump efficiency, suppose that a pump driven by an electric motor delivers 1,200 gpm at a total dynamic head of 45 ft. A torque transducer indicates that 68 ft-lb of torque are applied to the shaft which rotates at 1,750 rpm during pump operation. Calculate the power requirements and efficiency of the pump.

$$WP = \frac{1,200 \times 45}{3,960} = 13.6\,\text{hp}$$

$$BP = \frac{68 \times 1,750}{5,252} = 22.7\,\text{hp}$$

$$E_p = \frac{13.6\,\text{hp}}{22.7\,\text{hp}} = 0.60 \text{ or } 60\%$$

11.3.3 Affinity Laws

Total dynamic head, discharge, and brake power of a pump are related to size and speed of impeller, and the pump curve retains its characteristic features over a range of impeller size and speed (Stepanoff 1957; Wheaton 1981). Changing the size or speed of the impeller modifies the operational characteristics of a pump. This allows pump manufacturers or users to alter the performance of a single pump to match the system needs. Relationships among total dynamic head, discharge, brake power, and NPSH are described by the following affinity laws, in which subscripts 1 and 2 refer to before and after changes in the pump variables, respectively:

$$\frac{Q_1}{Q_2} = \frac{RPM_1}{RPM_2} = \frac{D_1}{D_2} \tag{11-9}$$

$$\frac{TDH_1}{TDH_2} = \left(\frac{RPM_1}{RPM_2}\right)^2 = \left(\frac{D_1}{D_2}\right)^2 \tag{11-10}$$

$$\frac{BP_1}{BP_2} = \left(\frac{RPM_1}{RPM_2}\right)^3 = \left(\frac{D_1}{D_2}\right)^3 \tag{11-11}$$

$$\frac{(NPSH_r)_1}{(NPSH_r)_2} = \left(\frac{RPM_1}{RPM_2}\right)^2 = \left(\frac{D_1}{D_2}\right)^2 \tag{11-12}$$

where RPM = pump speed and D = impeller diameter.

Example 11-3: Application of Affinity Laws of a Pump

Performance data for a pump operated at 1,750 rpm are $Q = 2,120$ gpm, TDH = 54.0 ft, $BP = 35.2$ hp, and $NPSH_r = 12.0$ ft at a peak efficiency of 82%. Affinity laws permit calculation of pump characteristics at 3,450 rpm:

$$\frac{2,120}{Q_2} = \frac{1,750}{3,450} = 0.507; \qquad Q_2 = 4,180 \text{ gpm}$$

$$\frac{54.0}{TDH_2} = \left(\frac{1,750}{3,450}\right)^2 = 0.507^2; \qquad TDH_2 = 210.0 \text{ ft}$$

$$\frac{35.2}{BP_2} = \left(\frac{1,750}{3,450}\right)^3 = 0.507^3; \qquad BP_2 = 270.0 \text{ hp}$$

$$\frac{12.0}{(NPSH_r)_2} = \left(\frac{1,750}{3,450}\right)^2 = 0.507^2; \qquad (NPSH_r)_2 = 46.7 \text{ ft}$$

The water power (WP) of the new pump is

$$WP_2 = \frac{4,180 \text{ gpm} \times 210.0 \text{ ft}}{3,960} = 221.0 \text{ hp}$$

The revised pump efficiency (E_p) is

$$E_p = \frac{WP_2}{BP_2} = \frac{221.0\,\text{hp}}{270.0\,\text{hp}} = 0.82 \text{ or } 82\%$$

Note that the efficiency of the pump at 3,450 rpm stays the same as that at 1,750 rpm. However, the new pump requires a larger power unit to supply the increased break power.

11.3.4 Pump Performance Curves

Pumps have well-defined operating characteristics, and these vary among type of pumps and manufacturers. Pump manufacturers operate pumps over a range of heads and determine discharge, brake power, and efficiency. The findings usually are plotted on a graph paper to prepare pump performance curves. There are various types of pump performance curves, and some are shown in Figure 11-13. NSPH$_r$ is also provided in some performance curves (Fig. 11-13b). Manufacturers normally provide a set of performance curves for each pump model. These curves represent the average performance of a specified pump.

Figure 11-13a shows performance curves of a typical single-stage mixed-flow pump. As shown in this figure, the peak efficiency for this pump is approximately 82%. At this point, the pump discharges 1,000 gpm at a total dynamic head of 14 ft and requires 45 hp as a brake power. The head decreases as discharge increases, but brake power stays fairly constant. A deep-well turbine pump is composed of multistages of mixed-flow pumps. Some pump performance curves are given for different impeller diameters (Fig. 11-13b) and others for different impeller speeds (Fig. 11-13c). Figure 11-13b shows pump curves for a typical radial-flow pump; pump curves for an axial-flow pump are shown in Figure 11-13c. For a given discharge, increasing the impeller size increases the head. For example, for an impeller diameter of $7\frac{5}{16}$ in. whose performance is illustrated in Figure 11-13b, the pump discharges 135 gpm at TDH = 50 ft, but with a 9.0-in. impeller it discharges at the same rate against a TDH of 84 ft. Of course, brake power increases from near 2.7 hp to about 4.5 hp. The pump efficiency decreased from 68% to 66%. Likewise for a given head, increasing impeller size will increase discharge. At 50-ft TDH, the pump discharges 135 gpm with a $7\frac{5}{16}$-in. impeller, and a 9.0-in. impeller discharges 350 gpm at the same TDH, but requires higher BP.

Shapes of pump performance curves vary with pump types, as is shown in Figure 11-14. The top part of Figure 11-14 shows general shapes of pump performance curves for three impeller types. For a radial-flow

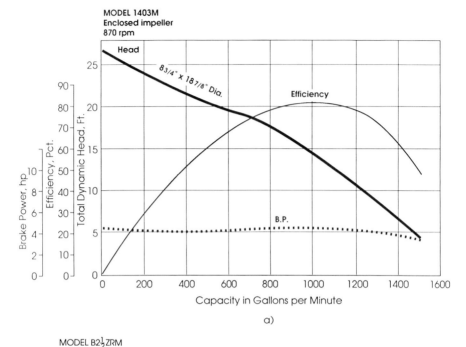

a)

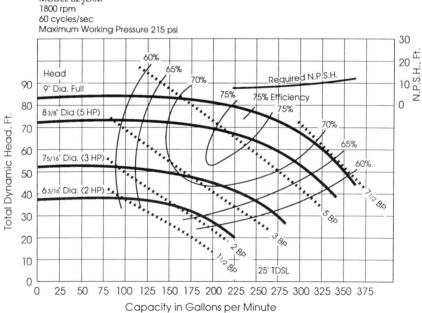

b)

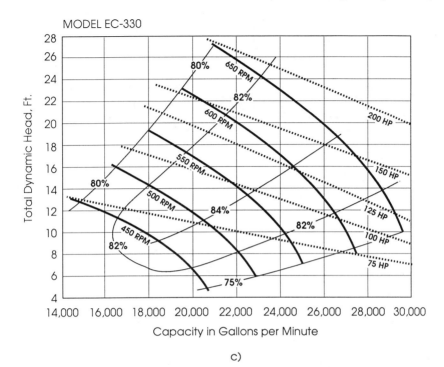

FIGURE 11-13. Examples of pump performance curves: (a) mixed-flow pump, model 1403M of Berkeley Pump Co.; (b) radial-flow pump, model B2½ZRM of Berkeley Pump Co.; and (c) axial-flow pump, model EC-330 of M&W Pump Corp.

pump the head changes slightly and then sharply drops after the maximum efficiency as discharge increases. However, the head continuously decreases as discharge increases for mixed- and axial-flow pumps. The head at the zero discharge is called the shutoff head, which is the maximum head a pump produces. At the shutoff head the pump efficiency is, of course, zero since the energy is used only to turn the pump without discharge. For all cases, the pump efficiency steadily increases to the maximum and rapidly decreases after the maximum as discharge continues to increase. Only one maximum pump efficiency exists for a given pump.

It is important to observe that brake power steadily increases as discharge increases for radial- and mixed-flow pumps. Since brake power is a minimum when discharge is zero, the discharge side should be closed when these pumps are started. This will minimize the power demand for the initial load. In contrast, brake power decreases as discharge increases

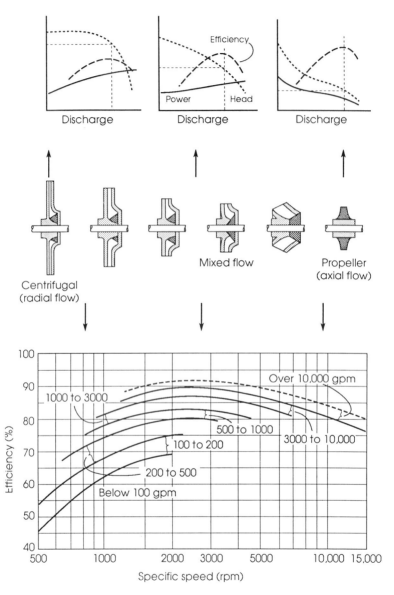

FIGURE 11-14. Shapes of pump performance curves and specific speeds for different types of pump impeller.

for axial-flow pumps. Thus, the discharge side of the pump should be open to start an axial-flow pump for the minimum initial load to the power unit.

11.3.5 Specific Speed

Specific speed (N_s) is an index that represents discharge, total dynamic head, and pump speed at peak efficiency of a pump. Mathematically it is expressed as

$$N_s = \frac{\text{RPM} \times Q^{0.5}}{\text{TDH}^{0.75}} \qquad (11\text{-}13)$$

As shown in equation (11-13) specific speed at the same pump speed increases as the pumping head decreases or as the discharge increases. Thus, low-specific-speed pumps deliver small discharges against high heads, and high-specific-speed pumps discharge large volumes at low head.

Pumps of identical impeller shapes but of different sizes have the same specific speeds. The specific speed varies as a function of impeller shape. The relationships of specific speed to pump type, discharge, and pump efficiency are summarized in the chart on the bottom of Figure 11-14. Mixed- and axial-flow pumps have higher specific speeds than do radial-flow pumps.

Example 11-4: Specific Speed of a Pump

For a pump with the pump curves shown in Figure 11-13a, at maximum efficiency (82%) Q = 1,000 gpm, and TDH = 21 ft at the pump speed of 870 rpm. The specific speed is

$$N_s = \frac{870 \, \text{rpm} \times (1,000 \, \text{gpm})^{0.5}}{(21 \, \text{ft})^{0.75}} = \frac{27,512}{9.81} = 2,800 \, \text{rpm}$$

Reference to Figure 11-14 indicates the pump is a mixed-flow pump. Head and discharge for the same pump at half the pump speed are Q = 500 gpm and TDH = 5.25 ft, calculated by the affinity laws. The specific speed for the new pump speed is then

$$N_s = \frac{435 \, \text{rpm} \times (500 \, \text{gpm})^{0.5}}{(5.25 \, \text{ft})^{0.75}} = 2,800 \, \text{rpm}$$

This example illustrates that changes of pump speed do not affect the specific speed of the pump. The same is true for changes in impeller size.

In other words, this pump remains a mixed-flow pump at the slower pump speed. However, the maximum efficiency of the pump falls to 80% (Fig. 11-14).

11.3.6   Multiple Pumps

Two or more pumps may operate in series or parallel to meet the changing characteristics of a pump system (i.e., seasonal changes in head or discharge requirement). To select and operate multiple pumps properly, one must develop combined performance curves of the new multiple-pump system.

Pumps in Series
More than two pumps are connected in series to obtain larger pumping heads than individual pumps can produce. The discharge from the first pump is passed into the successive pumps. Therefore, the same discharge passes through all pumps connected in series while more energy is added

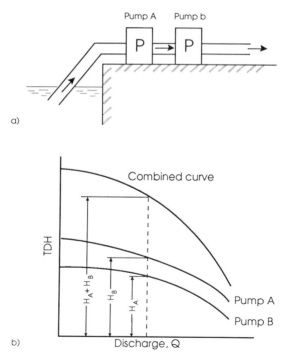

FIGURE 11-15.   Pumps connected in series, and individual and combined head-discharge curves for two pumps operating in series.

to the water to increase the TDH. A typical serial connection of two pumps is shown in Figure 11-15a. The combined TDH, BP, and E_p for a series pump are derived as follows:

$$TDH(Q)_s = \sum_{i=1}^{n} TDH(Q)_i \qquad (11\text{-}14)$$

$$BP(Q)_s = \sum_{i=1}^{n} BP(Q)_i \qquad (11\text{-}15)$$

$$E_p(Q)_s = \frac{\displaystyle\sum_{i=1}^{n} WP(Q)_i}{\displaystyle\sum_{i=1}^{n} BP(Q)_i} = \frac{Q\displaystyle\sum_{i=1}^{n} TDH(Q)_i}{3{,}960\displaystyle\sum_{i=1}^{n} BP(Q)_i} \qquad (11\text{-}16)$$

where n = total number of pumps in series
$TDH(Q)_s$ = combined TDH
$BP(Q)_s$ = combined BP
$E_p(Q)_s$ = combined pump efficiency in decimal
$WP(Q)_i$ = water power for pump i
$BP(Q)_i$ = brake power for pump i
$TDH(Q)_i$ = TDH for pump i

Performance curves for a series pump are developed by calculating these equations at various discharges and plotting the results for each parameter against Q on an arithmetic scale graph. As shown in equations (11-14) and (11-15), TDH and BP of a series pump are sums of TDH and BP of the individual pumps connected in series, respectively. The combined TDH curve of two pumps in series is shown in Figure 11-15b. The NPSH$_a$ of the pump next to the water source is only necessary to examine the cavitation. The combined E_p in series is the same as for the original pumps if the connected pumps are identical.

Pumps in Parallel

Two or more pumps may be operated in parallel to pump water from a single source or multiple sources into a single pipeline (Figure 11-16a). This type of operation is useful where a pump system requires various discharges at approximately the same head. The combined Q, BP, and E_p for a parallel pump are calculated as follows:

$$Q(TDH)_p = \sum_{i=1}^{n} Q(TDH)_i \qquad (11\text{-}17)$$

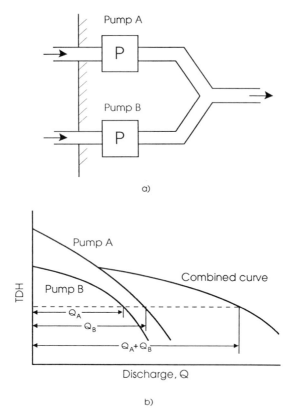

FIGURE 11-16.   Pumps connected in parallel, and individual and combined head-discharge curves for two pumps operating in parallel.

$$BP(\text{TDH})_p = \sum_{i=1}^{n} BP(\text{TDH})_i \qquad (11\text{-}18)$$

$$E_p(\text{TDH})_p = \frac{\sum\limits_{i=1}^{n} WP(\text{TDH})_i}{\sum\limits_{i=1}^{n} BP(\text{TDH})_i} = \frac{\text{TDH}\sum\limits_{i=1}^{n} Q(\text{TDH})_i}{3{,}960\sum\limits_{i=1}^{n} BP(\text{TDH})_i} \qquad (11\text{-}19)$$

where n = total number of pumps in parallel
$Q(\text{TDH})_p$ = combined Q
$BP(\text{TDH})_p$ = combined BP
$E_p(\text{TDH})_p$ = combined pump efficiency in decimal

$WP(\text{TDH})_i$ = water power for pump i
$BP(\text{TDH})_i$ = brake power for pump i
$Q(\text{TDH})_i$ = Q for pump i

To develop performance curves of a parallel pump, these equations are repeated at various TDHs, and data for each parameter are plotted on an arithmetic scale graph against the combined Q. As shown in equations (11-17) and (11-18), Q and BP are sums of Q and BP of the individual pumps connected in parallel, respectively. The combined head-discharge curve of two pumps in parallel is shown in Figure 11-16b. If individual pumps are taking water from different sources, the system is more complicated because the pumps may develop different pumping heads. The NPSH_a of each pump must be evaluated against the NPSH_r of the pump unless pumping conditions are identical for the pumps in parallel. The combined E_p is the same as for the original pumps if the conditions of the connected pumps are identical.

Example 11-5: Developing Combined Pump Curves for a Series Pump

Two identical pumps will be connected in series. Readings for Q, BP, and TDH of the original pumps are given in Table 11-2. These data are obtained from published pump curves. The calculations in Table 11-2 are needed to develop the combined performance curves. Data for TDH_s, BP_s, and $(E_p)_s$ versus Q are plotted for developing new pump curves for the serial pump system.

Example 11-6: Developing Combined Pump Curves for a Parallel-Pump System

If the pumps of Example 11-5 are connected in parallel, the calculations shown in Table 11-3 are needed to develop the combined performance curves. Data for TDH, BP_p, and $(E_p)_p$ versus Q_p are plotted for developing new pump curves for the serial pump.

11.3.7   System Curve and Pump Operating Point

The system curve of a pump system may be developed by plotting the TDH required by the pump system at various flow rates. Equation (11-4c) may be used to determine this curve with consideration of any drawdown of the water source due to pumping. In equation (11-14c), elevation and pressure heads remain constant in case of zero drawdown, because they are independent of discharge. All other variables increase due to increased flow velocity as discharge increases. Drawdown for a large

TABLE 11-2   Solution for Example 11-5: Developing Combined Pump Curves for Pumps in Series

Q	BP_1	BP_2	BP_p	TDH_1	TDH_2	TDH_s	$(E_p)_p$
0	36	36	72	106	106	212	0
800	34	34	68	88	88	176	52
1,600	36	36	72	68	68	136	76
2,000	35	35	70	56	56	112	81
2,400	34	34	68	44	44	88	78
2,800	28	28	56	27	27	54	68

Note: Subscripts 1 and 2 denote pump 1 and pump 2, which are identical in this example.

TABLE 11-3   Solution for Example 11-6: Developing Combined Pump Curves for Pumps in Parallel

TDH	BP_1	BP_2	BP_s	Q_1	Q_2	Q_s	$(E_p)_s$
27	28	28	56	2,800	2,800	5,600	68
44	34	34	68	2,400	2,400	4,800	78
56	35	35	70	2,000	2,000	4,000	81
68	36	36	72	1,600	1,600	3,200	76
88	34	34	68	800	800	1,600	52
106	36	36	72	0	0	0	0

Note: Subscripts 1 and 2 denote pump 1 and pump 2, which are identical in this example.

surface of supply water is usually low and may be neglected. However, drawdown in a well or a small pond is often noticeable when the pumping rate is higher than the source can supply. Any noticeable drawdown in the water source should be considered in the equation. A pump operates at a combination of TDH and Q. The TDH–Q relationship developed for a pump system is plotted on the TDH curve of the pump's performance curves, and the intersection of the two curves gives an operating point of the system (Figure 11-17). Other pump parameters can be obtained from the performance curves once the operating point has been determined. The most desirable operating point of a pump is where maximum pump efficiency is achieved. When the efficiency is lower than the maximum at the operating point, changing impeller speed of the pump may be considered to improve the efficiency. If this is not practical, another type or size of pump should be selected. The system curve is important in properly selecting and installing a pump (Schwab et al. 1981).

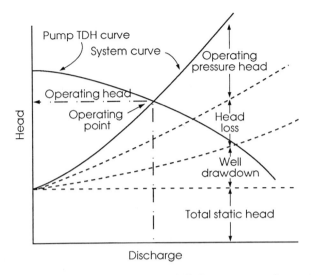

FIGURE 11-17.   Typical pump and system head-discharge curves and operating point.

11.4 POWER UNITS

Stationary pumps for use in aquaculture are powered by electric motors or by internal combustion engines fueled by gasoline, diesel, liquid propane (LP) gas, or natural gas. Nonstationary or portable pumps are also in common use. Small portable pumps usually have gasoline engines, but large portable pumps normally are driven by power takeoffs (PTOs) of farm tractors.

11.4.1 Electric Motors

Electric motors convert electrical energy into the mechanical energy of a rotating shaft. An electric motor is a generator run in reverse. In a generator, a rotating metal loop cuts through a magnetic field and produces an electric current as the loop cuts magnetic flux lines. In an electric motor, electric current is fed into the metal loop, and the magnetic force that develops on the loop causes it to rotate. Because electric motors are popular for driving pumps and other fish-farming equipment, the aquaculturist should know something about them.

If a loop of wire is connected between the terminals of a battery, electric current is forced through the circuit by the electric pressure (electromotive force or voltage). This electric pressure is measured in volts. The rate of current flow is measured in amperes, and the resistance

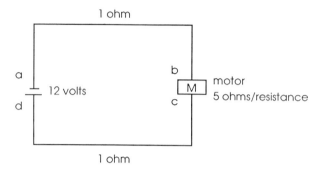

FIGURE 11-18.   A diagram to describe electromotive force of a DC-supplied device.

to current flow in the wire is measured in ohms. Resistance to current flow is a function of the wire material, length, and cross-sectional area. Resistance to flow occurs primarily at the wire surface, so the larger the cross-sectional area the less resistance per unit length for a given type of wire. Some materials are better conductors of electricity than others. For example, for wires of the same length and diameter, copper wire has less resistance than aluminum wire, and hence is a better conductor. Of course, doubling the length of a wire doubles its total resistance. The following symbols are normally used: I is current in amps (A), E is voltage in volts (V), and R is resistance in ohms (Ω). Ohm's law expresses the relationship among them:

$$I = \frac{E}{R} \tag{11-20a}$$

By definition, 1.0 V causes 1.0 A to flow against a resistance of 1.0 Ω. The voltage across any part of a circuit, according to equation (11-20a), is

$$E = IR \tag{11-20b}$$

and the resistance is thus

$$R = \frac{E}{I} \tag{11-20c}$$

Ohm's law applies to an entire circuit. Suppose that a small motor with a resistance of 5.0 Ω is connected by two leads, each with a resistance of 1.0 Ω to a 12-V battery (Figure 11-18). Total resistance in the circuit is thus 7.0 Ω, and the current is $I = 12.0\,V \div 7.0\,\Omega = 1.71\,A$. A voltmeter connected from a to b in the circuit would read 1.71 V (1.71 A $\times$ 1.0 Ω).

From c to d the voltmeter would also read 1.71 V, but from b to c it would read 8.55 V (1.71 A × 5.0 Ω). Of the 12.0 V applied, then, 1.71 V are used in the upper lead, 8.55 V are used by the motor, and 1.71 V are also used in the lower lead. Thus, the 12.0 V are "used up" in voltage "drops around" the circuit.

Power is the rate at which work is done. In an electrical circuit, power (P) is the product of current and voltage and is measured in watts (W). For a DC circuit,

$$P = EI = I^2R \tag{11-21}$$

Power for a motor that draws 5.0 A in a 24-V DC circuit is thus 120 W. The energy use in a circuit depends on the length of time that work is done at a particular rate. If this motor is operated for 10 hr, the energy consumed is 120 W × 10 hr = 1,200 watt-hour. Since consumption of electricity is usually measured in kilowatt-hours (kWh), we write 1,200 watt-hour = 1.2 kWh.

Alternating Current (AC) Motors

Alternating current (AC) is the major source of electric power used in aquaculture systems. Different from DC motors, most AC-powered motors add inductance to circuits. The common AC induction motors used in most aquacultural applications have the alternating current supplied to stationary field coils (stator) rather than to the metal loop. The armature (rotor) has the metal loops wrapped around it (windings), and current is induced in the rotor by electromagnetic induction from the time-varying field of the stator current. Induction of current in the rotor causes it to rotate.

The product of voltage and current overestimates true power in inductance circuits; this product is called *apparent power* and is measured in volt-amps (VA) or kilovolt-amps (kVA). Volts and ohms in an AC circuit may be measured with a common inexpensive instrument known as a multimeter. A more expensive and less common instrument known as a wattmeter may be used to measure *true* power. See Lloyd (1969) or Pansini (1989) for a detailed discussion of motor circuit. Here, we need only recognize that power for circuits including electric motors must be calculated differently from power in DC or pure-resistance AC circuits. The power factor (pf) is the ratio of true power to apparent power in a AC circuit. For a single-phase circuit (1-φ), the power factor is

$$\text{pf} = \frac{P}{E_L I_L} \tag{11-22}$$

where E_L = line voltage and I_L = line current. The power in watts in an AC circuit is

$$P = E_L I_L \times \text{pf} \tag{11-23}$$

In a three-phase circuit (3-φ), the appropriate equation is

$$\text{pf} = \frac{P}{E_L I_L \sqrt{3.0}} \tag{11-24}$$

and

$$P = E_L I_L \sqrt{3.0} \times \text{pf} \tag{11-25}$$

The power factor for AC motors at full load ranges from 0.6 to 0.9. The power factor of most motors declines when the motor is underloaded. A three-phase system may be supplied by a wye or delta connection, as shown in Figure 11-19. These names were derived from the shape of the connections. The delta connection supplies 230 V, and the wye connection gives 208 V. For example, the power factor for a 10-hp AC motor might be 0.91 at full load, 0.89 at three-fourths load, and 0.81 at one-half load. For a three-phase circuit where E_L = 230 V (delta connection), I_L = 26.5 A, and pf = 0.84, the power is

$$P = 230\,\text{V} \times 26.5\,\text{A} \times \sqrt{3.0} \times 0.84 = 8{,}870\,\text{W or } 8.87\,\text{kW}$$

The synchronous rotor speed depends on the number of poles in the motor and the frequency of the current. Motor speed is measured in rpm:

$$\text{Motor speed (rpm)} = \frac{60\,\text{sec/min} \times \text{frequency (Hz)}}{\text{no. of pairs of poles/phase}} \tag{11-26}$$

In the United States, electrical frequency is 60 cycle/sec, or, more commonly, 60 Hertz (Hz). Therefore, synchronous speeds for 60-Hz motors are 3,600 rpm for one pair per phase; 1,800 rpm for two pairs per phase; 1,200 rpm for three pairs per phase; and 900 rpm for four pairs per phase. Under load, a motor runs slower than its synchronous speed. The percentage difference between synchronous speed and full-load speed is called *percentage slip*; typical induction motors have 2–5% slip. Thus, common four-pole (two pairs) single-phase motors have full-load speeds of 1,710–1,760 rpm instead of 1,800 rpm.

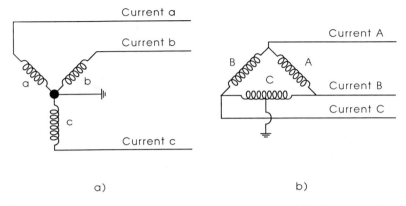

FIGURE 11-19.   Wire connection of a three-phase AC system: (a) wye-connected system, and (b) delta-connected system.

Common line voltages for AC motors used in aquaculture are 115 V, 230 V, and 460 V. Motors less than 1 or 2 hp usually are single-phase and require line voltages of 115 V or 230 V. Larger motors usually are three-phase and operate at 230 V or 460 V. Single-phase motors of 5–10 hp are available, but they are more expensive than three-phase motors of the same horsepower. Line voltage varies from time to time because of power load conditions; most motors are designed to run on variations of ±10% of their rated voltage. Where electrical power is available, one can depend on having access to 115-V and 230-V single-phase electricity. Three-phase electricity is not always available, and even where it is, 230-V electricity is more common than 460 V.

A motor is designed to produce a specific torque and speed at its output shaft. Motor power supplied at the output shaft is called brake power (*BP*) and is related to torque (*T*) in foot-pounds and speed in rpm. Equation (11-6) is used to determine motor brake power in horsepower (hp). Motors are rated in horsepower delivered at the output shaft for the full-load speed. When a motor is operating at full load at the rated voltage, it uses a specific amount of current, often termed the *rated amperes*. For example, a typical 10-hp, three-phase motor uses about 27 A at full load. The current increases under increasing load, so a motor using more than the rated current is overloaded. Many motors have a service factor of 1.10 to 1.20. The service factor is a coefficient that may be multiplied by the rated power or current to indicate the amount of permissible overload. A service factor greater than unity is desirable, because it prevents damage to the motor when, for some reason, the load increases above the design load. However, for long-term operation, a

TABLE 11-4   Full-Load Currents in Amps for Single-Phase and Three-Phase Induction-Type AC Motors (recommended fuse sizes in amps are given in parentheses)

Motor Size (hp)	Single-Phase[a]		Three-Phase[b]		
	115 V	230 V	115 V	230 V	460 V
0.25	5.8 (20)	2.9 (15)			
0.50	9.8 (30)	4.9 (15)	4.0 (15)	2.0 (15)	1.0 (15)
0.75	13.8 (45)	6.9 (25)	5.6 (25)	2.8 (15)	1.4 (15)
1	16 (50)	8 (25)	7.2 (30)	3.6 (15)	1.8 (15)
1.5	20 (60)	10 (30)	10.4 (40)	5.2 (15)	2.6 (15)
2	24 (80)	12 (40)	13.6 (50)	6.8 (25)	3.4 (15)
3	34 (110)	17 (60)		9.6 (30)	4.8 (15)
5	56 (175)	28 (90)		15.2 (45)	7.6 (25)
7.5	80 (250)	40 (125)		22 (70)	11 (35)
10	100 (300)	50 (150)		28 (80)	14 (45)
15				42 (125)	21 (60)
20				54 (175)	27 (80)
25				68 (200)	34 (100)
30				80 (250)	40 (125)
40				104 (350)	52 (175)
50				130 (400)	65 (200)
60				154 (450)	77 (225)
75				192 (600)	96 (300)
100				248 (800)	124 (400)
150					180 (600)
200					240 (800)

Reprinted with permission from NFPA 70-1990, *National Electrical Code*®, Copyright© 1989, National Fire Protection Association, Quincy, MA 02269. This reprinted material is not the complete and official position of the National Fire Protection Association on the referenced subject which is represented only by the standard in its entirety. *National Electrical Code*® and *NEC*® are registered trademarks of the National Fire Protection Association, Inc., Quincy, MA 02269.

[a]Voltages are rated motor voltages. The currents listed are considered valid for voltage ranges of 110–120 V and 220–240 V.
[b]For motors with 80% and 90% power factors, multiply above numbers by 1.25 and 1.1, respectively. Voltages listed are rated motor voltages. The currents listed are considered for voltage ranges of 110–120, 220–240, and 440–480 V.

motor should not be loaded beyond its rated power. An electric motor will last longer if it is not operated continuously at more than 90–95% of full load. On the other hand, motors should not be underloaded to a large extent. A motor operated at little or no load may use as much as 50% of the rated current. Full-load currents for AC motors are provided in Table 11-4. Note that doubling the voltage reduces the current by 50%, and at the same voltage the current for a single-phase motor is greater than the current for a three-phase motor. Load currents and their required fuse sizes for AC motors are also provided in the table. When a motor starts,

TABLE 11-5   Typical Motor Efficiencies in Percent (%) for 60-Hz, Single- and Three-Phase AC Motors for Nominal rpm at Full Load

Motor Size (hp)	Single-Phase		Three-Phase	
	1,750 rpm	3,450 rpm	1,750 rpm	3,450 rpm
0.50	64.0	61.0		
0.75	68.0	65.0	69.0	
1	70.0	66.0	79.5	76.0
1.5	72.0	72.0	82.0	79.5
2	76.0	73.0	84.5	83.0
3	76.0	75.0	84.5	85.0
5	76.0	78.0	86.0	84.5
7.5			87.0	86.5
10			87.5	85.5
15			88.5	87.5
20			89.5	87.5
25			89.5	87.5
30			90.5	89.5
40			90.5	90.0
50			91.0	90.5
60			91.5	90.5
75			92.0	91.0
100			92.0	92.0
125			92.5	91.5
150			92.5	92.5

Source: Berkeley Pump Co. (1987).

a current much larger than the full-load current is required to get the rotor off dead center (locked rotor position) and accelerate it to full load speed. Thermal overload heaters in a starter protect the motor; heaters are selected from the motor and starter specifications. Fuses for motor circuits are recommended only to protect the wiring. Recommended fuse sizes for various electric motors are given in Table 11-4. Details of AC motor control and protection are given by Smeaton (1987).

Motor efficiency is the percentage of the power input in electrical energy that is delivered by the output shaft in mechanical energy. Typical motor efficiencies are given in Table 11-5; large motors usually are more efficient than small motors. The efficiency of a motor usually declines when it is not fully loaded. When fully loaded, the energy consumption of a motor roughly is the rated energy output divided by motor efficiency. For example, the power consumption of a fully loaded 10-hp motor with a motor efficiency of 89% running for 10 hr is

$$(10\,hp \times 0.746\,kW/hp \times 10\,hr) \div 0.89 = 83.8\,kWh$$

Three-phase motors generally are more efficient and less expensive than single-phase motors. Therefore, their use is recommended for applications requiring 2 hp or more where three-phase electricity is available.

Standard motors are designed to operate at ambient temperatures not higher than 104°F. Obviously, a motor does not need to be operated in a tightly enclosed space or it will overheat, because the motor generates heat. Electric motors are rated for allowed temperature rises caused by resistance. The allowed temperature rise above the ambient temperature for Class A insulated motors is 117°F. The allowed temperature rises depend on the types of insulation and motor enclosures. Advances in the development of insulating materials raised the allowed temperature rise, which provided advantages for smaller motor-space requirements (Smeaton 1987). Class B insulated motors have an allowed temperature rise of up to 162°F above the ambient temperature. The temperature rise of a motor at the rated load under recommended operational conditions will not exceed the rated temperature rise for the motor guaranteed by the manufacturers (F.E. Myers & Bro. Co. 1966). A motor must be operated under the following conditions to meet the rated temperature rise: (1) specified voltage and frequency (no tolerance allowed), (2) power rating no higher than specified, (3) free-flow of air to the motor, (4) no exposure to moisture, dirt, or vapor, and (5) ambient temperature no higher than the rated temperature rise.

Sizing Electric Wires

Wire size is important in motor installation. *Voltage drop* is the term used to describe the decrease in voltage between the power supply and the motor or other load. Voltage drop occurs because the wire resists the flow of current. The voltage drop in a wire depends on wire type, diameter, length, and amount of current. Ampacity and voltage drop must therefore be considered when selecting wire. *Ampacity* refers to the current that a wire can handle without overheating and damaging the insulation.

Wire is classified according to the American Wire Gauge (AWG); actual diameters of bare wire of different AWG sizes are depicted in Figure 11-20. The National Fire Protection Association (1990) publishes the *National Electrical Code®*, which provides ampacities for all types of insulated wire conductors. Selected ampacity values for some insulation types of copper and aluminum wires are provided in Table 11-6. The table shows that a AWG size 8 copper wire is rated at 40 A, but AWG size 6 aluminum wire is also rated at 40 A. Ampacity is determined as

$$\text{Ampacity} = 1.25 \times \text{rated motor current} \qquad (11\text{-}27)$$

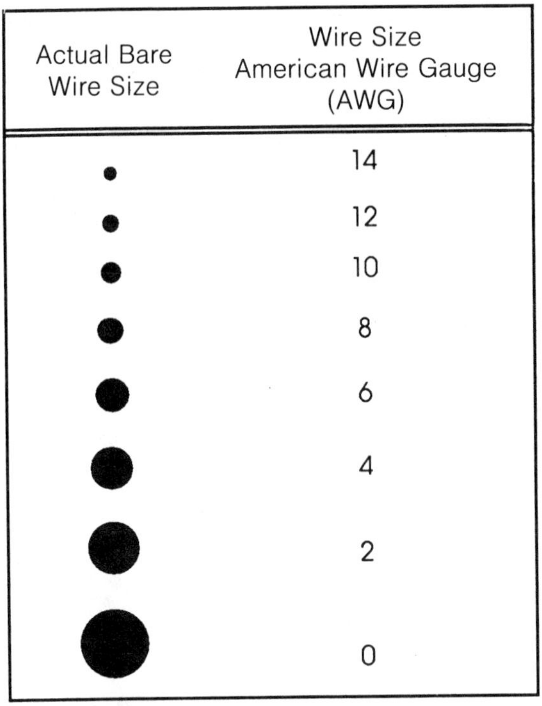

Actual Bare Wire Size	Wire Size American Wire Gauge (AWG)
•	14
•	12
•	10
•	8
•	6
•	4
•	2
•	0

FIGURE 11-20.   Cross-sectional area of electrical wires and American Wire Gauge (AWG) wire sizes.

The factor of 1.25 is commonly used for short wire length (<60 ft) to determine the current for selecting a wire size. This is due to the 80% rule on overcurrent device protection (National Fire Protection Association 1990). For example, if we want to select a short wire for a 20-hp motor rated at 54 A and 230 V (Table 11-4), the ampacity is 1.25 × 54.0 A or 67.5 A. From Table 11-6, a copper wire of AWG size 4 or an aluminum wire of AWG size 2 is required. For long wires, larger wire sizes than that for ampacity are usually required to minimize voltage drop, as discussed next.

Cross-sectional areas of wires are measured in circular mils (Cm). A mil is 0.001 in., and 1.0 circular mil is the area of a circle having a diameter of 1.0 mil. Area in circular mils for round wire is obtained by simply squaring the diameter measured in mils. Hence, a single strand of wire with a diameter of 0.015 in. (15.0 mil) has an area of 225.0 Cm. Based on this calculation, areas in circular mils of selected wire sizes are provided in Table 11-6. The equations for estimating required cross-

TABLE 11-6   Cross-Sectional Areas in Circular Mils (Cm) and Square Inches for American Wire Gauges and Their Ampacities for Insulated Copper and Aluminum Conductors (types TW and UF)[a] Rated 0–2000 Volts with Motor Temperature Ratings of 140–194°F Based on Ambient Temperatures of 84°F[b]

AWG Size	Diameter (mils)	Ampacity (amps)		Cross-Sectional Area	
		Copper	Aluminum	Cm	in.^{2}
14	64.0	20	—	4,110	0.00323
12	81.0	25	20	6,530	0.00513
10	102.0	30	25	10,380	0.00815
8	128.0	40	30	16,510	0.0130
6	162.0	55	40	26,240	0.0206
4	204.0	70	55	41,740	0.0328
3	229.0	85	65	52,620	0.0413
2	258.0	95	75	66,360	0.0521
1	289.0	110	85	83,690	0.0657
0	325.0	125	100	105,600	0.0829
00	365.0	145	115	133,100	0.105
000	410.0	165	130	167,800	0.132
0000	460.0	195	150	211,600	0.166

Reprinted with permission from NFPA 70-1990, *National Electrical Code®*, Copyright© 1989, National Fire Protection Association, Quincy, MA 02269. This reprinted material is not the complete and official position of the National Fire Protection Association on the referenced subject which is represented only by the standard in its entirety.

[a] Conductor application and insulations: TW, moisture-resistant thermoplastic; UF, underground feeder and branch-circuit single-cable conductor.
[b] Multiply the ampacities by ampacity correction factors: 1.08 for 70–77°F, 0.91 for 88–95°F, 0.82 for 97–104°F, and 0.71 for 106–113°F.

sectional area for a wire to satisfy permissible voltage drop along the wire are

$$A_w = \frac{2.0KLI}{VD} \quad \text{for single-phase (1-}\phi\text{)} \tag{11-28a}$$

$$A_w = \frac{KLI\sqrt{3.0}}{VD} \quad \text{for three-phase (3-}\phi\text{)} \tag{11-28b}$$

where A_w = wire cross-sectional area (Cm)
L = wire length (ft)
K = resistance factor (ohm-Cm/ft), 10.4 for copper wire and 17 for aluminum wire at 68°F
V = voltage (volts)
D = permissible voltage drop

Example 11-7: Determining Wire Size for a AC Motor

A 20-hp single-phase motor rated at 54 A and 230 V is 500 ft from the power source, and a voltage drop of 5% is acceptable. What size copper wire is needed? The required area in circular mils for a copper wire is

$$A_w = \frac{2 \times 10.4 \, \Omega\text{-Cm/ft} \times 500 \, \text{ft} \times 54 \, \text{A}}{230 \, \text{V} \times 0.05} = 48{,}830 \, \text{Cm}$$

An AWG size 3 copper wire has a cross-sectional area of 52,620 Cm (Table 11-6). Thus, this wire meets the voltage drop and ampacity requirements. If the same motor used three-phase power,

$$A_w = \frac{10.4 \times 500 \times 54 \times \sqrt{3.0}}{230 \times 0.05} = 42{,}290 \, \text{Cm}$$

The motor requires the same size wire as for the single-phase motor.

Motor Nameplate

All electric motors have a nameplate that contains information about the motor so that it can be properly selected, installed, operated, and maintained. Standard size frames and shaft heights have been established by the National Electrical Manufacturers Association (NEMA) for integral motors. This allows motors from different manufacturers to be interchanged according to application. The motor nameplate (Fig. 11-21)

```
┌──────────────────────────────────────────────────────────┐
│ DAYTON                         CAPACITOR START            │
│                                A.C. MOTOR                 │
│                                                           │
│  MODEL  5K658    H.P.  3/4    R.P.M.  3450                │
│  VOLTS  115/208-230  AMPS 9.6/4.8 HZ  60                  │
│  TIME RATING CONT   TEMP RISE 40 °C PH  1                 │
│  FRAME  56C    BRGS  BALL    CODE  L                      │
│  S.F.  1.5    SFA  12.8/6.4    AMB.  40 °C                │
│  NO.  KS55CXPA-619    INSU. CLASS  A                      │
│  THERMALLY PROTECTED  AUTO                                │
│                                                           │
│  DAYTON ELECTRIC MFG. CO. CHICAGO 60648 USA               │
└──────────────────────────────────────────────────────────┘
```

FIGURE 11-21.   Typical information on an electric motor nameplate.

TABLE 11-7   Motor Code Letters on the Motor Nameplate for Starting Current Requirement on Full Load

Code Letter	Locked Rotor Current (kVA/hp)
D	4.0–4.5
E	4.5–5.0
F	5.0–5.6
G	5.6–6.3
H	6.3–7.1
J	7.1–8.0
K	8.0–9.0
L	9.0–10.0
M	10.0–11.2
N	11.2–12.5
P	12.5–14.0

Reprinted with permission from NFPA 70-1990, *National Electrical Code*®, Copyright © 1989, National Fire Protection Association, Quincy, MA 02269. This reprinted material is not the complete and official position of the National Fire Protection Association on the referenced subject, which is represented only by the standard in its entirety.

TABLE 11-8   NEMA Index Letter Code for Motor Efficiency

Efficiency Index Letter	Minimum Efficiency[a] (%)	Normal Efficiency[b] (%)
D	91.7	93.0
E	90.2	91.7
F	88.5	90.2
G	86.5	88.5
H	84.0	86.5
K	81.5	84.0
L	78.5	81.5
M	75.5	78.5
N	72.0	75.5
P	68.0	72.0
R	64.0	68.0
S	59.5	64.0
T	55.0	59.5

Source: Gustafson (1988).

[a] All rated motors will meet or exceed the minimum efficiency.

[b] Average efficiency of the most motors of the same make and model.

contains the NEMA designation: manufacturer's name, serial number and suitable identification, frame designation number, power output, motor code for starting current requirement (Table 11-7), temperature rise, motor speed, frequency, phrase, voltage, rated load current, type of electricity, time rating, service factor, insulation class, lubrication, and pump efficiency (Table 11-8).

11.4.2 Internal Combustion Engines

Pumps and other fish-farming equipment may also be powered by internal combustion engines. The price of fuel for internal combustion engines may be less expensive than electricity and it may be more convenient to operate than an electric motor. Power outages can be a problem with electric motors, but internal combustion engines can malfunction. Therefore, one cannot say that internal combustion engines are any more reliable than electric motors, or vice versa.

In an internal combustion engine, fuel is exploded in a cylinder above a reciprocating piston. There usually are two or more cylinders and pistons, and the pistons are connected to a crankshaft. When the fuel explodes, the piston is forced downward, and the downward movement of the piston is converted to a rotating motion by the crankshaft. The rotating motion of the crankshaft causes the piston to move upward in the cylinder. In this process, gases resulting from fuel combustion are forced out of the cylinder. More fuel is introduced to the cylinder and exploded in order to keep the engine operating. The crankshaft is attached through a suitable drive train to an output shaft, and a portion of the energy resulting from the combustion of fuel is converted to mechanical energy in the rotating output shaft. Those interested in details of internal combustion engines should consult Arcoumanis (1988) and Heywood (1988).

There are two basic types of internal combustion engines. In the spark-ignition engine, a carburetor prepares a proper mixture of fuel and air that is introduced to the cylinder. The fuel mixture is ignited by a spark created by a spark plug. Of course, this process is not as simple as it sounds, for the fuel must be introduced and ignited at the proper time in order for the engine to run smoothly. Spark-ignition engines normally use gasoline as a fuel. With proper carburetion systems, liquid propane gas, natural gas, and other hydrocarbon fuels may be used. The compression-ignition engine relies on the heat of compression of air in the cylinder to spontaneously combust the fuel when injected into the cylinder. Compression-ignition engines often use diesel fuel, and they commonly are called diesel engines.

The efficiency of an internal combustion engine is called the brake

TABLE 11-9   Energy Content and Power Output of Hydrocarbon Fuels and Electricity

Fuel	Energy Content (Btu/unit)[a]	Power Output[b] (hp-hr/unit)	Consumption Rate
Diesel	140,000	16.7	0.06 gal/hp·hr
Propane	94,500	9.2	0.11 gal/hp·hr
Gasoline	124,000	11.5	0.087 gal/hp·hr
Natural gas	950,000–1,000,000	82.0–89.0	12.2–11.2 ft^3/hp·hr
Electricity[c]	3,410	1.34	0.75 kWh/hp·hr

[a] Units are diesel, propane, and gasoline—gallons; natural gas—1,000 ft^3; electricity—kWh.

[b] Power output is the work being accomplished by the power unit with losses considered (Pair et al. 1983).

[c] Assume 100% motor efficiency. For lower motor efficiency: multiply by the efficiency for power output and divide by the efficiency for consumption rate.

thermal efficiency, which is the ratio of its brake power (the power supplied at the output shaft) to the fuel equivalent power or energy content of the fuel. Factors affecting the efficiency are friction losses in moving parts and losses involved in the induction of fuel and the exhaust of combustion heat. Typical efficiencies of internal combustion engines range from 25% to 35%. The energy content of fuel is expressed in British thermal units (Btu) per unit volume of fuel. The representative energy content of fuels and their power output per unit volume with losses considered are given in Table 11-9.

Any engine operating against a light load uses more fuel per horse-power-hour than at a heavier load. The best use of fuel is normally achieved at 90–100% of full load, but the engine will have a shorter service life if it runs continuously near full-load conditions. Good fuel use and service life may be obtained by operating an engine at 70% of full load. As with electric motors, internal combustion engines should be selected for a particular load. A large engine should not be selected for a small load, or vice versa; an engine too small for a particular job will be continuously overloaded.

11.4.3 Power Transfer

The power from an electric motor or an internal combustion engine must be transferred to the shaft of a pump or other equipment. The pump shaft and electric motor or engine shaft may be coupled directly for high-speed pumps. Pulley and belt systems may be used where direct connection is not possible. The efficiency of a power-transfer mechanism may affect the operating cost of a pumping system.

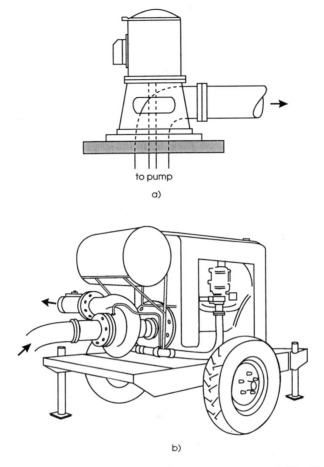

to pump

a)

b)

FIGURE 11-22.   Direct-driven pumps: (a) vertical turbine pump, and (b) trailer-mounted pump. (*Source: U.S. Soil Conservation Service 1970.*)

Direct Drives

The direct drive is the most efficient power-transfer mechanism because there is little loss of power. It is also the least expensive and most efficient drive, especially for vertical turbine pumps using an electric motor (Fig. 11-22a). The limitation of the direct drive is that the speed of the pump is the same as that of the power unit. Therefore, the pump speed cannot easily be varied. Figure 11-22b shows a direct connection of a centrifugal pump to an internal combustion engine. This system is convenient because it can be relocated easily.

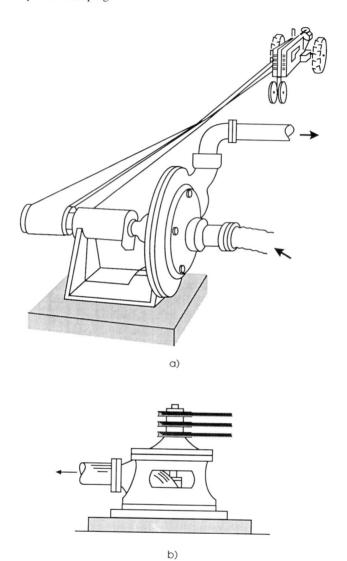

a)

b)

FIGURE 11-23. Pulley-driven pumps: (a) flat-face pulley, and (b) V-belt pulley. (*Source: U.S. Soil Conservation Service 1970.*)

Pulley Drives

Pulleys may be used to change the speed and torque output of shafts (Fig. 11-23), and are made in a variety of styles. The most common in aquaculture applications are flat-face pulleys (Fig. 11-23a) and V-belt sheaves (Fig. 11-23b); sheaves may be single or multiple. Flat belts are used on flat-face pulleys, V-belts on sheaves. The contact surface of V-belts may be smooth or cogged, where cogs on the belts prevent slipping. Belts are selected based on pulley type, pulley diameter, distance between pulleys, speed, and power transferred between shafts. Pulley and belt manufacturers provide detailed information on pulley characteristics. The efficiency of energy transfer with flat belts is 80–90% and is 90–95% for V-belts. The losses are due mainly to belt slippage.

The product of the rotational speed and diameter of a driving pulley must equal the product of the rotational speed and diameter of the driven pulley:

$$D_1 \times RPM_1 = D_2 \times RPM_2 \qquad (11\text{-}29)$$

where D = diameter of flat-face pulley or pitch diameter of V-belt pulley, RPM = rotational speed of pulley (rpm), and subscripts 1 and 2 denote driving and driven pulleys, respectively.

Suppose the output shaft of an electric motor rotates at 1,750 rpm. A 12-in.-diameter pulley will be used on the output shaft. A pulley diameter for a pump shaft that will produce a shaft rotation of 900 rpm is

$$D_2 = \frac{D_1 \times RPM_1}{RPM_2} = \frac{12.0 \, \text{in.} \times 1{,}750 \, \text{rpm}}{900 \, \text{rpm}} = 23.3 \, \text{in.}$$

Notice, to slow the driven shaft, a larger pulley must be attached on the driven shaft than on the driving shaft. Likewise, to speed up the driven shaft, the larger pulley must be on the driving shaft. This type of drive requires considerable attention, because of temperature and humidity, which affect the tension and running position of the pulley. The power generated at the motor shaft and the pump shaft are approximately the same even though they have different speeds. The same procedure used for pulleys may be applied to chain drives and gears. One simply uses the number of teeth in the sprocket of the chain or gear rather than the pulley diameter in equation (11-29) to determine the transfer ratio. The efficiency of this type is 95% or more and is rarely affected by temperature and humidity changes.

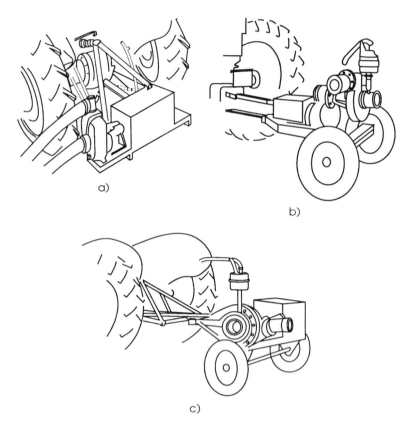

a)

b)

c)

FIGURE 11-24.   PTO-driven pumps: (a) drawbar-mounted PTO drive; (b) trailer-mounted PTO drive with spur-gear-type speed increaser; and (c) trailer-mounted PTO drive with bevel-gear-type speed increaser. (*Source: U.S. Soil Conservation Service 1970.*)

PTO-Driven Pumps

Trailer-mounted pumps powered by the power takeoff (PTO) of a farm tractor often are useful in aquaculture (Fig. 11-24) for small centrifugal pumps. The pumps are mobile, sturdy, and highly dependable and can be operated over a range of TDH–Q combinations at various pump speeds. The standard PTO speed of farm tractors is 540 or 1,000 rpm, depending on the tractor size. Desired pump speed is obtained through the use of a gear-type speed increaser. Care should be taken to be sure that the PTO shaft is properly aligned.

11.5 PUMP SELECTION

Pump performance curves provided by pump manufacturers are the basis of selecting a pump. A pump should be selected to meet head and discharge requirements for the expected operating conditions at or near maximum efficiency. The conditions of an aquacultural project often limit the size and type of pumps to be selected.

Example 11-8: Determining Brake Power and Wire Size for a Pump Powered by an AC Motor

A pump is needed to supply at least 150 gpm against 70 ft TDH. The pump will be operated at 2,000 ft above m.s.l. for 70°F water. The calculated $NPSH_a$ is 10 ft. Single-phase, 230-V AC power is available, and a 1,750-rpm electric motor will power the pump by a direct drive. The electrical service is 200 ft from the motor. The specific speed of the pump is

$$N_s = \frac{1,750 \times 150^{0.5}}{70^{0.75}} = \frac{21,433}{24.2} = 885 \, \text{rpm}$$

According to Figure 11-14, a radial-flow pump is required. The next task is to examine pump performance curves in pump manufacturers' catalogs to find a pump that has high efficiency at the desired discharge and head and that can be satisfactorily installed at the proposed location. The performance curve in Figure 11-13b is for a pump that could be used for this application; a 5-hp pump with $8\frac{3}{8}$-in.-diameter impeller discharges approximately 175 gpm at 70 ft with an efficiency of about 72%. Other pumps may have higher efficiency. Reference to Figure 11-13b shows that the pump has an $NPSH_r$ of 8 ft, which is lower than the $NPSH_a$. Therefore, cavitation is not a problem. The brake power required is approximately 4.3 hp, from the figure.

The pump shaft is coupled directly to the motor output shaft, so there is little power loss in the transfer. A 5-hp motor will suffice in this application. It will operate at about 86% (4.3 hp/5 hp) of full load, at 230 V will use 28 A at full load (Table 11-4). An AWG size 10 copper wire has the necessary ampacity; however, voltage should not decrease by more than 5% along the wire between the electrical service and the motor. Hence, the cross-sectional area of the wire is

$$A_w = \frac{2 \times 12 \times 200 \times 28}{0.05 \times 230} = 11,690 \, \text{Cm}$$

Table 11-6 indicates that an AWG size 8 copper wire is needed to meet ampacity and voltage drop requirements.

Example 11-9: Determining Brake Power Requirement of a Pump System

A vertical turbine pump similar to the one shown in Figure 11-7b will be installed to lift 20,000 gpm from a creek and discharge it into a water supply canal. The elevation of the pipe outlet will be 18 ft above the lowest water level expected in the creek. The pump will be placed beneath the water surface of the creek. There will be a 30-ft-long suction pipe with a strainer bucket entrance with a foot valve and a 120-ft-long discharge pipe with two 45° elbows and an ordinary free-flow exit. No pressure head is needed at the discharge point because water will freely discharge into the canal.

Solution: Considering the high volume and low head, we select an axial-flow pump. A suitable axial-flow pump would have performance features such as those in Figure 11-13c. Because large steel pipe will be used to convey the large volume and distance is not great, the total dynamic head (TDH) will not differ greatly from the elevation head. The pump will have a 36-in.-diameter steel pipe for both suction and discharge. The friction head loss for 36-in.-diameter steel pipe is 0.34 ft per 100 ft according to reference data (Berkeley Pump Co. 1987). Total friction loss of the 150-ft-long steel pipe is

$$H_f = 0.34\,\text{ft}/100\,\text{ft} \times 150\,\text{ft} = 0.5\,\text{ft}$$

Minor losses due to pipe fittings are (Fig. B-2)

Fitting	Number	Equivalent Length of 36-in.-diameter pipe (ft)
45° elbow	2	2 × 45 = 90
Free-flow exit[a]	1	0
Strainer bucket entrance with foot valve[b]	1	230
Total		320

[a] Free-flow exit has no head loss.
[b] The head loss of this entrance is approximately the same as for a full open check valve.

$$H_m = 0.34\,\text{ft}/100\,\text{ft} \times 320\,\text{ft} \times 1.1\,\text{ft}$$

The total head loss is

$$H_L = H_f + H_m = 0.5\,\text{ft} + 1.1\,\text{ft} = 1.6\,\text{ft}$$

The velocity head of the discharge is calculated as follows:

$$A = 0.785D^2 = 0.785(3.0\,\text{ft})^2 = 7.0\,\text{ft}^2$$

$$Q = \frac{20{,}000\,\text{gpm}}{450\,\text{gpm/cfs}} = 44.4\,\text{cfs}$$

$$v = \frac{Q}{A} = \frac{44.4\,\text{cfs}}{7.0\,\text{ft}^2} = 6.3\,\text{ft/sec}$$

$$H_v = \frac{v^2}{2g} = 0.0155 \times v^2 = 0.0155 \times 6.3^2 = 0.6\,\text{ft}$$

The total dynamic head is

$$\text{TDH} = H_e + H_l + H_v + H_p = 18.0\,\text{ft} + 1.6\,\text{ft} + 0.6\,\text{ft} + 0 = 20.2\,\text{ft}$$

No priming is needed since the pump is located below the supply waterline. This pump has a specific speed of

$$N_s = \frac{650 \times 20{,}000^{0.5}}{20.2^{0.75}} = \frac{91{,}920}{9.53} = 9{,}645\,\text{rpm}$$

which indicates an axial-flow propeller pump (Fig. 11-14).
    Reference to the pump performance curves (Fig. 11-13c) indicates that at 20.2 ft of TDH and 20,000 gpm discharge, the pump efficiency is about 81% at 580 rpm. The brake power requirement at the pump shaft is

$$BP = \frac{Q \times \text{TDH}}{3{,}960 \times E_p} = \frac{20{,}000\,\text{gpm} \times 20.2\,\text{ft}}{3{,}960 \times 0.81} = 126.0\,\text{hp}$$

The brake power may be directly read from the pump performance curves. A diesel engine will be connected by a flat pulley to operate the pump. Assuming a drive efficiency of 90% for the pulley, the required engine power to provide a brake power of 126.0 hp at the pump shaft is

$$\frac{126.0\,\text{hp}}{0.9} = 140.0\,\text{hp}$$

Assuming the output shaft of the engine turns at 1,200 rpm. An 18-in.-diameter pulley is selected for the engine shaft. The pulley diameter for the pump shaft to provide the 580 rpm is

$$\frac{18\,\text{in.} \times 1{,}200\,\text{rpm}}{580\,\text{rpm}} = 37.0\,\text{in.}$$

Of course, there should be some adjustment in selecting the pulley size because the exact size as calculated may not be available.

In summary, the axial-flow pump in Figure 11-13c will be powered by a 140-hp diesel engine. An 18-in. pulley on the engine shaft will be connected to a 36-in. pulley on the pump shaft with a belt. This connection provides a pump speed of 600 rpm, which will discharge 22,000 gpm at 20-ft TDH with 82.5% efficiency.

11.6 COST OF PUMPING WATER

A pumping system has both fixed and variable costs. Fixed cost is the annual initial investment cost, which includes the construction and development costs of a pump system. Table 11-10 shows estimated service life of various components of a pump system to calculate depreciation. Other

TABLE 11-10   Service Life of Pumping System Components

Item	Estimated Service Life
Well and casing	20 yr
Housing	20 yr
Pump turbine	
Bowl	16,000 hr or 8 yr
Column, etc.	32,000 hr or 16 yr
Pump, centrifugal	32,000 hr or 16 yr
Power transmission	
Gear head	30,000 hr or 15 yr
V-belt	6,000 hr or 3 yr
Flat-face, rubber	10,000 hr or 5 yr
Flat-face, leather	20,000 hr or 10 yr
Electric motor	50,000 hr or 25 yr
Diesel engine	28,000 hr or 14 yr
Gasoline engine	
Air-cooled	8,000 hr or 4 yr
Water-cooled	18,000 hr or 9 yr
Propane engine	28,000 hr or 14 yr

Source: U.S. Soil Conservation Service (1970).

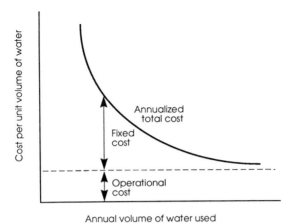

FIGURE 11-25.   Typical relationship of annualized pumping cost and pumping volume.

fixed costs paid annually, such as taxes, insurance, and energy hookup, must be included to determine the total annual fixed cost. Several methods can be used to determine an annual cost, and these are covered in many introductory economics books. The annual variable cost includes mostly costs of power and personnel. The total annual pumping cost is the sum of the annual fixed cost and variable cost. Annual pumping cost per unit volume of water depends on the total volume used per year. As shown in Figure 11-25, annual pumping cost per unit volume of water pumped decreases as the total volume of annual water use increases while operational cost stays constant.

It is not possible to calculate exactly how much it will cost to pump water in a particular application, but close estimates of power cost for various power units may be made with Anderson's (1984) method.

For electric motors,

$$C = \frac{Q \times \text{TDH} \times 0.746 \times R}{3,960 \times E_P \times E_D \times E_M} \tag{11-30}$$

where C = pumping cost per hour of operation ($/hr)
   R = cost of electricity ($/kWh)
   E_P = pump efficiency (decimal)
   E_D = drive train efficiency in decimal; use 1.0 if the pump is directly driven by electric motor
   E_M = motor efficiency (decimal)

For internal combustion engines;

$$C = \frac{Q \times \text{TDH} \times R}{3{,}960 \times E_P \times E_D \times K} \tag{11-31}$$

where R = cost of fuel ($/unit volume)
$\quad\ K$ = average power output per unit volume of fuel

Example 11-10: Fuel Cost for a Pump Powered by a Diesel Engine

Consider again Example 11-9. In operation the pump discharges 22,000 gpm at 20.0-ft TDH with 82.5% efficiency. The power output of diesel fuel is 16.7 hp-hr/gal (Table 11-9). What is the cost of fuel per hour operation at $1.05/gal diesel cost?

Solution: The hourly cost of operating the pump is

$$C = \frac{22{,}000\,\text{gpm} \times 20.0\,\text{ft} \times 1.05\,\$/\text{gal}}{3{,}960 \times 0.81 \times 0.90 \times 16.7\,\text{hp-hr/gal}} = 9.4\,\$/\text{hr}$$

If this pump operates 10 hr per day for six months each year, the annual fuel cost is

$$9.4\,\$/\text{hr} \times 10\,\text{hr} \times 6\,\text{months} \times 30\,\text{days/month} = 16{,}920\,\$/\text{yr}$$

Example 11-11: Power Cost for an AC Motor-Driven Pump

Suppose a 200-gpm discharge is to be pumped against a 75.0-ft head with an electrically powered pump. Assuming a pump efficiency of 60%, a motor efficiency of 85%, direct drive, and a price of $0.075/kWh for electricity, find the power cost.

Solution: The power output of an AC motor is 1.34 hp-hr/kWh $\times$ 0.85, or 1.14 hp-hr/kWh (Table 11-9). The hourly operating cost is

$$C = \frac{200\,\text{gpm} \times 75.0\,\text{ft} \times 0.746 \times 0.075\,\$/\text{kWh}}{3{,}960 \times 0.60 \times 1.0 \times 1.14\,\text{hp-hr/kWh}} = 0.31\,\$/\text{hr}$$

If this pump operates 12 hr/day for eight months each year, the annual fuel cost is

$$0.31\,\$/\text{hr} \times 12\,\text{hr} \times 8\,\text{months} \times 30\,\text{days/month} = 893.0\,\$/\text{yr}$$

Equations (11-30) and (11-31) show that pumping cost is directly proportional to discharge and TDH and inversely proportional to efficiencies. Increasing discharge or TDH will increase pumping costs. Excessive discharge or TDH must be avoided to obtain minimum pumping cost.

11.7 INSTALLATION OF PUMPS

A horizontal pump should be located as close to the supply water level as possible on a solid foundation. This will allow a short suction pipe to be used. The pump should also be located where good drainage is provided to protect it from leakage and flooding water. Figure 11-26 illustrates recommended methods of pump installation and problems to be avoided. Suction pipe should be kept free of air leaks. A foot valve is recommended to maintain water in the suction side for priming and a strainer to prevent suspended or floating debris from entering the pump. To maximize $NPSH_a$, the fewest number of fittings, bends, and joints should be used along with the shortest length of suction pipe. An eccentric, rather than concentric or straight taper, reducer should be used to connect the pump and suction pipe. The latter reducers can form air pockets at the top of the suction pipe. The section of suction pipe on the ground should be well supported independently so that no strain is transmitted to the pump casing or other pipe joints. Properly installed pumps prevent some common pump problems that might develop during operation.

11.8 OPERATION AND MAINTENANCE OF PUMPS

A centrifugal pump should not be operated until the suction side of the pump is filled with water, unless the pump is located below the supply water level, because the impeller cannot create enough suction to allow water to enter the pump impellers unless the suction side is full of water or pressure is reduced below atmospheric pressure by removing air. Some pumps are self-priming or equipped with priming devices. Foot valves are used to hold water in the suction pipe after the pump is stopped, but over time water may leak from the pump in spite of the foot valve. Priming was covered earlier in the chapter. Details of pump priming are usually provided by pump manufacturers.

Centrifugal pumps, like other machines, are subject to malfunction and failure. Pump operators should be alert to locate any problems that pumps may develop in their early stages. Such vigilance will help avoid further serious problems. If a pump fails to deliver water or if it exhibits a

RECOMMENDED

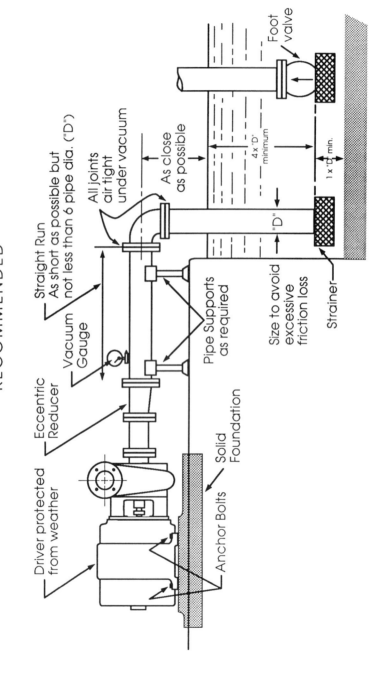

Driver protected from weather

Eccentric Reducer

Straight Run
As short as possible but
not less than 6 pipe dia. ("D")

Vacuum Gauge

All joints
air tight
under vacuum

As close
as possible

Foot valve

Pipe Supports
as required

Anchor Bolts

Solid Foundation

"D"

4 x "D" minimum

1 x "D" min.

Size to avoid
excessive
friction loss

Strainer

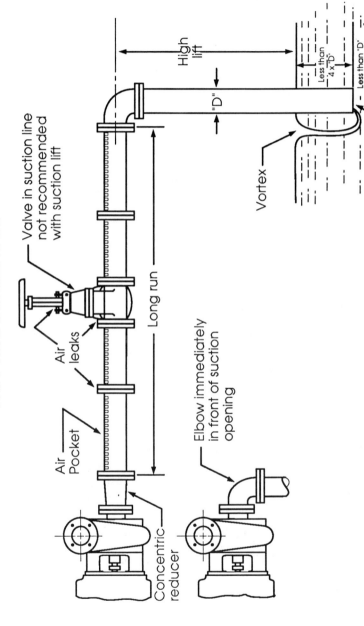

NOT RECOMMENDED

Valve in suction line
not recommended
with suction lift

High
lift

"D"

Vortex

Less than
4 x "D"

Less than "D"

No strainer

Air
leaks

Air
Pocket

Long run

Concentric
reducer

Elbow immediately
in front of suction
opening

FIGURE 11-26. Installation guidelines for pump stations. (*Reprinted with permission of Sta-Rite Industries, Inc. © 1987 Sta-Rite Industries, Inc.*)

reduction in discharge or pressure, one or more of the following problems may exist (Pair et al. 1983).

1. No water or not enough water would be delivered if
   a. Pump not primed
   b. Speed too low
   c. Not enough head
   d. Suction head too high
   e. Impeller clogged
   f. Impeller rotating in wrong direction
   g. Suction entrance clogged
   h. Air leak on suction side
   i. Mechanical damage to rings, seals, or impeller
2. Failure of pump to supply enough pressure may be caused by
   a. Air leak in suction line
   b. Impeller too small
   c. Impeller speed too slow
   d. Valve settings incorrect
   e. Mechanical damage to pump components
3. Erratic action may result from
   a. Leaks in suction line
   b. Pump shaft misalignment
   c. Air entrained in water
4. When the pump overloads the power unit, the problem might be
   a. Impeller speed too high
   b. Pump too small for required TDH and Q
   c. Water temperature too high
   d. Mechanical damage to pump components

Vibration and noise from pumps are usually caused by cavitation due to insufficient $NPSH_a$. In this case either lower the pump close to the water surface or replace the suction pipe with a larger pipe. However, improperly mounted pumps and motors may also vibrate, and worn or damaged bearings and pump parts may cause noise. Good maintenance, even with the additional cost, will maintain high pump efficiencies, reduce pumping cost, and, most of all, provide extended pump life. Regular pump maintenance should include lubricating pump and drive units and checking for excessive drawdown of the supply water level, varied discharge or TDH, and any unusual noise during operation.

12

Ground Water and Wells

12.1 INTRODUCTION

Ground water is the largest and most widely available supply of high-quality freshwater for use in aquaculture. Many fish farmers rely on well water to fill and maintain ponds, and aquaculture often is most profitable in regions where shallow wells yield large volumes of water, such as along floodplains of major rivers. This chapter discusses features of ground-water, aquifers, and function and design of wells.

12.2 AQUIFERS

At some depth beneath most places on the land, soil and rock are saturated with water that has percolated down after infiltrating the ground surface. The saturated volume of geological material is known as an *aquifer*. There are two kinds of aquifers, confined and unconfined. Figure 12-1 shows typical formation of these aquifers. In an unconfined aquifer, the bottom of the aquifer is an impervious confining stratum, and the top of the aquifer is open to the atmosphere through voids in the overlying geological material (Fig. 12-1). The top of the saturated layer is called the *water table*, so an unconfined aquifer is known as a *water table aquifer*.

Water in an unpumped well in an unconfined aquifer stands at the level of the water table. In a confined aquifer, water is trapped between two impervious layers. Water at any point in a confined aquifer is under pressure (Fig. 12-1). This happens because the confining strata act like a pipe with one end higher than the other. Water pressure within the aquifer is a function of elevation difference between a point and the level

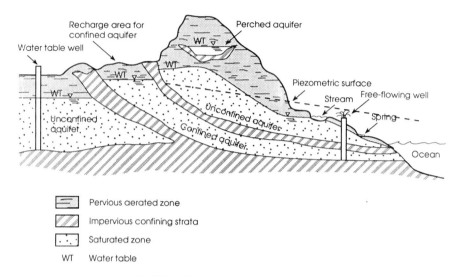

FIGURE 12-1.   An aquifer formation.

of the water table beneath the recharge area. The land area between outcrops of the two confining strata of a confined aquifer is the recharge area, because water infiltrating here enters the confined aquifer. Water stands above the upper confining stratum when a well is drilled into a confined aquifer. The level at which water stands in an unpumped well is the piezometric level. Piezometric levels mapped over the entire aquifer would form a piezometric surface. In some cases, water may flow above the ground surface from a well drilled into a confined aquifer—the piezometric level is above the ground surface.

Confined aquifers often are called artesian aquifers, and free-flowing wells frequently are said to exhibit artesian flow. The piezometric surface slopes downward; that is, water in a well drilled into the aquifer does not rise to the same elevation as that of the water table in the recharge area between the two confining layers (Fig. 12-1). This results primarily from reduction in hydraulic head caused by friction losses between water and solid particles as water moves through the aquifer. Note that a well drilled from the recharge area into a confined aquifer will behave as a water table well. Also, a water table aquifer almost always exits above an artesian aquifer.

As illustrated in Figure 12-1, several water tables may occur beneath a particular place on the ground surface. Small perched aquifers are common on hills where water stands above a small hardpan or impervious layer. Water may leak from an artesian aquifer into aquifers above or below. One or more artesian aquifers may occur below the same point, or artesian aquifers may be absent.

Water in aquifers moves in response to decreasing hydraulic head. Ground water may seep into streams where their bottoms cut below the water table, flow from springs, or seep into the oceans. Ocean water may seep into aquifers, and the saltwater may interface with the freshwater. The position of the interface depends on the hydraulic head of freshwater relative to that of saltwater. As the hydraulic head of an aquifer decreases due to excessive pumping, the saltwater moves inland into the aquifer. Saltwater intrusion in coastal regions, where excessive pumping reduces the hydraulic head of the aquifer, often causes water-quality problems in existing wells.

Aquifers vary greatly in size and depth beneath the land. Some water table aquifers may be only 4 or 5 ft thick and a few acres in extent. Others may be many feet thick and cover many square miles. Artesian aquifers normally are rather large, at least 15–20 ft thick, a few miles wide, and several miles long. Recharge areas may be many miles away from the location of a well drilled into an artesian aquifer. The depth of aquifers below the land varies from a few feet to several thousand feet. For obvious reasons, aquifers near the ground surface are more desirable. The water table is not level but tends to follow the surface terrain, being higher in recharge areas in hills than in discharge areas in valleys (Fig. 12-2). However, the vertical distance from the ground surface to the water table is often less under valleys than under hills. Water table depth also changes in response to rainfall; it falls during prolonged periods of dry weather and rises following heavy rains. The water table also is higher during a cool, wet season than during a warm, dry season.

12.2.1   Ground-Water Storage and Well Yield

Aquifers serve to store and convey water. Contrary to popular belief, water beneath the land does not occur in large flowing streams. The old statement that ground water is mostly rock is a better analogy, because ground water exists in voids that occur between individual grains or between aggregates of grains in soils, in solution channels within limestone, and in fractured rock (Fig. 12-3). A measure of the relative volume of voids is porosity (η), which is defined as the ratio of void volume to bulk volume:

$$\eta = \frac{V_v}{V_b} \qquad (12\text{-}1)$$

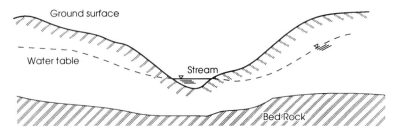

FIGURE 12-2.   Typical location of water table in hills and valleys.

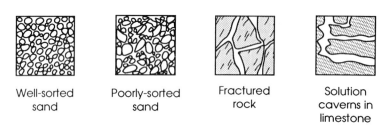

| Well-sorted sand | Poorly-sorted sand | Fractured rock | Solution caverns in limestone |

FIGURE 12-3.   Examples of aquifer materials.

TABLE 12-1   Typical Values of Porosity and Specific Yield of Aquifer Materials. Hydraulic Conductivities of Some of These Soils are Given in Table 1-4

Material	Porosity (% by volume)	Specific Yield S_y
Coarse, repacked gravel	28	0.23–0.25
Coarse sand	39	0.23–0.27
Silt	46	0.08
Clay	42	0.03
Fine-grained sandstone	33	0.21–0.27
Limestone	30	0.14
Loess	49	0.18

Source: Morris and Johnson (1967).

where V_v = void volume and V_b = bulk volume. For example, if $1\,ft^3$ of sand contains $0.35\,ft^3$ of voids, the porosity is 0.35 or 35%. Porosity is a fundamental property of aquifers because it determines how much water can be stored in a unit volume of an aquifer. Porosity values for various types of aquifer material are provided in Table 12-1. The porosity of

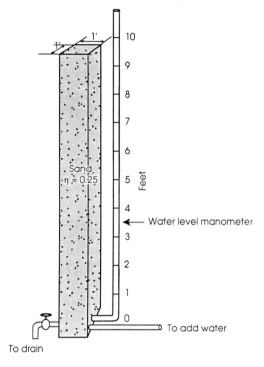

FIGURE 12-4.   A test setup to determine volume–height relationships of aquifer materials.

aquifer materials depends on particle shape and aggregation, degree of compaction and cementation, and particle size distribution. To illustrate the concept of porosity, consider a 1-ft³ container filled with water. Suppose that 1 ft³ of dry sand is poured into the container, and 0.57 ft³ of water is displaced. The solid volume is equal to the volume of water displaced, or 0.57 ft³; bulk volume equals container volume, or 1 ft³; void volume is bulk volume minus solid volume, or 1 ft³ − 0.57 ft³ = 0.43 ft³; porosity is void volume divided by bulk volume, or 0.43 ft³ ÷ 1.0 ft³ = 0.43.

To visualize how an aquifer stores, takes in, and yields water, consider a prismatic section with a 1-ft by 1-ft base from top to bottom of a 10-ft-thick aquifer and think of it as a tank as shown in Figure 12-4. If the tank is filled with water only, the volume (V) of water is simply cross-sectional area (A) times height (h):

$$V = Ah \tag{12-2a}$$

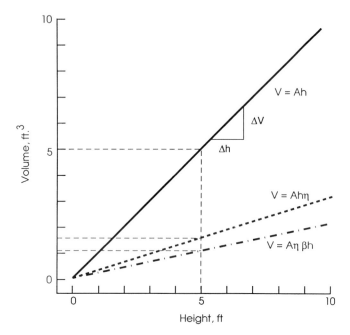

FIGURE 12-5.   Typical volume–height relationships of aquifer materials.

If a small amount of water is added or removed from the tank, height changes slightly, and the change in volume is

$$\Delta V = A \, \Delta h \tag{12-2b}$$

where ΔV = change in volume and Δh = change in height. Volume depends only on h and A, and for the tank the volume may be read directly from the solid line in Figure 12-5 if the height is known.

Suppose the tank is filled with sand of porosity η. Fit the tank with a pipe at the bottom through which water can be added or removed, and attach a manometer to the side of the tank so that the water level within the sand can be read (Fig. 12-4). If a certain amount of water is added to the tank, the relationship between volume and height of water is

$$V = Ah\eta \tag{12-3a}$$

Adding water in increments, measuring the height after each increment, and plotting volume versus height again gives a straight line—the dotted line in Figure 12-5. The relationship of the straight line is expressed as

$$\Delta V = A\eta \, \Delta h \tag{12-3b}$$

The slope of this line is

$$A\eta = \frac{\Delta V}{\Delta h} \qquad (12\text{-}3c)$$

The volume of water in the tank is still a function of height, but the volume for a particular height depends on A, h, and η. Volume can be directly read from height, but now the volume for a particular height is less than before the tank was filled with sand.

When water is drained from the tank, we get less volume of water per unit change in height than was required to fill the tank. This occurs because the sand retains a fraction of the water by adhesion and capillarity. The ratio of water drained to water added is called β, where

$$\beta = \frac{\text{volume of water drained}}{\text{volume of water added}} \qquad (12\text{-}4)$$

In other words, β is the fraction of water in the tank that can be drained by gravity. The fraction of water retained by the sand is $1 - \beta$; this quantity is known as the *specific retention* (S_r), which is the ratio of the volume of water retained against gravity after saturation to its own volume:

$$S_r = \frac{V_r}{V_b} \qquad (12\text{-}5)$$

where V_r = volume of retained water and V_b = bulk volume of the soil. It follows that porosity minus specific retention equals *specific yield* (S_y), the ratio of the volume of water that can be drained by gravity to its own volume:

$$S_y = \frac{V_y}{V_b} \qquad (12\text{-}6)$$

where V_y is the volume of water drained. Then

$$\eta = S_r + S_y \qquad (12\text{-}7)$$

Some representative values for the specific yield (S_y) are given in Table 12-1.

The amount of water that must be added to the tank to restore the height to the level before draining is

$$V = A\eta\beta h \qquad\qquad (12\text{-}8\text{a})$$

Draining increments of water, measuring the value of h each time, and plotting V versus h yields a straight line, semidotted in Figure 12-5. The relationship of the straight line is expressed as

$$\Delta V = A\eta\beta\ \Delta h \qquad\qquad (12\text{-}8\text{b})$$

The slope is

$$A\eta\beta = \frac{\Delta V}{\Delta h} \qquad\qquad (12\text{-}8\text{c})$$

Therefore, change in storage in the empty tank is a function of height only, but change in aquifer storage depends on A, h, η, and β.

Specific yield (S_y) is usually reported as a decimal fraction instead of a percentage. It is a constant property describing an unconfined aquifer's ability to store and release water. Specific yield may be defined as the volume of water that may be drained by gravity from a vertical prism of unit base area extending through an unconfined aquifer in response to a unit lowering of the saturated level or hydraulic head. Recall the tank analogy used above, and think of water as being drained from the pipe in the bottom of the tank. Specific yields depend on grain size, shape and distribution of the pores, degree of compaction of the aquifer materials, and duration of pumping. Specific yields are low for fine materials and high for coarse materials. They also decrease with aquifer depth, reflecting the effect of compaction.

For unconfined aquifers, specific yield is essentially the same as the storage coefficient (S), which is defined as the volume of water that an aquifer releases or stores per unit surface area per unit change in hydraulic head normal to the ground surface.

$$S_y = S = \frac{V}{Ah} \qquad\qquad (12\text{-}9\text{a})$$

or

$$V = AhS_y = AhS \qquad\qquad (12\text{-}9\text{b})$$

The size of the storage coefficient depends on whether an aquifer is confined or unconfined (Heath 1984). In a confined aquifer, the water is trapped between two impervious layers and is under pressure. Water released from storage when the head declines comes from expansion of the water and from compression of the aquifer. Remember that a head decline does not reduce the saturated volume of the aquifer as long as the piezometric surface is above the upper confining stratum. Therefore, if head decreases, pressure is less, and water in a confined aquifer expands. For a confined aquifer with a porosity of 0.2 and water temperature of 59°F (15°C), expansion amounts to $3 \times 10^{-7} \, \text{ft}^3$ of water per cubic foot of the aquifer for each foot of decline in head of the aquifer. To determine the value of storage coefficient (S) of an aquifer because of water expansion, multiply aquifer thickness by 3×10^{-7}. For a 100-ft-thick aquifer the storage coefficient would be 3×10^{-5}. Storage coefficients for confined aquifers range from about 10^{-5} to 10^{-3}, much less than the range of about 0.1 to 0.3 for unconfined aquifers. This indicates that large head changes in a confined aquifer are required to yield substantial volumes of water.

The total load on the top of the confined aquifer is supported partly by the skeleton of the solid material in the aquifer and partly by the water itself. When the water pressure declines in response to reduced head, more of the load must be supported by the aquifer material. The aquifer material is compressed (particles are distorted and pressed together) and porosity is reduced. Water forced out from pores when porosity is reduced represents the part of the storage coefficient resulting from aquifer compression. Continued removal of water from a confined aquifer will cause the piezometric surface to drop below the upper confining stratum. The storage coefficient will then increase drastically to that of a confined aquifer. Permanently reducing the piezometric surface below the confining bed by continuous withdrawal of water from wells can lead to compression of the aquifer and land subsidence, which may cause *sink holes*.

In summary, the source of water from unconfined aquifers is gravity drainage through the aquifer materials. The volume of water obtained from the expansion of water and from compression of the aquifer is insignificant. Hence, the storage coefficient has essentially the same value as the specific yield. In a confined aquifer with the piezometric surface above the upper confining stratum, water is derived from expansion of water and compression of the aquifer. Therefore, the storage coefficient is much smaller than that of an unconfined aquifer. Once the piezometric surface falls below the upper confining stratum, the source of water from the aquifer is gravity drainage, and the storage coefficient increases to take essentially the same value as the specific yield.

Example 12-1: Water Storage and Water Yield of a Water Table Aquifer

A water table aquifer is 25 ft thick and composed of coarse sand. From Table 12-1 its porosity is 0.39; the specific yield S_y is 0.25. Determine water storage and yield.

Solution   The storage volume per square mile of the aquifer is

$$V = Ah\eta$$
$$= (1\,\text{mi}^2 \times 640\,\text{acres/mi}^2 \times 43{,}560\,\text{ft}^2/\text{acre}) \times 25\,\text{ft} \times 0.39$$
$$= 271{,}814{,}400\,\text{ft}^3$$

The amount of water that can be removed from the aquifer by wells is

$$V = AhS_y$$
$$= (1\,\text{mi}^2 \times 640\,\text{acres/mi}^2 \times 43{,}560\,\text{ft}^2/\text{acre}) \times 25\,\text{ft} \times 0.25$$
$$= 174{,}240{,}000\,\text{ft}^3$$

The aquifer would yield 5,000 gpm continuously to wells within the square-mile area for about six months without recharge, as shown below:

$$174{,}240{,}000\,\text{ft}^3 \times 7.48\,\text{gal/ft}^3 = 1{,}303{,}315{,}200\,\text{gal}$$
$$1{,}303{,}315{,}200\,\text{gal} \div (5{,}000\,\text{gpm} \times 1{,}440\,\text{min/day}) = 181\,\text{days or 6 months}$$

12.2.2   Hydraulic Head of Ground Water

Total hydraulic head of flowing water consists of elevation head, pressure head, and velocity head, as discussed in Chapter 9. Velocity head may be ignored in movement of ground water because its velocity is slow, and the total hydraulic head of an aquifer is

$$h_t = h_e + h_p \qquad (12\text{-}10)$$

where
h_t = total head (ft)
h_e = elevation head measured at the top of aquifer above a reference plane (ft)
h_p = pressure head in the aquifer (ft)

It is convenient to report hydraulic head of an aquifer in terms of height. The total head for an unconfined aquifer is determined by measuring the elevation of the surface of the water table in an unpumped well relative to the elevation of a reference plane because no pressure head exists in the aquifer. The mean sea level (m.s.l.) is usually used as the reference plane. In practice, the elevation of the top of the well casing is known, and the distance from the top of the casing to the unpumped water level in the well is measured. Total head is measured by subtracting the depth to the water level from the elevation of the top of the casing. The water depth in shallow-water-table wells may be measured by attaching a weight to the end of a steel tape and coating about 0 to 3 ft of the tape with carpenter's chalk. The tape is lowered until the weight strikes water. An even foot mark on the tape is aligned with the top of the casing. The tape is then withdrawn and the wetted distance is subtracted from the total measured distance. For deeper wells, an electrical device is available for measuring water depth. When the electrode at the end of the tape strikes water, a current flows through a circuit and is recorded on an ammeter.

In confined aquifers, the water level may be above the ground surface. In this case, the water pressure at the top of the casing can be measured. The pressure in a capped well casing in a confined aquifer capable of producing free-flow is termed the *shut-in head*. To illustrate total head of an aquifer, consider a confined aquifer in which the top of a well casing is 200.75 ft above the mean sea level. The unpumped water level in the well stands 10.11 ft below the top of the casing. The elevation of the top of the aquifer is 104.15 ft above m.s.l. The elevation of the piezometric surface or the total head is 200.75 ft − 10.11 ft = 190.64 ft above m.s.l. The pressure head in the well is thus 190.64 ft − 104.15 ft = 86.49 ft.

12.2.3 Movement of Ground Water

As stated earlier, aquifers store and convey water. Water is stored in voids and flows through the voids in the direction of decreasing hydraulic head. The rate of ground-water movement is determined by the hydraulic conductivity of the aquifer material and the hydraulic gradient. The normal range of ground-water movement is from less than a few feet per year to several feet per day (Todd 1980). The hydraulic conductivity of an aquifer with known porosity may be measured by using a tracer between two measuring points of an aquifer. Tracer materials may include dyes, salts, and low-level radioactive materials. However, the most reliable method is by pumping test wells (Wenzel 1942). In consideration of ground-water movement, hydraulic conductivity values are usually expressed in feet per day or gallons per day per square foot of aquifer cross-sectional area. Typical values of hydraulic conductivity of aquifer materials are presented in Table 12-1.

The Darcy equation (see Chapter 1) is used to estimate the amount of water moving through an aquifer:

$$Q = KA\frac{\Delta h}{\Delta L} = KiA \qquad (12\text{-}11a)$$

or

$$v = \frac{Q}{A} = K\frac{\Delta h}{\Delta L} = Ki \qquad (12\text{-}11b)$$

where Q = aquifer flow rate (ft³/day)
   K = hydraulic conductivity of aquifer material (ft/day)
   A = flow direction cross-sectional area of aquifer (ft²)
   Δh = hydraulic head change between two test wells (ft)
   ΔL = distance between two test wells (ft)
   v = aquifer flow velocity (ft/day)
   i = hydraulic gradient between two test wells in aquifer

Equation (12-11) states that the flow velocity is equal to the product of the hydraulic conductivity (K) and the hydraulic gradient. For example, consider a 30-ft-thick unconfined aquifer. The aquifer material consists of coarse sand with a hydraulic conductivity of 13.7 ft/day. Well A at the upstream has a total head of 125.12 ft, and well B at the downstream has a total head of 113.38 ft. The distance between the two wells is 1,250 ft. The hydraulic gradient between the two wells is

$$\frac{125.12\,\text{ft} - 113.38\,\text{ft}}{1,250\,\text{ft}} = 0.0094\,\text{ft/ft, or } 50\,\text{ft/mile}$$

The flow rate in the aquifer per unit width is

$$Q = (13.7\,\text{ft/day})(30\,\text{ft} \times 1\,\text{ft})(0.0095\,\text{ft/ft}) = 3.9\,\text{ft}^3/\text{day/ft}$$

12.3 WELLS

Wells are constructed to permit withdrawal of water from aquifers. The first wells were holes dug deep enough into the ground to intercept and extend into an aquifer. Water filled the part of the hole that was below the water table or piezometric surface, and water was withdrawn with a bucket attached to a rope. Today, most wells consist of a borehole cased with pipe. A well screen on the end of the pipe allows water from the aquifer to enter the well. A pump provides energy to remove and pressurize water from the well. Aquifer characteristics determine the

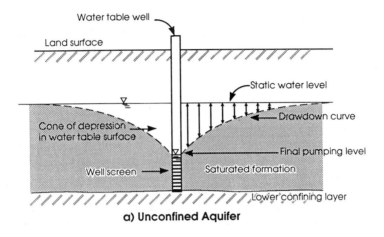

a) Unconfined Aquifer

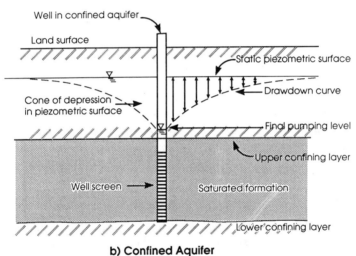

b) Confined Aquifer

FIGURE 12-6. Illustration of drawdown, pumping level, and cone of depression for wells in unconfined and confined aquifers.

amount of water that can be withdrawn by wells at a site. Some aquifers are more suitable than others for supplying water to wells. Well design depends to a large extent on aquifer characteristics.

This section presents information on relationships between well and aquifer, well hydraulics, well design and construction, and well testing. An aquaculturist will seldom design and construct a well; a contractor would usually be hired. Nevertheless, knowledge of well function and design is valuable when one negotiates with a well-drilling firm to install a well. The water well manuals by Driscoll (1986) and the U.S. Bureau of Reclamation (1981) and other references are recommended to readers wanting further details.

12.3.1 Well Hydraulics

Wells in a water table aquifer and in a confined aquifer are depicted in Figure 12-6. The water table well, when not being pumped, will have water standing in the well at the elevation of the water table. However, the water level in an unpumped well in a confined aquifer will stand above the top of the aquifer; in some cases, water will flow freely from a well in a confined aquifer where the piezometric level is above the top of the well. (The elevation to which water will rise in a well casing penetrating a confined aquifer is called the piezometric level.) If piezometric levels are plotted over an area, the levels at various wells may be connected to depict the piezometric surface.

Well Drawdown and Cone of Depression
When pumping is initiated in a well, the pump must impart enough energy to the water to raise it to the ground surface. This lowers the water level in the well casing. The hydraulic gradient created by drawdown of water in the casing permits water from the aquifer to flow through the well screen. The well screen is cylindrical in cross section, so in the vicinity of the well water will flow laterally to the well screen from all directions. If the well is discharging at a constant rate, the flow of water through the aquifer and into the well screen must equal the well discharge. Consider water flowing through a saturated layer of aquifer surrounding the well screen, as shown in Figure 12-7. Flow converges toward the well through successive cylindrical surfaces that become progressively smaller in area. In Figure 12-7, let $R_1 = 5$ ft, $R_2 = 10$ ft, $h_1 = 8$ ft, and $h_2 = 10$ ft. The area of a cylindrical surface is $A = \pi D h$ and $D = 2R$; hence, the areas of flow for cylindrical radii of 5 ft and 10 ft are 251 ft^2 and 628 ft^2, respectively. Discharge is area times velocity, so for the same discharge the velocity must be about 2.5 times as great at 5 ft from the well as at 10 ft. According to the Darcy equation [equation (12-11)], velocity is the product of the hydraulic conductivity and the hydraulic gradient. The hydraulic conductivity of an aquifer does not change, and the length of flow is fixed by distance from the well. The only way water can accelerate as it flows toward the well screen is for drawdown to continually increase, which increases the hydraulic gradient as the well is approached. Pumping results in dewatering of a cone-shaped

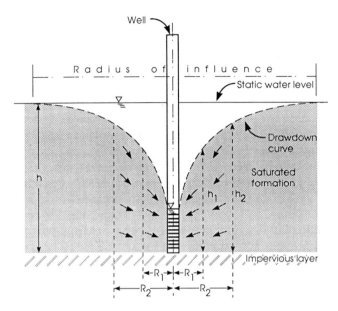

FIGURE 12-7. Illustration of the drawdown curve and the radius of influence from a water table well.

region around a water table well. Confined aquifers will not show a cone-shaped region if the pumping level in the well stays above the top of the upper confining layer. However, if the piezometric surface is mapped in the vicinity of a pumped well in a confined aquifer, the piezometric surface will also show a cone-shaped region (Fig. 12-6b).

The shape of the cone of depression depends on the pumping rate, duration of pumping, aquifer characteristics, and amount of recharge occurring within the area of drawdown. The maximum radius of the area in which drawdown of the static level occurs is called the radius of influence (Fig. 12-7). During continuous pumping, the cone of depression will increase for a while, but equilibrium finally will be reached when the cone of depression expands to intercept recharge that is equal to the pumping rate. Recharge in an aquifer results from natural flow of water in the aquifer into the cone of depression, seepage into the cone from surface water, leakage from other aquifer formations, and infiltration of precipitation in the recharge area.

The specific yield of an aquifer is defined as the quantity of water per unit volume of saturated material that can be drained by gravity. In the absence of recharge during pumping, the volume of the cone of depression may be estimated as the total volume of water pumped divided by the specific yield. Suppose a well pumping 500 gpm for 6 hr discharges 24,061 ft³ of water. If the specific yield of the aquifer is 0.05, then the volume of the cone of depression is 24,061 ÷ 0.05 = 481,220 ft³. The cone of depression for a comparable well in an aquifer with a specific yield of 0.1 is only half as large.

Water in an aquifer flows from higher hydraulic head to lower, so when pumping stops in a well the water will continue to converge into the well and the water level throughout the cone of depression will rise until the effects of drawdown are no longer apparent. Of course, if the rate of water removal by wells from an aquifer exceeds the recharge rate, the static water level in an aquifer will eventually decline. This is why wells in small aquifers may dry up during droughts, and the water table may steadily fall in a large aquifer subjected to excessive pumping.

The flow rate to a well from an unconfined aquifer is determined from the equation

$$Q = \frac{K(H^2 - h^2)}{62 \ln(R/r)} \tag{12-12}$$

where
Q = rate of flow (gpm)
K = hydraulic conductivity (ft/day)
H = initial water level above impervious layer (ft)
h = final water level in the well above impervious layer (ft)
R = radius of influence (ft)
r = radius of well (ft)

In a confined aquifer the flow rate is

$$Q = \frac{Kd(H - h)}{31 \ln(R/r)} \tag{12-13}$$

where
d = thickness of confined layer (ft)
H = initial piezometric level above top of aquifer (ft)
h = final piezometric level above top of aquifer (ft)

Example 12-2: Computing the Flow Rate from an Unconfined Aquifer

A well is installed in an unconfined aquifer with a 55-ft-thick saturated zone of coarse sand. The well is 12 in. in diameter, and the final

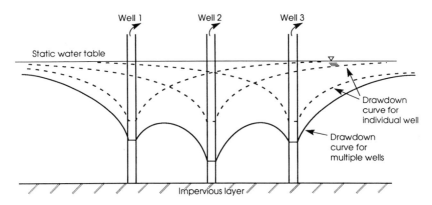

FIGURE 12-8.   Illustration of well interference by multiple wells.

drawdown in the well is 40 ft below the static water table. Find the flow rate from the well.

Solution   Hydraulic conductivity of the aquifer is 13.7 ft/day (Table 12-1). Assume that the radius of influence is 1500 ft. The final water level in the well above the impervious layer is $h = 55\,\text{ft} - 40\,\text{ft} = 15\,\text{ft}$, and the well radius is 0.5 ft. The flow rate from the well is

$$Q = \frac{13.7(55^2 - 15^2)}{62 \times \ln(1500/0.5)} = 77\;\text{gpm}$$

Because the flow rate is inversely proportional to the logarithm of the radius of influence, an error in estimating the radius of influence results in very small error. In the example if the radius of influence is assumed to be 3000 ft, the flow rate would be 71 gpm.

A well may be drilled within the radius of influence of other wells. If all wells in a given aquifer field are pumping simultaneously, their cones of depression will overlap and the wells are said to *interfere* (Fig. 12-8). Well interference reduces available drawdown and well yield. A common problem of multiple wells is to determine the effect of interference on the drawdown and yield of each well. The drawdown at any point in the area of influence due to pumping of multiple wells is equal to the sum of the drawdown caused by each well individually (Todd 1980).

$$D_t = D_1 + D_2 + \cdots + D_n \tag{12-14}$$

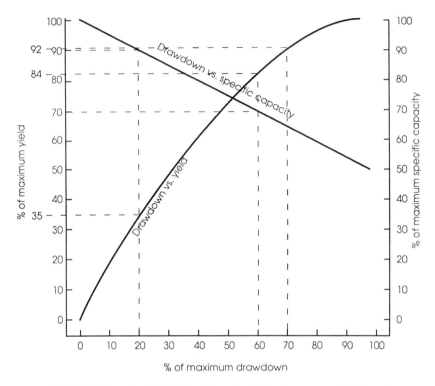

FIGURE 12-9.   Typical relationships of well drawdown vs. well yield and specific capacity of an unconfined aquifer.

where D_t = total drawdown at a given point
D_i = drawdown at the point by well i
n = total number of wells

The number of wells and distance between wells are important in determining the drawdown in multiple wells. Well interference is reduced by increasing the spacing between wells and by locating wells along a line rather than a circle or grid.

Well Drawdown and Well Yield
Specific capacity of a well is determined by dividing well yield by drawdown in the well casing. The specific capacity of a well is constant at any pumping rate as long as the aquifer is not dewatered. Hence, in a confined aquifer with the pumping level above the top of the aquifer, doubling the drawdown will double the yield, as shown in equation (12-13). For a water table aquifer, dewatering of the aquifer during pumping

influences the ratio of drawdown to well yield. Doubling drawdown will not double yield, and specific capacity decreases in direct proportion to drawdown (Fig. 12-9). Maximum drawdown occurs when the pumping level is lowered to the bottom of the well; maximum yield occurs at maximum drawdown. Of course, at 0% drawdown there is no yield. Suppose that an aquifer is 40 ft thick, and a well extended to the bottom of the aquifer has a drawdown of 8 ft and a yield of 200 gpm. The drawdown is 20% of maximum, so the yield is 35% of maximum. Then the maximum yield of the well is 200 gpm ÷ 35% = 571 gpm. If drawdown is increased to 24 ft or 60% of maximum, 84% of maximum yield will be attained. The yield at 60% drawdown is 571 gpm × 84% = 480 gpm.

Maximum specific capacity corresponds to 0% of maximum drawdown, for with no drawdown there is no reduction in the saturated thickness of the aquifer. Minimum specific capacity is 50% of maximum, and it occurs at 100% of maximum drawdown. In the example, specific capacity would be 90% of maximum at 20% of maximum drawdown. Increasing drawdown to 60% of maximum would reduce specific capacity to 70% of maximum. When drawdown is 70% of maximum, the yield is about 92% of maximum. Therefore, it is not economical to increase drawdown an additional 30% just to increase yield by 8%. Increased drawdown requires more pumping head and, consequently, pumping energy. This means that well screens should be positioned in the bottom 30–40% of the saturated thicknesses of water table aquifers.

12.3.2   Well Design

For the typical well, a borehole is drilled to the bottom of the aquifer, a well screen is positioned in the aquifer, and the well casing extends upward to the ground surface (Fig. 12-10). The space between the wall of the borehole and the well screen and well casing must be filled. Sand or gravel is applied around the screen and the casing to extend above the saturated thickness of the aquifer. If it is desired to seal the aquifer, a layer of bentonite pellets are added. The pellets absorb moisture and swell to seal off the opening around the casing. Next, clay backfill may be used to close the opening within a few feet of the ground surface. The top layer of the hole is filled with cement grout. The information needed to design a well includes (a) thickness, character, and sequence of materials above the aquifer; (b) properties of the aquifer, including thickness, transmissivity, specific yield, and size and gradation of aquifer material; (c) aquifer depth and fluctuation; and (d) water quality. This information may often be obtained from wells in the vicinity that penetrate the aquifer

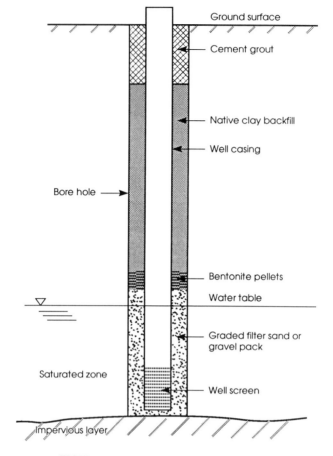

Ground surface

Cement grout

Native clay backfill

Well casing

Bore hole

Bentonite pellets

Water table

Graded filter sand or gravel pack

Saturated zone

Well screen

Impervious layer

FIGURE 12-10.   Features of a typical well.

in question. If not, a test well may be constructed; a small-diameter test well would be inexpensive when compared with the cost of the production well being designed. Other factors to consider are desired yield; pump size; well diameter and depth; length, diameter, and type of casing; length, diameter, slot size, and placement of well screen; and need for gravel pack.

Well Depth, Diameter, and Casing

The well should be completed to the bottom of the deepest saturated layer from which water will be withdrawn. As mentioned, well diameter has no appreciable effect on well yield, and diameter is regulated by

TABLE 12-2   Nominal Sizes of Pump Bowls and Minimum Casing Diameters for Wells with Different Yields

Expected Well Yield (gpm)	Nominal Diameter of Pump Bowl (in.)	Pump Chamber	
		Diameter of Casing (in.)	Surface Casing Diameter (in.)[a]
<100	4	6 ID	8–10
70–175	5	8 ID	10–12
150–400	6	10 ID	12–14
350–650	8	12 ID	14–16
600–900	10	14 OD	18–20
850–1,300	12	16 OD	18–22
1,200–1,800	14	20 OD	22–24
1,600–3,000	16	24 OD	26–28
3,000–5,000	20	28 OD	30–32

Source: Driscoll (1986).

[a] For naturally developed wells. Surface casing must be larger for gravel-packed wells.

desired yield and depth. The two factors, yield and depth, determine the pump size necessary to produce the desired yield, and the well casing must be large enough to accommodate the pump and discharge. Information on well yield, pump diameter, and casing diameter are provided in Table 12-2. The pump chamber casing should have a nominal diameter at least 2 in. larger than the nominal diameter of the pump bowls.

All wells have at least one casing, the pump chamber casing or pipe-flow casing, which furnishes a conduit between the aquifer and the ground surface. The casing also prevents water from seeping into the aquifer from strata above, and it supports the borehole. Wrought-iron pipe with wall thickness of at least 0.25 in. is normally used for well casing. Small pipe sections may be joined with couplings, but it is best to weld casing joints for diameters of 6 in. or larger. Plastic pipe also may be used as well casing; this practice is popular for wells of diameter 6 in. or smaller. Other types of casings include pipe of stainless steel and other alloys, sheet steel, asbestos cement, fiberglass, copper alloys, ceramic clay, and concrete.

The casing extends to the depth where the pump will be set. The pump depth must be selected after consideration of the minimum expected water level in the well over a long period, expected drawdown, possible interference from other wells, and required pump submergence. The casing must be straight and plumb; deviation from the vertical should not exceed two-thirds of the inside diameter of the casing per 100 ft of depth.

A second casing, called a surface casing, is sometimes placed around the pump chamber casing. Minimum diameters of surface casings are given in Table 12-2.

Well-Screen Selection

The well screen provides the open area through which water from the aquifer flows into the well. The well screen also supports the borehole and prevents unconsolidated material from the aquifer from entering the well. The term *open area* refers to the total open space in the screen through which water can flow.

Perforated casing or slotted pipe may be used as a substitute for well screen. Slots are cut into pipe with a saw or torch. The size of the slots varies, and even with large slots the open area of the completed slotted pipe do not exceed 10–12% of the total screen area. The open area of slotted pipe is less than the porosity of most aquifers, so drawdown is often increased by use of slotted pipe. A similar method uses screens formed by machine punching or stamping openings in pipe. The slots are uniform in size, and the edges of openings can be smoothed to reduce corrosion. With large slots, the open area may be as high as 20%. Most of these screens are made of lightweight material and are useful only in shallow wells.

Pipe-base well screens are made by winding wire around perforated pipe. A wide range of slot size is possible, and slot size can be controlled. Suitable open area can be achieved, and the pipe base contributes strength. However, pipe base screens clog badly in aquifers with fine materials. The best well screens are made by wrapping continuous lengths of triangular wire around a base of vertical rods (Fig. 12-11). Such screens are called continuous-slot screens. The triangular-shaped wire permits

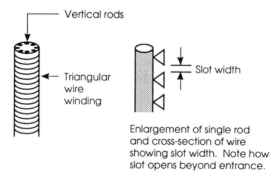

FIGURE 12-11.   Illustration of a well screen. (*Source: Driscoll 1986.*)

TABLE 12-3   Acid and Corrosion Resistance of Well-Screen Material

Material	Acid Resistance	Corrosion Resistance
Low-carbon steel	Poor	Poor
Toncan and Armco iron	Poor	Poor
Admiralty red brass	Good	Good
Silicon red brass	Good	Very good
304 stainless steel	Good	Very good
Everdune bronze	Very good	Very good
Monel metal	Very good	Very good
Super nickel	Very good	Very good
Plastic	Very good	Very good

Source: U.S. Bureau of Reclamation, former Water and Power Resources Service, U.S. Dept. of the Interior (1981).

sand grains that enter the openings to pass through without clogging the screen. Slot sizes of 0.006 to 0.250 in. generally are available. Satisfactory well screens are also manufactured from plastic pipe by cutting slots into the pipe.

Ground waters that contain more than 2 mg/L dissolved oxygen, 50 mg/L carbon dioxide, 0.5 mg/L hydrogen sulfide, 50 mg/L chloride, and 500 mg/L total dissolved solids or have pH values less than 7 are potentially corrosive to some well-screen materials. The acid and corrosion resistances of common materials used for manufacturing well screens are provided in Table 12-3. Unfortunately, material cost increases as resistance to acidity and corrosion increases. Of course, plastic well screens are comparatively inexpensive and highly resistant to corrosion and low pH. Mineral deposition on the well screen and in the slots of the screen is called *encrustation*. Ground waters with carbonate hardness above 300 mg/L, iron in excess of 2 mg/L, manganese concentrations greater than 1.0 mg/L, or pH values above 7.5 have the potential for encrustation of well screens.

Entrance velocity through the well screen should not exceed 6 ft/min (Driscoll 1986). A high velocity of water entering the well screen encourages corrosion and encrustation. High entrance velocity also increases drawdown necessary for a given well yield. The expected well yield (in ft^3/min) divided by the maximum entrance velocity, 6 ft/min, gives the necessary open area of the screen (in ft^2). Many times, 50–100% more open area than the minimum is used to avoid effects of clogging and encrustation. Slot width is expressed in thousandths of an inch; a number 10 slot screen has a slot width of 0.010 in., a number 20 slot screen has a slot width of 0.020 in. and so on. The amount of open

TABLE 12-4   Minimum Opening Areas of Continuous Slot Screens (in²/ft) and the Percentage of Opening Area of Screen (in parentheses) with Continuous Slot

Nominal Screen Diameter (in.)	Slot size[a] (1/1,000 in.)						
	20	30	40	50	60	95	125
4	44	58	72	78	90	106	100
	(25)	(33)	(41)	(45)	(52)	(61)	(58)
6	45	61	77	88	100	127	127
	(18)	(25)	(31)	(35)	(40)	(51)	(51)
8	58	80	98	114	135	165	166
	(18)	(25)	(30)	(35)	(41)	(51)	(51)
10	72	100	122	143	135	179	207
	(18)	(25)	(30)	(35)	(33)	(44)	(51)
12	69	77	99	117	135	182	214
	(14)	(16)	(21)	(24)	(28)	(38)	(45)
16	68	97	124	146	169	228	268
	(11)	(16)	(21)	(24)	(28)	(38)	(45)

Source: Driscoll (1986).

[a] Screens with other types of slot are found in the reference.

TABLE 12-5   Recommended Minimum Diameters of Well Screens for Various Well Capacities

Discharge (gpm)	Minimum Nominal Diameter (in.)
<50	2
50–125	4
126–350	6
351–800	8
801–1,400	10
1,401–2,500	12
2,501–3,500	14
3,501–5,000	16
5,001–7,000	18
7,000–9,000	20

Source: U.S. Bureau of Reclamation, former Water and Power Resources Service, U.S. Dept. of the Interior (1981).

area per linear foot of screen depends on screen diameter, slot width, and screen type. Open areas per linear foot in continuous-slot screens of different slot widths are given in Table 12-4. Well-screen manufacturers provide similar data for their screens.

Screen diameters are selected on the basis of desired well yield and aquifer thickness. Recommended minimum diameters for different well yields (Table 12-5) may have to be adjusted upward for aquifers that necessitate screens with small slot widths. The necessary open area in square feet divided by the open area per linear foot of screen indicates the number of linear feet of well screen needed for the desired yield without exceeding the recommended maximum entrance velocity of 6 ft/min.

Example 12-3: Determining Well Screen Length

Suppose a yield of 300 gpm is expected from a number 40 slot screen of a well. Reference to Table 12-5 suggests that the screen should be no less than 6 in. in diameter. Determine the length of the well screen.

Solution: The flow rate is

$$300 \, \text{gpm} \div 7.48 \, \text{gal/ft}^3 = 40.0 \, \text{ft}^3/\text{min}$$

The open area of screen is

$$40.0 \, \text{ft}^3/\text{min} \div 6 \, \text{ft/min} = 6.68 \, \text{ft}^2 \text{ open area}$$

From Table 12-4, a number 40 slot screen of 6 in. diameter has 77 in.^{2} or 0.534 ft^2 open area per linear foot. The required screen length is

$$6.68 \, \text{ft}^2 \div 0.534 \, \text{ft}^2/\text{linear foot} = 12.5 \, \text{ft}$$

Drawdown in a water table aquifer should not exceed two thirds of the aquifer thickness, so screen length should be about one third the aquifer thickness. In an aquifer where a screen of small slot width is necessary, the required length of minimum diameter screen may exceed the permissible screen length for the aquifer. In this event, a shorter length of larger-diameter screen should be selected.

The slot width of a screen is determined by the particle size of aquifer material. Information on particle size is often obtained by sieve analysis of samples from the aquifer. In a sieve analysis, the sample is dried, gently pulverized, and separated into particle-size classes by allowing particles to pass through a set of nested sieves. A set of five or six sieves that covers the particle-size range of the sample is nested with the mesh size increasing from bottom to top of the sieve stack, and the stack of sieves is placed in the bottom pan. A sample weighing 200–400 g (0.5–1.0 lb) is placed on the top sieve, and the sieve stack is shaken and

TABLE 12-6   The Tyler and U.S. Standard Sieve Classifications and Actual Sieve Openings

Aquifer Material	Tyler Sieve No.	U.S. Standard Sieve No.	Actual Sieve Opening (in.)
Fine gravel	4-mesh	No. 4	0.187
Very fine	6-mesh	No. 6	0.132
gravel	8-mesh	No. 8	0.0937
Very coarse	9-mesh	No. 10	0.0787
sand	10-mesh	No. 12	0.0661
	12-mesh	No. 14	0.0555
	14-mesh	No. 16	0.0469
Coarse	16-mesh	No. 18	0.0394
sand	20-mesh	No. 20	0.0331
Medium	28-mesh	No. 30	0.0234
sand	32-mesh	No. 35	0.0197
	35-mesh	No. 40	0.0165
	42-mesh	No. 45	0.0139
	48-mesh	No. 50	0.0117
Fine	60-mesh	No. 60	0.0098
sand	65-mesh	No. 70	0.0083
	80-mesh	No. 80	0.0070
	100-mesh	No. 100	0.0059

vibrated to cause the sample particles to pass through the sieves. Shaking can best be done with a mechanical sieve shaker, but shaking by hand is acceptable. Particles larger than the mesh opening of a particular sieve are retained on the sieve. Usually, some fine particles will pass through the finest-mesh sieve and accumulate in the bottom pan. The amount of material retained on each sieve is removed and weighed. The series designations for the Tyler and U.S. standard sieve classifications are presented in Table 12-6. The classification of soil particles based on their size and used by the U.S. Geological Survey is presented in Table 12-7.

Data analysis consists of adding the weight of material in the largest-mesh sieve to that of the second-largest-mesh sieve, and so forth, and plotting the cumulative percentage retained on each sieve versus particle size. Suppose a 250-g sample from an aquifer is air-dried at 140°F in an oven. The material is then separated by sieving with 8-, 10-, 14-, 20-, and 32-mesh Tyler series sieves. Results are summarized in Table 12-8, and the percent cumulative retention is plotted against opening size (particle size), as shown in Figure 12-12. The uniformity coefficient is defined as the 40% size of the sample divided by the 90% size of the sample. In this sieve analysis, the 40% size is 0.064 in., and the 90% size is 0.039 in.; the uniformity coefficient is 1.64. The uniformity coefficient of unity means

TABLE 12-7   Particle Size Classification of Soils

Material	Particle Size (mm)
Clay	<0.004
Silt	0.004–0.062
Very fine sand	0.062–0.125
Fine sand	0.125–0.25
Medium sand	0.25–0.5
Coarse sand	0.5–1.0
Very coarse sand	1.0–2.0
Very fine gravel	2.0–4.0
Fine gravel	4.0–8.0
Medium gravel	8.0–16.0
Coarse gravel	16.0–32.0
Very coarse gravel	32.0–64.0

Source: Morris and Johnson (1967).

TABLE 12-8   Results of a Sieve Analysis Illustrated in Section 12.3.2

Mesh	Opening (in.)	Weight Retained (g)	Cumulative Weight (g)	Cumulative Retention (%)
8	0.0937	15	15	6.0
10	0.0661	77	92	36.9
14	0.0469	103	145	78.3
20	0.0331	44	239	96.0
32	0.0197	8	247	99.2
Pan	0.0000	2	249	100.0

Note: 1 g of material was lost in this analysis. This results because some particles may get trapped in the sieve openings.

the material is practically all of the same size, and a large coefficient indicates a large range in sizes.

Where the uniformity coefficient of a sample is 5 or less, a well screen slot size is selected that will retain 40–50% of the aquifer material. If the uniformity coefficient of a sample is greater than 5, a slot size that will retain 30–50% of the aquifer material is selected. Of course, during initial pumping, fine particles from the aquifer enter the well and will be pumped out. After a while, the well water will be free of these particles. The process of removing from the aquifer those particles smaller than openings in the well screen is known as *well development*. To illustrate selection of a well screen, consider Figure 12-12. For this aquifer the

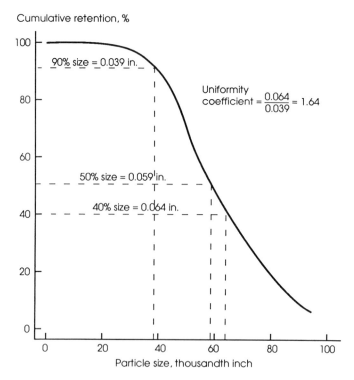

Cumulative retention, %

90% size = 0.039 in.

Uniformity
coefficient $= \dfrac{0.064}{0.039} = 1.64$

50% size = 0.059 in.

40% size = 0.064 in.

Particle size, thousandth inch

FIGURE 12-12.   Results of sieve analysis of an aquifer material illustrated in Section 12.3.2.

uniformity coefficient is 1.64, so a screen that will retain 40–50% of the aquifer material is suitable. A screen with a slot width of 0.064 in. would retain 40% of the particles, and a 0.059-in. slot would retain 50%. The choice is quite narrow, and from Table 12-4 a number 60 slot screen (0.060-in. slot size) is suitable.

An aquifer may have different particle-size distributions at different depths and at different locations. Therefore, it is best to have particle-size data from several depths in the aquifer at the location where a well will be constructed. It is possible to join well screens of different slot sizes when the particle-size distribution of an aquifer is layered. This practice is not worth the effort unless particle sizes in the two layers are considerably different and the different layers are more than 5 ft thick. If screens of different slot sizes are to be joined and fine material lays over coarse material, the screen for the layer of fine material should extend 2 ft into the layer of coarse material. This prevents the finer material that may settle downward during well development from entering the well through

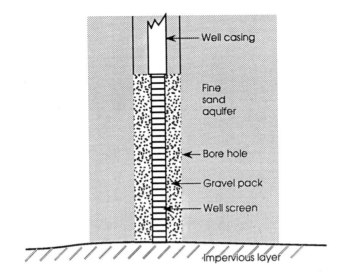

FIGURE 12-13.   Illustration of gravel packing around a well screen.

the coarse screen. When joining screens of different slot sizes, the slot size of coarse screen should not be more than twice the slot size of the fine screen. If the slot size difference is too great, a 1- or 2-ft length of intermediate screen may be inserted between the fine and coarse screens.

Gravel Pack

The gravel pack procedure is used to replace material in the vicinity of the well screen with coarser material of greater hydraulic conductivity (Fig. 12-13). A gravel pack stabilizes the aquifer and minimizes sand pumping. Because the gravel pack is next to the screen and tends to hold back fine particles, it permits a larger screen slot to be used. This increases the open area per linear foot of screen. The gravel provides an annular zone of high hydraulic conductivity that increases the effective radius of the well and, consequently, the well yield. A gravel pack would be useful in a formation consisting of very fine sand. Gravel packs are not used unless they are absolutely necessary, for they are difficult and expensive to install. Procedures for selecting gravel size and well screen slot size for a gravel-packed well in a specific aquifer are found in Anderson (1984) and Driscoll (1986).

Screen Length and Placement

Aquifers may be put into one of four categories for screen length and placement: homogeneous confined, nonhomogeneous confined, homo-

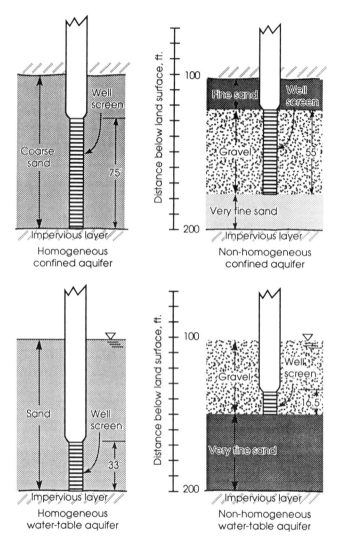

FIGURE 12-14.   Examples of well-screen placement in different types of aquifer formation. (*Source: Driscoll 1986.*)

geneous water table, and nonhomogeneous water table. For a homogeneous confined aquifer, the lower three fourths of the aquifer thickness should be screened. The most permeable portions of a nonhomogeneous confined aquifer should be screened. As shown earlier, the optimal drawdown for a homogeneous water table aquifer is 67% of the saturated thickness of the aquifer. Therefore, the bottom one third of a homogeneous water table aquifer should be screened. For a nonhomogeneous water table aquifer, the most permeable layer should be screened. The different possibilities are illustrated in Figure 12-14 for aquifers 100 ft thick. For the homogeneous confined aquifer, the lower 75 ft is screened. In the nonhomogeneous confined aquifer, the bottom 25 ft is very fine sand and the top 20 ft is clay. The 55-ft layer between the fine sand and very fine sand comprises gravel. The hydraulic conductivity of the gravel layer will be greatest, so the screen is extended throughout the gravel layer. For the homogeneous water table aquifer, the screen is placed in the bottom 33 ft. The nonhomogeneous aquifer has a 50-ft layer of gravel above a 50-ft layer of very fine sand. Because sand is less permeable than gravel, a screen placed in the sand layer would not be efficient. Hence, only the lower one third (16.5 ft) of the gravel stratum is screened.

12.4   PUMPING OF GROUND WATER

A pump transfers energy from a power unit to water in order to lift and pressurize water from a well. The centrifugal pump is by far the most common in pumping ground water. In shallow wells, the pump may be positioned on the ground surface. It produces a partial vacuum at its intake, and atmospheric pressure acting on the water surface in the well forces water up the casing and out of the well. More often the pump is placed in the well. In the vertical turbine pump, the power unit is mounted on the ground surface and connected to the pump in the well by a shaft running through the well casing. Alternatively, in the popular submersible pump, an electric motor is located just below the pump intake and the pump and motor assembly are positioned in the well casing (see Chapter 11).

12.4.1   Pumping Power

The basic equation to determine the energy requirement of a pump given in Chapter 11 is

$$BP = \frac{\text{TDH} \times Q}{3,960E_p} \tag{11-8}$$

where BP = brake power (hp)
Q = well yield (gpm)
TDH = total dynamic head (ft)
E_p = pump efficiency in fraction

The total dynamic head (equation (11-4)) is the head a pump needs to lift and deliver the required water yield from a well:

$$\text{TDH} = H_s + H_d = (H_e + H_f + H_v)_s + (H_e + H_f + H_v + H_p)_d$$
(11-4)

where H_s = suction head (ft)
$\quad\quad H_d$ = discharge head (ft)
$\quad\quad H_e$ = elevation head (ft)
$\quad\quad H_f$ = friction head (ft)
$\quad\quad H_v$ = velocity head (ft)
$\quad\quad H_p$ = pressure head (ft)

and subscripts s and d denote the suction and discharge sides of the pump, respectively.

Example 12-4: Pump Design for a Ground-Water Well

A well to yield 2,000 gpm is to be constructed in an area where there is a good water table aquifer capable of yielding 50-gpm/ft drawdown. No interference from other wells is expected. The aquifer material is a homogeneous mixture of very coarse sand and fine gravel that has a uniformity coefficient of 5.23. The saturated thickness is 90 ft, and the water table stands 30 ft below the ground surface during drier times of the year. A vertical turbine pump will be inserted in the well, and a continuous-slot type well screen will be used. Only one permanent casing, the pump chamber casing, will be installed, but a temporary surface casing will be needed to facilitate drilling operations. Water from the well casing will be lifted 6 ft above the ground surface and delivered to a pond. Assume a 5-ft pressure head is sufficient at the pipe outlet. A summary of this example is illustrated in Figure 12-15.

Solution

Well Diameter   Reference to Table 12-2 suggests that a 24-in. O.D. pump chamber casing with 16-in. pump bowl is suitable for 2,000 gpm. The temporary surface casing should be 26–28 in. in diameter, and the borehole must be at least as great in diameter as the surface casing.
Well Depth   The well should be completed to the bottom of the aquifer. The total depth will be 120 ft below the ground surface.
Well Screen Slot Size and Placement   The bottom one third of the homogeneous water table aquifer will be screened. The saturated thickness is 90 ft, so the well screen should be 30 ft long. The uniformity

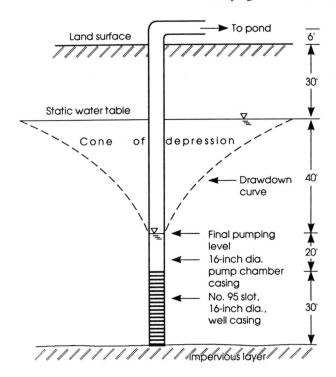

FIGURE 12-15.   Illustration of a well in Example 12-4.

coefficient of the aquifer material is 5.23, so the screen should retain 30–50% of the formation. Assume that a sieve analysis of the aquifer material showed that the relationship between cumulative retention and particle size was 0.077 in. at 30% retention and 0.137 in. at 50% retention. Hence, any slot width between 0.077 in. and 0.137 in. would suffice. A number 95 slot screen will be selected. A 16-in. diameter, number 95 slot screen has an open area of 228 in². or 1.58 ft² linear ft (Table 12-4). The flow rate in ft³/min is

$$2{,}000 \text{ gpm} \div 7.48 \text{ gal/ft}^3 = 267.4 \text{ ft}^3/\text{min}$$

The required screen open area is

$$267.4 \text{ ft}^3/\text{min} \div 6 \text{ ft/min} = 44.6 \text{ ft}^2$$

The minimum length of screen necessary to maintain entrance velocity at 6 ft/min or less is

$$44.6 \, \text{ft}^2 \div 1.58 \, \text{ft}^2 \, /\text{ft} = 28.2 \, \text{ft}$$

The 30-ft length of 16-in.-diameter well screen will exceed the minimum requirement. The extra length may be desirable in case some slots become clogged over time.

Drawdown   The anticipated drawdown is 1 ft per 50-gpm pumping rate, so at a yield of 2,000 gpm drawdown should be 40 ft. To account for fluctuations in the water table, the pump will be set 45 ft below the top of the saturated level.

Pump Capacity   The total dynamic head is the discharge head, and no suction head is needed since the pump is submerged. Elevation head (H_e) is the distance from the final pumping level in the well to the point of discharge. The final pumping level is 40 ft below the static level, and the static level is 30 ft below the ground surface. It follows that $H_e = 30 \, \text{ft} + 40 \, \text{ft} + 6 \, \text{ft} = 76 \, \text{ft}$. The required pressure head (H_p) is 5 ft. Velocity head is ignored in this example. The casing and pipe are made of steel; there is one standard elbow and one ordinary exit. Assume the total head loss in the system is 0.8 ft. Computing pipe friction head loss was discussed in Chapter 9. Thus, TDH = 76 ft + 5 ft + 0.8 ft = 81.8 ft.

Power Demand   A pump with an efficiency of 72% is used for this application. The power applied to the pump shaft is

$$BP = \frac{2,000 \, \text{gpm} \times 81.8 \, \text{ft}}{3,960 \times 0.72} = 57.4 \, \text{hp}$$

12.4.2 Pumping Costs

If the pump shaft horsepower requirement is known, the cost of pumping water may be calculated if the motor efficiency and cost of power are known. The cost of pumping well water will be illustrated for wells fitted with electric pumps. Consumption of electricity (in kW) may be calculated by equation (11-30):

$$C = \frac{Q \times \text{TDH} \times 0.746 \times R}{3,960 \times E_p \times E_D \times E_M} \tag{11-30}$$

where C = pumping cost per hour of operation ($/hr)
   R = cost of electricity ($/kWh)
   E_P = pump efficiency in decimal
   E_D = drive train efficiency in decimal; use 1.0 if pump driven directly by electric motor

E_M = motor efficiency in decimal

0.746 = factor for converting hp to kW

Large electric motors usually have high efficiency (greater than 90%) when properly loaded.

Example 12-5: Determining Pumping Cost

Consider the pump system of Example 12-4. Assume a 90% efficient electric motor is directly connected to the pump and that the power rate is $0.08/kW-hr. Find the operating cost of the pump.

Solution

$$C = \frac{2,000 \text{ gpm} \times 81.8 \text{ ft} \times 0.746 \times \$0.080 \text{ kW-hr}}{3,960 \times 0.72 \times 1.0 \times 0.90} = 3.80 \text{ \$/hr}$$

The pumping will cost $3.80 for one hour of operation. The cost of pumping water per acre-foot may be estimated from the total operating hours required to pump 1 acre-ft of water. One acre-foot is equal to 325,782 gal. The operating hours for pumping 1 acre-ft of water are

$$\frac{325,782 \text{ gal/acre-ft}}{2,000 \text{ gpm} \times 60 \text{ min/hr}} = 2.71 \text{ hr/acre-ft}$$

The pump must operate for 2.71 hr to deliver 1 acre-ft of water. The total electricity cost to pump 1 acre-ft of water is

$$3.80 \text{ \$/hr} \times 2.71 \text{ hr/acre-ft} = 10.30 \text{ \$/acre-ft}$$

It is obvious from Examples 12-4 and 12-5 that pumping cost is directly proportional to well yield and to TDH. Hence, doubling either well yield or TDH would double pumping cost. Also, pumping cost is inversely proportional to pump efficiency and motor efficiency. Motors do not vary greatly in efficiency, but pumps do. Pumps should be selected and operated carefully to ensure high efficiency and, consequently, low pumping cost.

12.5 WELL TEST

A well test can provide information on the static water level, the pumping level and drawdown, and the specific capacity of the well. Sieve analysis of material collected from the aquifer and information from the well test

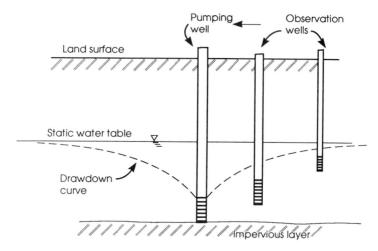

FIGURE 12-16.   Observation wells for use in a well test.

can provide data for selecting and placing the permanent well screen and pump. The well yield also may be determined. The second use of a well test is to obtain information on the aquifer, which may be used for planning and designing other wells. For example, suppose a large well is to be constructed at a particular location. A well test on an existing well in the vicinity or on a small well drilled nearby the site can provide valuable information on design features and expected performance of the production well.

In a typical single-well test, the static water level is measured and the well is then pumped at a known constant rate. The pumping level is measured at timed intervals. After some period of pumping, several hours to several days, the pump is stopped, and the depth of water in the well is measured as the well recovers from pumping. Even more information may be obtained if more than one observation well is located some distance from the pumping well (Fig. 12-16). A multiple-wells test is preferable to a single-well test. The depth of the water level may be measured with a tape or electrical sounder. Discharge of the well may be measured by a variety of techniques. Methods for measuring pump discharge are presented in Chapter 10.

12.6 WELL CONSTRUCTION

According to Heath (1984), seven different methods of well construction are commonly used. In areas where the soil is easily worked and the water table is near the ground surface, wells may be dug with pick and shovel, drilled with hand-turned or power-driven soil augers, constructed by driving a casing with screened drive point in the ground, and formed by jetting a hole into the ground with a high-pressure stream of water. These procedures are suitable to depths of 50–100 ft.

For harder materials and deeper wells, cable-tool (percussion) drilling procedures or rotary drilling techniques are employed. These procedures can be used to drill boreholes through hard rock down to 1,000 ft or more. The cable-tool techniques involve repeatedly raising and dropping a chisel bit that shatters the formation beneath the bit (U.S. Environmental Protection Agency 1976b). In unconsolidated material, the casing is driven ahead of the bit, but in hard material the bit must penetrate the formation ahead of the casing. Soil and rock particles are removed from the borehole with a stream of water. The well screen is placed inside the casing, and the casing is pulled back to expose the screen.

A bit attached to a rotating drill stem or pipe is used for rotary drilling. In the hydraulic rotary method, water and clay (drilling mud) are pumped down the drill pipe. This fluid passes into the borehole and serves to cool the bit and remove cuttings from the hole. The drilling mud is under pressure and it supports the borehole. When the hole is finished, screen and casing are lowered into it. In air rotary drilling procedures, air under high pressure is used to remove cutting from the hole. Many modern rotary drilling rigs have both air rotary and hydraulic rotary options. The procedures may be alternated, depending on the nature of the formation where the drill is working.

After the casing and screen have been installed in the borehole and the annular space has been filled with packing materials as shown in Figure 12-10, the well must be subjected to well development. During well development, compaction of the sides of the borehole caused by drilling is broken down and particles finer than the slot width of the well screen are removed from the aquifer in the vicinity of the screen. This allows the well to produce sediment-free water and enhances its yield. In unconsolidated aquifers, pumping the well at a rate higher than the design yield is sufficient for development. A variation of the overpumping technique is pumping and surging. A pump without a ratchet to prevent reverse rotation or a check valve is placed in the well. Pumping is conducted at 0.25, 0.5, 1, 1.5, and 2 times the design capacity. After pumping at the lowest rate until the water is free of sand, the pump is stopped and water in the casing surges back into the well. This procedure is repeated at each pumping rate; it is more reliable than the overpumping method. Surge blocks may be attached to a drill stem and moved up and down inside the well casing to force water back and forth through the screen and facilitate development. Techniques of well development employing compressed air, hydraulic jetting, or chemical treatment also are available (see Driscoll 1986; U.S. Environmental Protection Agency 1976; Anderson 1984).

13

Pond Design and Construction

13.1 INTRODUCTION

Ponds may be classified as watershed ponds, levee ponds, and excavated ponds. A *watershed* pond is created by constructing a dam across a watershed where the valley is depressed enough to permit storage of runoff. A *levee* pond is made by building levees around the area in which water is to be impounded. Water for filling levee ponds must be derived from wells, storage reservoirs, streams, or canals, because watersheds are absent. An *excavated* pond is constructed by digging a pit in which water is stored. Excavated ponds are filled by runoff, well water, or groundwater seepage. Pond construction is a broad subject. Construction techniques used for a particular pond depend on site characteristics, available equipment and material, cost, and preference.

Both levee and watershed ponds are well suited for commercial aquaculture. However, levee ponds have three distinct advantages over watershed ponds: (1) water may be better controlled; (2) levee ponds may be built side by side to consolidate management efforts; and (3) levee ponds afford a greater degree of control over size and slope, which are independent of site topographical characteristics. The main advantage of watershed ponds is that water is supplied from the watershed at no or minimum cost. Unfortunately, there is no control over the amount and timing of runoff. The runoff water might carry sediments and other materials that are eroded or washed out from the watershed. However, watershed ponds are most popular on farms where there is need to have water for multiple uses as well as for fish culture (e.g., livestock watering, irrigation, fire protection, recreation, domestic purposes).

13.2 SITE SELECTION

Selection of a suitable site for a pond or series of ponds is a critical consideration in aquaculture. Often, more than one site is available, and each site should be studied to determine which is the most practical and economical. If all sites are unsuitable, the project should be abandoned. Many aquacultural endeavors have failed because ponds were constructed at an unsuitable site. Terrain, soil characteristics, and water supply are obvious factors affecting site suitability. However, accessibility, proximity to markets for aquacultural products, availability of labor and supplies, security of facilities, and availability and cost of fuel and electricity must be considered. It is especially important to study market possibilities, for commercial aquaculture cannot succeed without adequate markets.

13.2.1 Terrain

Levee and excavated ponds should be built in relatively level areas. Common sites for such ponds are floodplains of streams and low-lying coastal areas. Flooding may be a problem, so it must be established that ponds are not likely to be inundated by floodwater.

Watershed ponds must be constructed on sloping terrain. Topographic features favorable for watershed ponds are a valley with moderate slopes and a depressed valley floor that permits a large area and volume of storage and a constricted section that allows construction of a short dam.

13.2.2 Water Supply

Most watershed and excavated ponds are filled by surface runoff or stream flow. Terrain interacts with climatic, topographic, and vegetative factors to regulate the quantity of runoff generated by a watershed. Hence, the allowable area and volume for a pond at a site depends on the area and runoff-producing characteristics of the watershed. Minimum permissible pond depth is regulated by climatic and topographic factors, for depth must be great enough to prevent the pond from drying up because of unreplaced evaporation and seepage losses during dry weather. Information on seasonal variation in stream discharges and on the frequency of droughts, which cause extremely low flow, is essential for planning ponds.

A guide for estimating the approximate area of watershed required for each acre-foot of storage in ponds in the United States is presented in Figure 5-4. Methods for making site-specific calculations for the pond volume that can be maintained by a watershed and recommended minimum depths of water for watershed ponds in different climatic regions are also presented in Chapter 5. Recommended minimum depths are for ponds with normal rates of seepage and evaporation losses. Where these losses are high, ponds must be deeper. Recommended watershed sizes from Figure 5-4 are based on average topographic and rainfall conditions. More reliable estimates may be made from consideration of runoff-producing characteristics of the specific watersheds as described in Chapter 4.

Suppose we are interested in knowing how large a pond can be maintained by a 20-acre watershed in central Georgia. Reference to Figure 5-4 suggests that 2.5 acres of watershed by interpolation are needed for each acre-foot of storage. The permissible pond volume is 20 acres ÷ 2.5 acres/acre-ft, or 8 acre-ft. The recommended minimum depth of water in a humid climate is 6–7 ft (from Chapter 5). Assuming a depth of 6 ft, pond area should be no greater than 1.25 acres (8 acre-ft ÷ 6 ft = 1.25 acres).

Watershed properties and management affect the quality of runoff and may impact aquaculture. Runoff from fertile soils or from fertilized pastures contains enough nutrients to favor high productivity of phytoplankton in ponds (Boyd 1976). Runoff from wooded watersheds generally has low concentrations of nutrients. Acidic soils yield acidic runoff, and calcareous or alkali soils yield basic runoff. Erosion on watersheds results in turbid runoff, which may lead to excessive turbidity in ponds. Agricultural, industrial, or domestic pollution may enter ponds in runoff from watersheds. Agricultural chemicals from a watershed may be particularly harmful to aquaculture species. Problems relating to water quality in aquaculture are discussed by Boyd (1990).

Levee ponds are filled by surface water from streams or by ground water from wells. Because ground water normally is unpolluted and free of wild fish, it is a preferred water supply for freshwater aquaculture. Consultation with a local well-drilling company or a ground-water geologist often can provide an estimate of the quantity and quality of water available from wells at a particular site. However, it would be a costly mistake to construct ponds at a site where well water is of low quality or to build more ponds than can be filled and maintained by the quantity of well water available. Questions about water quantity and quality may be answered by drilling small test wells. If water is of good quantity and quality, the production wells may be developed and tested for discharge so that permissible pond area and volume may be estimated. Chapter 12 discusses ground water and wells in detail.

13.3 SOILS

The soil that will form the pond bottom must be impervious enough to prevent excessive seepage. As discussed in Chapter 6, seepage rate is primarily a function of the particle size distribution of soil. Coarse-textured sands and sand–gravel mixtures transmit water easily and are not suitable for pond bottoms. Fine-textured soils that have a high clay content or are composed of a mixture of silt and clay resist seepage and make watertight pond bottoms. Sandy clays also are generally suitable as pond bottoms.

The entire depth of soil beneath a pond need not consist of impervious material, but a layer of relatively impervious material thick enough to prevent excessive seepage is essential. Sometimes, soils of the pond area are not homogeneous. Some places will be impervious, and other places may be more porous. Soil properties in the pond area should be investigated. Hand-operated soil augers are used to make several borings per pond in order to identify areas of porous soils. To reduce seepage, places with porous soils may be covered with a clay blanket or impermeable membrane or treated with bentonite or soil dispersant (see Chapter 6).

Soil texture analysis is the usual way of identifying high-seepage places in a proposed pond area. The hydraulic conductivity of a soil also may be estimated by digging or boring holes into the soil, filling holes with water, covering holes to prevent evaporation, and measuring seepage losses. If the hole penetrates the pervious layer, seepage will be erroneously high. Laboratory tests of hydraulic conductivity as shown in Figure 13-1 are another alternative for assessing potential seepage (Todd 1980). In Figure 13-1 hydraulic conductivity, K (ft/min), is

$$K = \frac{VL}{Ath} \tag{13-1}$$

where V = volume of water passed for time t (ft^3)
   L = length of sample (ft)
   A = cross-sectional area of sample (ft^2)
   t = time of flow (min)
   h = hydraulic head (ft)

13.3.1 Soil Analysis

Because of the importance of soil texture in pond construction, soil texture analysis merits discussion. The major classes where soils may be classified in terms of texture are gravel, sand, silt, and clay. The term *loam* is used for a certain mixture of sand, silt, and clay. The common

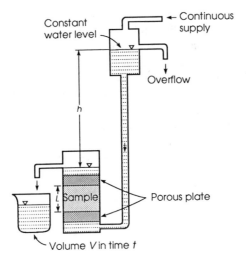

FIGURE 13-1.   Laboratory setup of constant-head permeameter.

FIGURE 13-2.   Feel test of soil textures.

field method for determining soil texture class is a "feel" test (Fig. 13-2). Soil is simply rubbed between the thumb and fingers. Wetting the soil is beneficial for obtaining a feel for plasticity. Sand feels gritty and is not sticky; it has no plasticity and cannot be rolled into a small ball. Silt has a floury or talcum-powder feel when dry. Wet silt is moderately plastic and sticky. Clay is sticky and plastic; it can be rolled into a ball or into a thin cylinder or "snake."

A simple test for estimating percentages of sand, silt, and clay particles in a soil may be conducted with a straight-sided, clear bottle and a ruler. Fill the bottle about one-third full with soil and finish filling the bottle with water. Shake the bottle vigorously for several minutes to suspend the soil in water. Set the bottle aside on a level shelf for 24 h so that the soil settles to the bottom. Sand particles settle first, silt particles next, and clay particles last. Measure the thicknesses of the different layers. For example, suppose the thicknesses of sand, silt, and clay layers are 2.1, 1.3, and 0.6 in., respectively. The percentages of sand, silt, and clay are

$$\text{Total thickness} = 2.1 + 1.3 + 0.6 = 4 \text{ in.}$$

$$\% \text{ sand} = \frac{2.1}{4} \times 100 = 52.5\%$$

$$\% \text{ silt} = \frac{1.3}{4} \times 100 = 32.5\%$$

$$\% \text{ clay} = \frac{0.6}{4} \times 100 = 15.0\%$$

Laboratory tests provide more accurate estimates of the proportions of sand, silt, and clay. Several procedures have been developed, but the following technique (Weber 1977) is simple and sufficiently accurate for practical application. Metric units are usually used in soil sampling and laboratory analysis of the samples.

A soil sample is air-dried and gently pulverized by hand. Next, the moisture content is estimated. Place 10–20 g of air-dried soil in a pre-weighed 50-mL aluminum weighing dish. Weigh the dish and soil and record the values. Dry dish and sample in a forced-draft oven at 105°C for 24 h. Remove dish and sample, cool in a desiccator, and reweigh. Calculate the percentage oven-dried weight of soil as follows:

Initial weight of dish and soil − dish weight
    = weight of air-dried soil   (13-2a)

Oven-dried weight of dish and soil − dish weight
    = weight of oven-dried soil   (13-2b)

$$\% \text{ oven-dried soil by weight} = \frac{\text{weight of oven-dried soil}}{\text{weight of air-dried soil}} \times 100 \quad \text{(13-2c)}$$

Next, a 50 g sample of air-dried soil is placed in a 500-mL bottle. Add 100 mL of hexametaphosphate solution [50 g $(NaPO_3)_6$ diluted to 1,000 mL with deionized water] and 300 mL of deionized water, stopper bottle, and shake on a mechanical shaker for 24 h. Wash bottle contents into a 1,000-mL graduated cylinder with deionized water. Add deionized water to within 5 cm of the 1,000-mL mark. Insert a soil hydrometer and fill the cylinder to the 1,000-mL mark. Remove the hydrometer, stopper the cylinder, turn the cylinder end over end 10 times, return the cylinder to an upright position, record the time, and set the cylinder on a laboratory bench. After 20 sec, insert the hydrometer into the soil suspension. Read the hydrometer exactly 40 sec after the cylinder was set upright. Repeat the mixing procedure and take another 40-sec hydrometer reading. Immediately record the temperature of the soil suspension. Allow the soil suspension to settle for 2 h and take hydrometer and temperature readings. Correct hydrometer readings for temperature by adding 0.36 g/L for every 1°C above 20°C; subtract 0.36 g/L for every degree below 20°C.

Calculate the percentages of sand, silt, and clay, using the temperature-corrected hydrometer readings and the following equations:

$$\text{Oven-dried weight of soil (g)} = \frac{(50\,\text{g})\,(\%\ \text{oven dry weight})}{100} \tag{13-3}$$

$$\%\ (\text{silt} + \text{clay}) = \frac{40\text{-sec hydrometer reading}}{\text{oven-dried weight of soil}} \tag{13-4}$$

$$\%\ \text{clay} = \frac{2\text{-h hydrometer reading}}{\text{oven-dried weight of soil}} \tag{13-5}$$

$$\%\ \text{silt} = \%(\text{silt} + \text{clay}) - \%\ \text{clay} \tag{13-6}$$

$$\%\ \text{sand} = 100\% - \%(\text{silt} + \text{clay}) \tag{13-7}$$

For general uses Figure 13-3 shows relationships between particle sizes and soil classification by the U.S. Department of Agriculture (USDA) and Unified Soil Classification Systems (USCS). To obtain the textural class of a soil sample, use a soil-texture triangle developed by the U.S. Army Corps of Engineers (Fig. 13-4). Suppose that a soil is 10% clay, 35% silt, and 65% sand. Reference to the soil-texture triangle identifies the soil as silty sand.

13.3.2 Fill Material

Earth fill for building embankments must be available at or near the pond site. Otherwise, a borrow pit must be secured to obtain the material. Earth fill must be structurally stable and impervious enough after com-

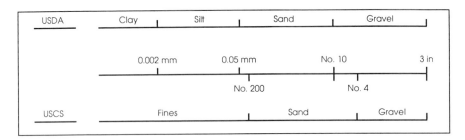

FIGURE 13-3. Relationships between particle size and soil classification. (*Source: U.S. Soil Conservation Service 1990.*)

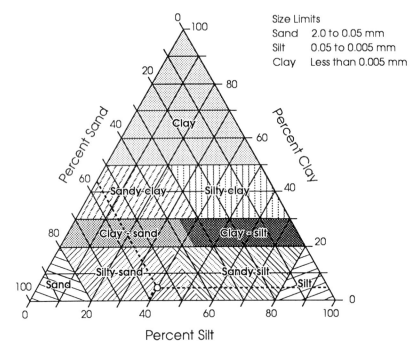

FIGURE 13-4. Triangle soil classification chart.

paction to prevent excessive seepage through the embankment. Soils for earth fill should contain particles of different sizes, ranging from small gravel or coarse sand to clay. A clay content of 20–30% is ideal. A very high clay content reduces soil stability. Also, when dried, soils with a high clay content may shrink and develop cracks. Pervious materials may be used for a homogeneous embankment as an earth fill if a clay cutoff is provided in the embankment and side slopes are flattened to provide a wider barrier to seepage.

13.4 EMBANKMENTS

In this discussion, *embankment* will be used as a general term referring to dams and levees. The single embankment for a watershed pond will be called a *dam*. The embankments used to form a pond in a level area by enclosing the area in which water will be stored will be termed *levees*. Levees often are longer than dams, but dams generally are taller and more massive. Nevertheless, principles of construction are the same for dams and levees.

13.4.1 Foundations

The weight of an embankment bears down on the soil beneath it, so the soil beneath an embankment is called the *foundation*. The place where the embankment joins the foundation is especially vulnerable to seepage. A thick layer of impervious, consolidated material just below the ground surface makes an ideal foundation because it is strong and watertight. Gravel, sand, or sand–clay mixtures will support the weight of an embankment, but seepage beneath the embankment must be controlled. Clays and silts resist seepage, but they are not as stable as coarser materials. However, embankments for aquaculture ponds are relatively small and can be constructed on foundations with high silt or clay contents. Peat and other organic materials are unsuitable for foundations; they must be removed and replaced with suitable foundation material.

Even when consolidated material serves for the foundation, seepage can occur through crevices. The foundation should be examined carefully, and any crevices filled with impervious material. Loose topsoil, roots, boulders, and other debris must be removed from the foundation. The top of the foundation should be scarified to provide a bond with the earth fill of the embankment. A cutoff is highly recommended to protect against seepage beneath the embankment.

13.4.2 Cutoffs and Clay Blankets

The most common means for reducing seepage is to dig a trench in the foundation and install a cutoff of impervious material. Alternatively, the wet side of the embankment and pond bottom can be covered with

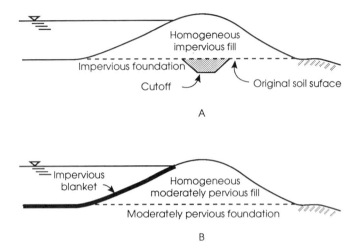

FIGURE 13-5.   Cutoff and impervious blanket structures on the wet side of embankments. (*Source: U.S. Soil Conservation Service 1979b*)

a clay blanket. Examples of a cutoff and a clay blanket are shown in Figure 13-5.

Cutoffs (also called cores) almost always are constructed of a soil with a high clay content. The cutoff runs along the centerline of the embankment for its entire length. The cutoff may extend only into the foundation, or it may extend into both the foundation and the embankment. The cutoff depth below the ground surface depends on the location of the impermeable stratum of the site.

To construct a cutoff, a trench at least 4 ft in bottom width is dug along the centerline. If possible, the trench should extend downward and key into an impervious layer. The side slopes of the trench should be no steeper than 1:1. Earth fill for the cutoff is applied in thin layers (a few inches), and each layer is compacted at optimal moisture conditions. If the cutoff is extended upward into the embankment, it must be raised along with the rest of the embankment. If proper materials for a cutoff are not available or are too far away to haul, certain plastic sheets or a concrete wall may be used.

13.4.3   Top Width

The top width of embankments depends on their height and uses. Minimum permissible top widths for earthen embankments are shown in Table 13-1. Tractors, trucks, and other equipment are used on embankments during

TABLE 13-1   Recommended Top Width of Embankment

Height of Embankment (ft)	Top Width (ft)
<10	8
10–15	10
15–20	12
20–25	14

Source: U.S. Soil Conservation Service (1979b).

operation of aquaculture projects. Therefore, minimum permissible top widths sometimes are not adequate. For example, the top of levees for channel catfish ponds should be at least 14 ft wide to accommodate equipment and vehicles.

13.4.4   Side Slopes

The sides of embankments are sloped to provide stability. Soil on steep slopes will slough off and slide down. Steeper side slopes may be allowed on the dry side of an embankment than on the wet side. Recommended side slopes for embankments made of clay, clayey sand, clayey gravel, sandy clay, silty sand, or silty gravel are 3:1 (horizontal:vertical) on the wet side and 2:1 on the dry side. Slopes of 3:1 should be provided on both sides of an embankment made of silty clay or clayey silt. Where well-graded soil has been compacted properly, the side slopes may be 1:1 or 2:1 on both sides. A stone layer, riprap rock, gabion, or concrete blocks are used to maintain stability below the normal waterline. Grass cover must be used above the normal waterline on the wet side and on the dry side to protect the embankment from damage by erosion.

13.4.5   Height

Embankments must have a freeboard extending above the anticipated maximum high water level in the pond. Freeboard is an added height provided as a safety factor to protect embankments from waves or overflow due to higher than expected storm events. Freeboard should be at least 1 ft, but for safety 2 ft of freeboard is recommended for most aquaculture ponds. Earth fill in an embankment will settle, and the height of an embankment must be determined with allowance for settlement. The amount of settlement usually will vary from 5% to 10%, depending on the material and initial compaction. For safety, a settlement value of 10%

is suggested. For example, suppose the high water level will be 5 ft above the base of a levee and 2 ft of freeboard is desired. The final levee height is 7 ft, but allowance must be made for 10% potential settlement. The height of the presettled levee must be 7 ft + (0.1)(7 ft) = 7.7 ft.

13.4.6 Compaction

Compaction increases the density of earth fill, which provides stability and reduces seepage for embankments. The density that can be achieved by a given amount of compaction varies with the moisture content of earth fill for a given fill material. If the moisture content of fill material is not near optimum when it is compacted, the embankment may not have the desired stability and resistance to seepage. Fill material that is too wet should be allowed to dry; water should be sprinkled over a layer of fill material that is too dry before compaction.

Laboratory tests are conducted to ascertain the optimal moisture content for compaction. Separate samples of fill material are compacted at different moisture contents. To accomplish this, successive layers of each sample are placed in a mold, and each layer is compacted under standardized conditions with a drop hammer (McCarthy 1988). The density of each compacted sample is determined. This test is called the Proctor density test, and the result is the Proctor density curve (Fig. 13-6). The energy used to compact soil in the laboratory test is comparable to the energy applied to the soil by standard compaction equipment.

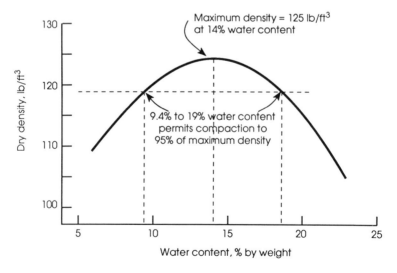

FIGURE 13-6.   Typical results of a Proctor soil compaction test.

TABLE 13-2   Equipment Commonly Used for Soil Compaction

Static rollers
    Smooth-wheel roller
    Sheep's-foot roller
    Rubber-tired roller
Vibrating rollers
    Smooth-wheel vibratory rollers
    Vibratory sheep's-foot or tamping rollers
    Rubber-tired vibratory roller
Tampers and vibrators
    Heavy-duty rammer
    Single-shoe vibrotamper
    Vibrotamper
    Vibratory compactor

It is a common field practice to compact soils to 95% of the maximum density achieved in laboratory tests. The moisture content that permits the 95% compaction may be determined from laboratory results.

Many contractors have equipment for making moisture and density determination of fill material during compaction. Laboratory compaction test results must be available for best use of the moisture-density relationship. McCarthy (1988) reported optimal moisture content (% by weight) for fill materials: 6–10% for sand, 8–12% for sand and silt mixture, 11–15% for silt, and 13–21% for clay. This information may be used as a guide when laboratory compaction tests cannot be conducted. Even at optimal moisture content, some soils compact better than others, and equipment recommended for compaction is not the same for all soils. Table 13-2 lists equipment commonly used for embankment compaction.

Pond construction often is done during dry periods when soil moisture normally is low, and most ponds are built with little attention to moisture content of fill material. Boyd (1982a) reported potential problems of seepage through the embankment due to lower than the allowable soil compaction achieved under this condition.

13.4.7   Volume of Fill

The volume of earth fill between two points on the centerline of an embankment is approximately equal to the mean of the two end areas multiplied by the distance between them. The appropriate formula for the volume of earth fill in a segment of embankment in cubic yards (yd^3) is

$$V = \frac{(A_1 + A_2)D}{2(27\,ft^3/yd^3)} \tag{13-8}$$

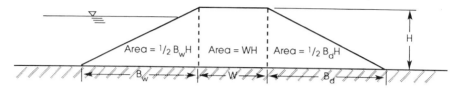

FIGURE 13-7.   Cross-sectional view of an embankment.

where
V = volume of earth fill in a segment (yd^3)
A_1 = cross-sectional area at one end of segment (ft^2)
A_2 = cross-sectional area at the other end of segment (ft^2)
D = length of segment along the embankment (ft)

The cross section of an embankment is illustrated in Figure 13-7; the formula for a trapezoidal cross-sectional area A is

$$A = WH + \frac{(B_w + B_d)H}{2} \qquad (13\text{-}9)$$

where W = width of middle section (ft)
H = height of embankment (ft)
B = width of triangle section (ft)

The subscripts w and d refer to the wet side and dry side of the embankment, respectively. The volumes of all segments are summed to obtain the earth-fill volume of an embankment:

$$\text{Total volume of embankment} = \sum_{i=1}^{n} V_i \qquad (13\text{-}10)$$

where i = segment number and n = total number of segments.

Materials for earth fill are cut (or borrowed) from the ground or a borrow pit and transported to the site for the embankment. Because the embankment is thoroughly compacted, it will take more than a cubic yard of cut material, often called *borrow material*, to give a cubic yard of final compacted fill. In practice a cut factor of 1.15 should be multiplied by embankment volume to determine the amount of borrow material required. The *cut factor* is the ratio of cut material volume to the volume of compacted fill before the final settlement. Details of determining the cut factor are given by Schwab et al. (1981).

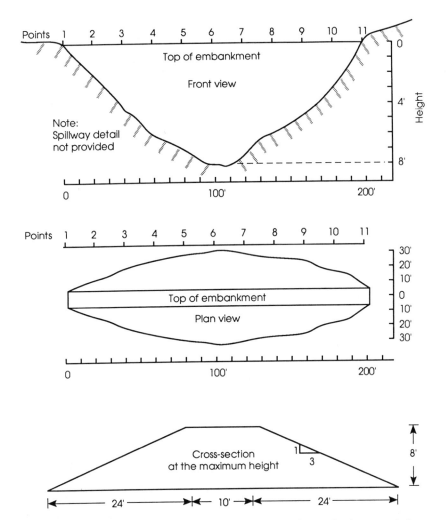

FIGURE 13-8. Plan view, front view, and cross section of an embankment and dimensions used in Example 13-1.

Example 13-1: Determining Embankment and Borrow Material Volumes

The volume of the embankment illustrated in Figure 13-8 and the amount of borrow material required for its construction will be computed. The design heights shown in Figure 13-8 have already been adjusted for the final settlement. The top width is 10 ft, the slope of the dry and wet sides is $3:1$, and the height (H) is found at each point from Figure 13-8. Hence,

TABLE 13-3   Summary of Embankment Calculation Illustrated in Example 13-1

Point	Cross-Sectional area (ft²)	Segment	Length of Segment (ft)	Volume of Segment (yd³)
1	0	1–2	20	11.9
2	32	2–3	20	44.4
3	88	3–4	20	94.8
4	168	4–5	20	142.6
5	217	5–6	20	181.1
6	272	6–7	20	181.1
7	217	7–8	20	142.6
8	168	8–9	20	108.5
9	125	9–10	20	67.4
10	57	10–11	20	21.1
11	0			0.0
Total				995.5

the value for B is $3H$. The cross-sectional area is $0\,\text{ft}^2$ at point 1. For point 2, $H = 2\,\text{ft}$, $W = 10\,\text{ft}$, and $B = 6\,\text{ft}$, and cross-sectional area is

$$A = (10\,\text{ft})(2\,\text{ft}) + \frac{(6\,\text{ft} + 6\,\text{ft})(2\,\text{ft})}{2} = 32\,\text{ft}^2$$

Distance between points 1 and 2 is 20 ft; the volume of this segment is

$$V = \frac{(0\,\text{ft}^2 + 32\,\text{ft}^2)(20\,\text{ft})}{2(27\,\text{ft}^3/\text{yd}^3)} = 11.9\,\text{yd}^3$$

Calculations for the entire embankment are summarized in Table 13-3. The calculated total volume is $995.5\,\text{yd}^3$. The volume needed before the settlement, assuming 10% settlement after compaction, would be $995.5\,\text{yd}^3 \times 1.1 = 1{,}095.0\,\text{yd}^3$. Next, multiply by the cut factor of 1.15 to determine the amount of borrow necessary for the embankment; the total volume of borrow material is $1{,}095.0\,\text{yd}^3 \times 1.15 = 1{,}259\,\text{yd}^3$.

13.4.8   Seepage through Embankment

The characteristics of the embankment material affect the seepage losses through the embankment. Understanding of the amount and location of the seepage on the dry side of the embankment allows adjustments or structures to be installed to control the seepage. The location and amount of seepage depend on the embankment material, compaction, type of foundation, cutoff, and drainage systems at the toe of embankment. It is important to know the amount and location of the seepage that intersects the dry-side slope. In many cases the amount of seepage is slight, but in some cases a large amount of seepage appears on the face of the embankment and will cause sloughing. Most aquaculture ponds made of a homogeneous material may be located on an impervious foundation. The seepage line of this type pond will appear at the embankment face above the base unless drainage is provided (Fig. 13-9). The seepage line is the upper line of seepage through the embankment. The location of the seepage line of a homogeneous embankment is approximately a third of the water depth in the pond from the impervious base of the embankment to the interception of the seepage line and the dry side of the embankment (Schwab et al. 1981). A toe drainage system (Fig. 13-10) is desirable to lower the seepage line for most homogeneous embankments. Toe drains may be constructed of sand, gravel, rock, or filter drain.

13.5 WATER-CONTROL STRUCTURES

The function of embankment spillways is to pass storm runoff around or under the embankment to prevent overtopping of the embankment by excessive inflow from heavy storm events. They are often equipped to serve as intake for water use at the downstream. Two types of spillways used in aquaculture ponds are emergency and principal spillways.

13.5.1 Emergency Spillways

Watershed ponds must be equipped with an emergency spillway to bypass excessive runoff after heavy rains. Without an adequate emergency spillway, water will overtop or flow around the lower end of the dam. The resulting erosion will cause serious damage to or failure of the dam.

Often, emergency spillways for aquaculture ponds are excavated waterways with vegetative cover designed to convey peak discharges of watersheds during severe storm events. The U.S. Soil Conservation Service (1982) recommends using the peak discharge from a rainfall expected over 24 h every 25 years for most aquaculture ponds. A 50-year or longer return period is recommended where the consequences of dam failure potentially are great. The methods to design peak runoff discussed in Chapter 4 may be used to estimate peak discharge where runoff data are not available.

An emergency earthen spillway (Fig. 13-11) often has a trapezoidal cross section with side slopes of 3:1. The approach channel should be

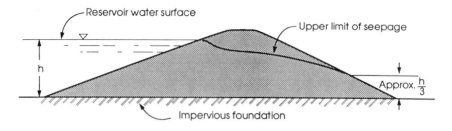

Completely homogeneous embankment

FIGURE 13-9.   Seepage line of a homogeneous embankment built on impervious foundation.

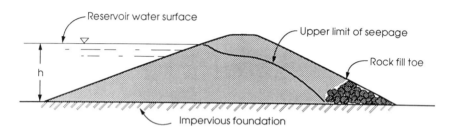

Modified homogeneous embankment

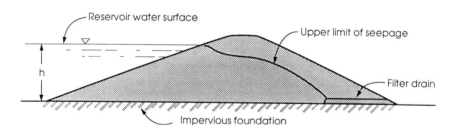

Modified homogeneous embankment

FIGURE 13-10.   Toe drainage systems of homogeneous embankments.

expanded and smooth to minimize turbulence. The middle or control section should slope slightly (0.5–1%) or be level, but the exit section may need a greater slope to expedite the flow. The dam should extend 2 ft above the highest water level expected in the spillway. A natural grass or

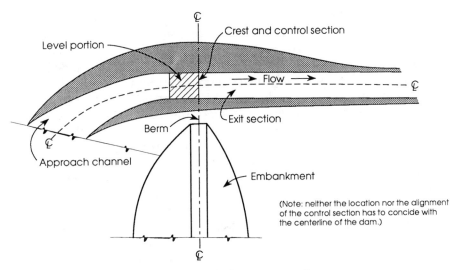

Plan view of excavated emergency spillway

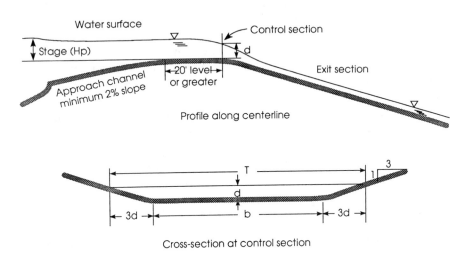

Profile along centerline

Cross-section at control section

FIGURE 13-11.   Design guidelines for an excavated emergency spillway of an embankment. (*Source: U.S. Soil Conservation Service 1979b.*)

riprap lining must protect the spillway from erosion. A flow depth of 1 ft is recommended in the control section (Schwab et al. 1981), and the selected bottom width of the spillway should not exceed 35 times the design depth of flow. The exit section must be constructed so that erosion will not occur in the spillway. Loose sands and other highly erodible material should be avoided. When discharge exceeds 300 cfs, two earthen spillways should be constructed or a concrete-lined spillway must be built.

An earthen spillway should be well protected against erosion by a good vegetative cover if soil and climate permit. The entire spillway should be prepared for seeding or sodding of adapted or native perennial grasses and treated until a good stand has been established. Mulching is usually necessary to protect the seeding on the spillway. Design of vegetated waterways is discussed in Chapter 8. Limitations of vegetated emergency spillways are that (a) topography for the waterway should be adequate to carry the design peak flow well away from the embankment along a relatively straight line; and (b) the velocity at the exit section should not exceed the permissible velocity for soil type and vegetation. Where steep slope is unavoidable due to topography, riprap rock or concrete lining should be used. Consideration should be given to select a natural spillway such as a natural saddle at one end of the embankment.

Drop-inlet spillways occasionally are installed in watershed ponds in conjunction with emergency spillways. A drop-inlet spillway essentially is a large, rectangular weir constructed in the dam or at one end of the dam. Water flows over the weir notch, and the force of the water is dissipated by a concrete or riprap rock apron. Equations for estimating the discharge of rectangular weirs are found in Chapter 10.

Natural spillways sometimes may be a suitable alternative to excavated spillways. Wherever there is a good vegetal cover on the spillway area and the topography, such as a natural saddle, is suitable, the use of a natural spillway should be strongly considered. One side of a natural spillway is the end of the dam; the other side is the natural ground beyond the end of the dam. Discharge from a natural spillway can be estimated by the Manning equation discussed in Chapter 8.

13.5.2 Principal Spillways

Principal spillways are mechanical spillways constructed of permanent materials. They are designed to provide flood protection and to reduce the frequency of operation of the emergency spillway to protect the vegetation from prolonged flooding during wet weather or from spring flows. The top of the entrance to the principal spillway of aquaculture ponds is usually 1 ft lower than the entrance to the control section of the emergency spillway. Two kinds of principal spillway are in common use: drop-inlet pipe spillway (Figs. 13-12 and 13-13) and hooded-inlet pipe spillway (Fig. 13-14).

A drop-inlet pipe spillway consists of pipe extending through the dam

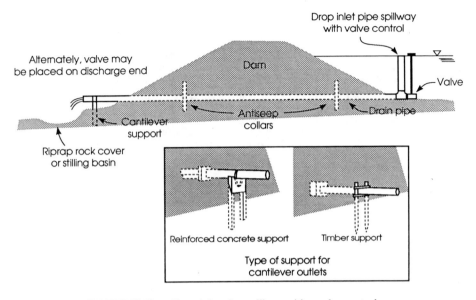

FIGURE 13-12.   Drop-inlet pipe spillway with a valve control.

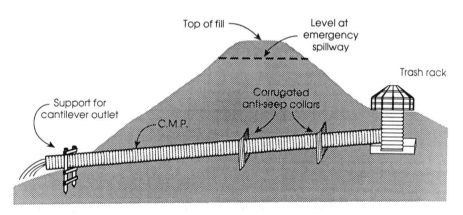

FIGURE 13-13.   Corrugated metal pipe (C.M.P.) spillway with a drop inlet. Note: Trash rack and antivortex plate on the top of the inlet and a cantilever-type outlet.

with a riser connected to the upstream end of the pipe. The top of the riser establishes the normal full-pool water level. The spillway may be used to drain the pond or to serve a supply-water use at the downstream if a suitable valve is attached near the pond bottom (Fig. 13-12). The valve may be placed on the upstream or downstream side of the spillway,

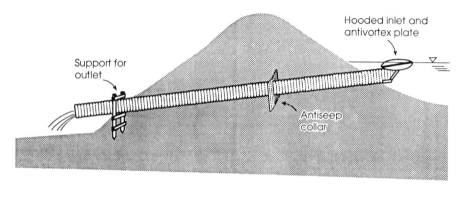

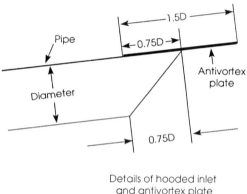

Details of hooded inlet
and antivortex plate

FIGURE 13-14.   Hooded-inlet pipe spillway. Note: Antivortex plate, antiseep collar, and cantilever-type outlet.

although it often is convenient to place the valve on the downstream side. This arrangement permits ready accessibility to the valve but increases the probability of the drain pipe clogging. The riser should be larger than the pipe diameter, and may be made of the same material as the pipe section. The capacity of a principal spillway must be great enough to discharge inflow from the watershed, springs or streams, and prolonged surface runoff following heavy rains. It is best to slightly oversize the spillway pipe for the safety of the embankment.

A hooded-inlet pipe spillway consists of a pipe extended through the embankment without a riser. Instead, the intake end of the pipe is cut at an angle to form a hood shape (Fig. 13-14). An antivortex plate should be installed at the pipe inlet. Details for the hood and antivortex device

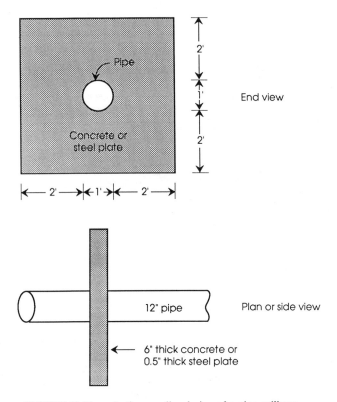

End view

Concrete or
steel plate

Pipe

12" pipe   Plan or side view

6" thick concrete or
0.5" thick steel plate

FIGURE 13-15.   Antiseep collar design of a pipe spillway.

are provided in Figure 13-14. Hooded inlets are not recommended for aquaculture ponds for they cannot be used to drain ponds.

Pipes for drop- or hooded-inlet spillways may be steel, concrete, plastic, or corrugated metal. The spillway pipe must extend through the dam, and the zone around the pipe is often vulnerable to seepage. Antiseep collars should be installed around pipes to prevent seepage along the pipe. With steel pipe, a plate can be welded around the pipe to form an antiseep collar (Fig. 13-15). Special collars are available for corrugated metal and plastic pipes. Of course, a collar may be improvised by installing a concrete structure around the pipe. Antiseep collars should extend at least 2 ft perpendicular to the pipe in all directions. One collar at the centerline of the dam is sufficient for dams up to 15 ft high; two or more collars equally spaced should be used for larger embankments.

A variation in drop-inlet spillway design is the deep-water release. Watershed ponds stratify during warm months, and nutrients and plankton

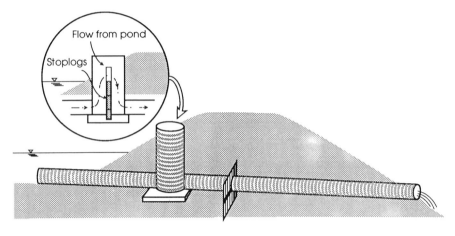

Full section of pipe riser with stoplogs

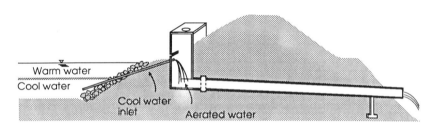

Cool water release for fish below structure

FIGURE 13-16.   Examples of deep-water release methods for drop-inlet pipe spillways. (*Source: U.S. Soil Conservation Service 1975.*)

are concentrated in surface water (Boyd 1990). The deep-water release structure prevents loss of plankton and nutrients following heavy rains as long as water does not exit through the emergency spillway. This method also prevents flow of oxygen-rich surface water from the ponds. Some structures used for this purpose are given in Figure 13-16.

Trash racks at the inlet are recommended to prevent debris from clogging in spillway pipes. Several types of trash racks are illustrated by the U.S. Soil Conservation Service (1979a), one of which is shown in Figure 13-17. Fish seldom escape through spillways. Wire-mesh screens sometimes are installed around spillways, but these fish barriers often clog with debris and reduce discharge capacity of spillways. Design and

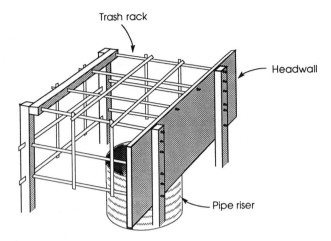

Trash rack

Headwall

Pipe riser

FIGURE 13-17.   An example of trash rack and headwall installed on the inlet riser of a drop inlet. (*Source: U.S. Soil Conservation Service 1975.*)

hydraulics of pipe spillways and methods to determine their flow capacity are discussed in Chapter 9.

13.5.3 Drains

A drop-inlet trickle tube can provide an adequate overflow and drain combination for a levee pond. Alternatively, a pipe extending through the levee with a swivel joint and standpipe attached (Fig. 13-18) may be used as a drain and overflow structure. The simplest type of swivel joint is a tee, which permits the standpipe to be rotated. To regulate water depth or drain the pond, the standpipe is rotated until its inlet end is at the desired elevation. A floating inlet described in Figure 13-19 is convenient for pond draining. A pipe is placed through the levee, and on the upstream side a flexible pipe is attached. A float is attached to the end of the flexible pipe and maintained at the desired elevation by a cable and anchor. For draining the pond, the end of the flexible pipe is lowered to the pond bottom. Fish seldom escape in the overflow through the inlet. However, when ponds are drained, it is safer to attach a wire-mesh or nylon-mesh basket, box, or bag to the outlet end of the drain pipe to capture any fish passing through.

A drop inlet structure fitted with removable crestboards or stoplogs (Fig. 13-20) is a popular means of regulating water levels and draining ponds in aquaculture. These structures often are called *flashboard risers*. They are also referred to as monks in many countries. Screens may be

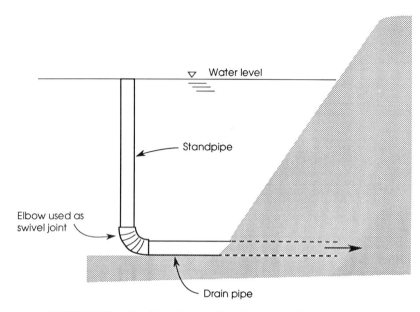

FIGURE 13-18.   Flexible-joint standing pipe used to drain a fish pond.

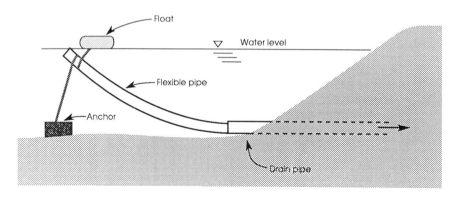

FIGURE 13-19.   Floating-inlet pipe used to drain a fish pond.

installed in or behind the structure to prevent escape of fish. Of course, the screens must be cleaned frequently or they will clog with debris and interfere with discharge. Some structures have a double set of crestboards between which mud may be packed to reduce seepage through spaces between them and the frame (Fig. 13-21). Discharge over crestboards may be estimated with the equation for discharge of rectangular weirs

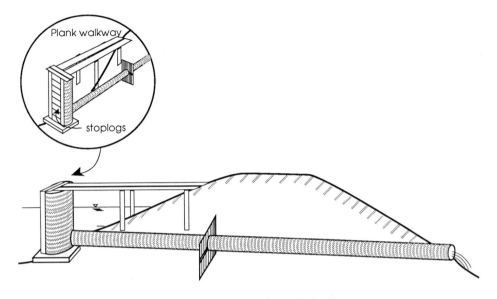

Half section of pipe riser with stoplogs

FIGURE 13-20.   Drop inlet with removable stoplogs (crestboards).

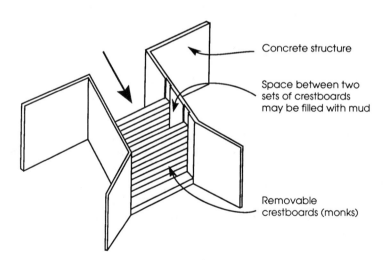

Concrete structure

Space between two
sets of crestboards
may be filled with mud

Removable
crestboards (monks)

FIGURE 13-21.   Crestboards (monks) to control water supply to a fish pond.

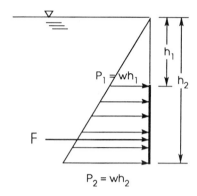

FIGURE 13-22.   Force diagram on the surface of a submerged structure.

discussed in Chapter 10. An alternative to crestboards is an adjustable headgate that may be raised or lowered with a spindle.

Lifting crestboards is often difficult due to the water force against them. Figure 13-22 shows the force diagram of a submerged surface. The total force against submerged surfaces such as crestboards or headgates due to the water depth is determined as follows:

$$F = \frac{P_1 + P_2}{2} = \frac{\omega h_1 + \omega h_2}{2} = \omega \frac{h_1 + h_2}{2} \qquad (13\text{-}11)$$

where F = total force (lb)
   P = pressure due to water depth (lb/ft^2)
   ω = unit weight of water (lb/ft^3)
   h = water depth (ft)

Most fish remain in the pond during draining, and are usually harvested from the shallow water with seines. Catch basins or kettles sometimes are installed around drains. These basins are concrete-lined depressions in which fish accumulate for easy capture. Because of their expense, catch basins are seldom used in commercial ponds. Catch basins have been widely used on research stations and government-operated fish hatcheries. They are especially useful for fingerling production ponds where small fish are to be concentrated.

13.5.4   Inlet Structures for Water Supply

Levee ponds filled by water from wells are supplied through pipes. Where a well supplies a single pond, a valve on the inlet pipe is not essential. More often, several ponds will be supplied water from the same well, and valves must be provided on inlet pipes. Water supply pipes can be above ground, but they are usually buried so that they will not interfere with traffic. When water from canals or streams supplies ponds, individual pumps may be used to pump water directly into a pond. Where there are several ponds on the same farm, a water supply canal may be constructed between the ponds, and water can be introduced into ponds through crestboards. Where the supplying water carries a large amount of suspended solids, the supply canal also serves as a sediment-settling basin, or else a sediment basin must be provided before sediment-laden water reaches the pond area. This feature is very important in ponds used for brackish-water shrimp culture, for water supplies often contain high concentrations of suspended sediments. Pipes extending through the levee between the supply canal and ponds are another means of introducing water into ponds. In low-lying coastal areas where tidal fluctuation is substantial, ponds are sometimes constructed so that water will enter them through flashboard risers at high tide. Obviously, inlet structures must be screened if wild fish and other animals are to be denied entry to ponds filled by stream or canal water.

13.6 POND BOTTOM

The area for the pond bottom must be cleared of trees, shrubs, and other vegetation. Stumps and large roots should be removed. Sometimes the impervious layer may be disturbed during removal of stumps and roots with a bulldozer. If so, earth fill should be placed in holes and compacted. If areas of sand or other porous material are found in the bottom area, they should be excavated and replaced with more suitable fill materials.

If ponds are to be drained for fish harvest, it is essential to smooth bottoms so that depressions that will hold water after draining are absent. To facilitate draining, the pond bottom must slope toward the entrance of the drain structure. Pond bottoms should have a slope of 0.05–0.1% or more along the draining direction, and near the drain some cross slope is necessary. Areas for levee ponds normally are fairly level. Therefore, earth fill for levees often is obtained from the pond bottom. If tractor-pulled scrapers are used to build levees, the earth fill may be cut uniformly from the entire pond bottom and the required slope provided. Of course, if levee ponds are small, earth fill from pond bottoms must be supplemented with fill material from outside bottom areas. When large levee ponds are built with bulldozers, it is not convenient to cut the bottom uniformly. Many times when fill for levees is pushed up from the bottom with bulldozers, the cut will be deeper near the inside toe of levees than in the central area of the pond bottom. The finished bottom is deepest around the inside perimeters of the levees, and the central area

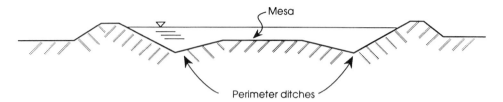

FIGURE 13-23.   Bottom shape of a levee pond. Note: Higher level on the central area.

is a mesa covered by shallow water (Fig. 13-23). If the perimeter ditch is sloped and smoothed adequately and the mesa is sloped toward the perimeter ditch and smoothed, the pond will drain readily. Shallow water in the central area is undesirable; nevertheless, this type of construction is widely used. When fill for levees is cut from within the pond area, the pond bottom will be of lower elevation than the surrounding ground. Therefore, ponds must be drained into a ditch that is excavated to a lower elevation than pond bottoms.

A technique for sloping pond bottoms is gaining popularity. The bottom is sloped toward the center and the drain is installed in the center. Aeration with paddle-wheel aerators is employed in the pond. A circular water movement is established by the aerators, and settleable solids tend to accumulate in the center of the pond by centrifugal forces where they can be removed by partially draining the pond. However, this technique may cause problems for ponds larger than 0.25–0.5 acre. The high velocity of water necessary to move solids to the center erodes the pond bottom. The eroded particles then settle over the central area of the pond and form a mound. Soil in the mound cannot be properly dried when ponds are drained between harvests.

The bottom shape of watershed ponds reflects the watershed topography. Excessive areas of shallow water are particularly undesirable because of the potential for weed infestation. The bottom area should be deepened so that the entire bottom will be filled with water at least 2 ft deep. When watershed ponds are used for aquaculture, it is desirable to limit maximum depth to 6 or 7 ft (Boyd 1985b). Soil may be cut from higher elevations in the bottom area and used to fill lower elevations in the bottom area. Soil cut from the higher elevation also may be used as fill for the dam.

13.7 OTHER FEATURES

Ponds for aquaculture must be accessible to service vehicles. Therefore, all-weather roads should be provided on levees and dams. Watershed ponds are not enclosed by levees, so perimeter roads should be constructed around them. Where it is not too expensive, gravel and sand should be applied to the roads. The roadway berm should be planted with grass to prevent erosion.

Levees and dams must be protected from erosion using grass covers. Areas on inside slopes of embankments that are subjected to strong wave action may be covered with riprap rocks. Wave action in watershed ponds may sometimes be reduced by planting trees around the perimeter areas as windbreaks. Wave height for moderate-size ponds may be determined by Hawksley's method, described by Schwab et al. (1981):

$$h = 0.025(D_f)^{1/2} \tag{13-12}$$

where h = wave height (ft) and D_f = fetch length (ft). Trees and shrubs should not be permitted to grow around pond edges or on embankments. Rooted aquatic vegetation around shallow edges of ponds is undesirable.

Watershed of a pond should be protected against soil erosion to prevent excessive turbidity and sedimentation in the pond. Well-maintained vegetative cover on a watershed will prevent erosion and sedimentation in the pond. Information on soil erosion control on natural or cultivated watersheds is found in Schwab et al. (1981), Norman and Hudson (1981), and Beasley et al. (1984).

13.8 EXCAVATED PONDS

Excavated ponds are made by digging a hole into the ground with a dragline, bulldozer, or other earth-moving equipment. Ponds to be filled with runoff must be built in a low-lying, fairly level area. The placement or disposal of the excavated material should be planned in advance of construction. Proper disposal will prolong the life of the pond as well as reduce the construction cost. The waste material may be stacked, spread, or removed from the site as condition warrants (Fig. 13-24). Grass cover must be established on the stacked material to prevent soil erosion. A spillway may be installed to bypass peak flow from surface runoff. Where the water table is near the surface, ground-water seepage will fill excavated ponds. If there is no high water table, pumped water from ground water or a nearby creek or stream may be used to fill the ponds. Seldom is it practical to install drain pipes in excavated ponds, and draining must be accomplished by pumping water from the ponds.

The sides of excavated ponds must be sloped in accordance with soil stability. Side slopes should be no steeper than 1:1, but 2:1 or 3:1 side

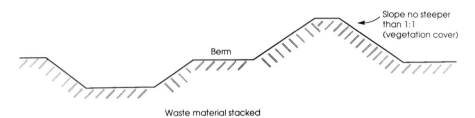

Berm

Slope no steeper
than 1:1
(vegetation cover)

Waste material stacked

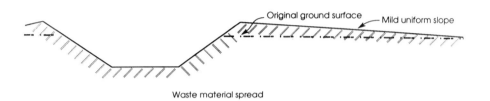

Original ground surface

Mild uniform slope

Waste material spread

Waste material removed

FIGURE 13-24.   Waste disposal methods of excavated ponds.

slopes are recommended. The volume of excavation in level ground may be estimated with the prismoidal formula

$$V = \frac{A + 4B + C}{6} \times D \times \frac{1}{27 \, \text{ft}^3/\text{yd}^3} \qquad (13\text{-}13)$$

where V = volume of excavation (yd^3)
A = area of excavation at ground surface (ft^2)
B = area of excavation at mid-depth (ft^2)
C = area of excavation at pond bottom (ft^2)
D = pond depth (ft)

13.9 SURVEYING AND STAKING

An initial survey is conducted and a topographic map is prepared for the proposed pond site. Where a pond will be filled by surface runoff, the survey must include the entire watershed. The proposed pond is designed in accordance with topographic features of the site, and information from the plan is transposed to the site by use of stakes. Stakes are set along the centerline of the embankment at intervals of 25–50 ft, depending on the variation of slope. Fill and slope stakes are set on each side of the centerline to mark the front and back edges of the bases of the embankment. These stakes also indicate embankment height. For a watershed pond, the centerline of the emergency spillway is staked and cut, and slope stakes are set where spillway side slopes intersect the ground surface. Cut and fill stakes are set in the pond bottom area if necessary. Locations of all water-control structures are marked with stakes. A traverse run around the valley at the same elevation as the top of the principal spillway establishes the position of the normal shoreline of a watershed pond. Stakes are used to identify the edge so that it may be deepened. For excavated ponds, stakes are installed to identify the outline of the excavation and to show slopes and depths. Where cut material will be used to build a levee around all or part of the excavation, appropriate stakes must be provided.

13.10 EXAMPLE PROBLEMS FOR POND DESIGN

Three example problems will be presented to illustrate design of different types of pond.

13.10.1 Levee Pond

A levee pond will be designed for construction on level land (Fig. 13-25). The inside dimensions measured from the bases of the levees will be 1,500 ft by 750 ft. Soil over the pond bottom is uniform and of high clay content. Therefore, fill material for the levees will be taken from the pond bottom. A cutoff in the foundation is not necessary. Side slopes will be 2:1 on wet and dry sides of the levees. A freeboard of 1.5 ft will be provided and the top width of the levees will be 12 ft. A swivel-joint standpipe will serve as the overflow and drain. Well water is available for filling the pond. Minimum water depth to the pond bottom will be 4 ft at the shallow end. For simplicity, cutting for the fill material will begin at one end and continue at a constant slope to the other end. The cross-sectional area of the levee at the shallow end is shown in Figure 13-25.

The height of the levee is determined based on the minimum water depth of 4 ft, assuming 10% final settlement. Then

Total height = (4 ft deep + 1.5 ft freeboard)(1.1) = 6.05 ft

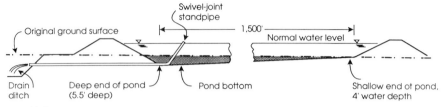

Note:
Shaded area will be excavated
to build embankment. At deep end
some cross slope must be provided
to facilitate complete draining.

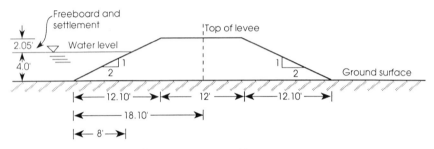

Cross-sectional area of levee

FIGURE 13-25.   Construction plan and cross section of a levee pond
design in Section 13.10.1.

The cross-sectional area is

$$A = (6.05)(12) + 2 \times 0.5(12.10)(6.05) = 145.8 \, \text{ft}^2$$

The centerline of the levee will be 18.10 ft from its inside edge, so the
perimeter of the levee centerline will be

$$2[1,500 \, \text{ft} + 2(18.10)] + 2[750 \, \text{ft} + 2(18.10)] = 4,645.2 \, \text{ft}$$

The total volume of the levee is

$$V = \frac{(145.8 \, \text{ft}^2)(4,645.2 \, \text{ft})}{27 \, \text{ft}^3/\text{yd}^3} = 25,084 \, \text{yd}^3$$

Assuming a cut factor of 1.15, the total amount of fill material
required is $25,084 \times 1.15 = 28,846 \, \text{yd}^3$.

The horizontal distance from the base of a levee to the place where the water surface strikes the levee is 8 ft. Hence, the surface area of the pond will be

$$[1,500 + (2)(8)] \times [750 + (2)(8)] = 1,161,256\,ft^2$$

which is 26.7 acres. The pond bottom slope of 0.1% may be use to estimate the amount of fill material that will be available from the pond bottom. Think of the uniform cut forming a triangle with maximum depth of $1,500 \times 0.001 = 1.50$ ft and a base of 1,500 ft (Fig. 13-25). The area of the triangle is the cross-sectional area of the cut or $1.5 \times (1,500 + 1.5 \times 2) \div 2.0 = 1,127\,ft^2$; the pond bottom width (750 ft) is multiplied by the cross-sectional area to obtain total available cut volume:

$$V = \frac{(1,127\,ft^2)(750\,ft)}{27\,ft^3/yd^3} = 31,305\,yd^3$$

Slightly more than the calculated fill volume will be available from the pond, for some cross slope must be provided near the drain to facilitate complete removal of water, and it would not be possible to cut the bottom exactly to the inside bases of levees in all places. However, the cut volume will be large enough to build the levees and other needs for the project. For most small levee ponds, fill material for levees must be obtained from places outside the pond area, for the cut required to obtain sufficient fill material for the levees will become excessively deep.

The elevation of the drain pipe will be 1.5 ft ($1,500$ ft $\times$ 0.1%) lower than the bottom of the pond at the shallow end. A minimum depth of 4.0 ft is desired, so the top of the standing drain pipe must be 5.5 ft above the bottom of the pond at the deep end. The end of the drain pipe should extend at least 5 ft beyond the inside bottom edge of the levee and at least 10 ft beyond the outside bottom edge of the levee. The pipe must slope downward from the horizontal and empty into a drain ditch. The diameter and number of the drain pipe may be selected based on how quickly it is desired to drain the pond. Techniques for determining drain pipe sizes are presented in Chapter 9.

13.10.2 Watershed Pond

A topographic map of a watershed is depicted in Figure 13-26a. Notice that the contour lines cross at the centerline of the proposed dam, which is indicated by the dashed line. The top of the embankment will be at 116-ft elevation. The control section of the emergency spillway will be 3 ft

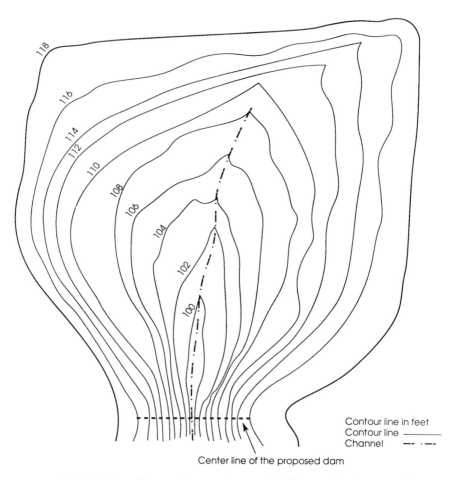

Contour line in feet
Contour line —————
Channel   —·· ·—··—

Center line of the proposed dam

FIGURE 13-26a.   Topographic map of a watershed for an embankment pond design in Section 13.10.2.

lower at 113 ft, and the top of the inlet of the principal spillway will be installed to establish the water level at 112 ft—1 ft lower than the bottom of the control section of the emergency spillway.

Total length of the embankment will be 180 ft. The height of the dam is indicated at intersections of contour lines with the centerline of the dam. The side slope on each side will be 3:1, and the top width will be 10 ft. The width of the base of the embankment is illustrated in the plan view (Fig. 13-26b). A cutoff with a bottom width of 10 ft and 1:1 side slopes will be installed to the embankment for a core trench. The cutoff will extend 8 ft below the ground surface. Assume plenty of soil with

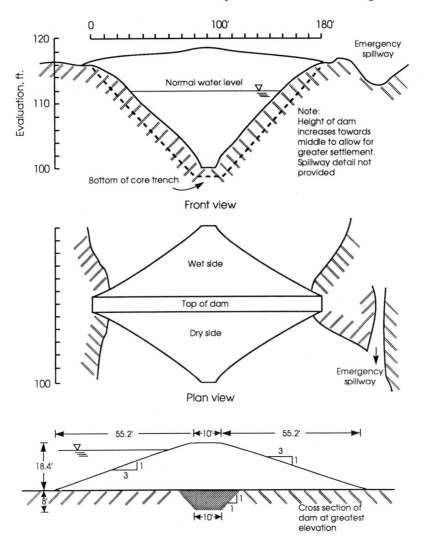

FIGURE 13-26b.   Front view, plan view, and cross section of an embankment pond design in Section 13.10.2.

a high clay content is available from a nearby borrow pit. The cutoff and embankment will be constructed from this material, so it will not be necessary to extend the cutoff above the ground level.

The cross-sectional area of the cutoff will be the same along its entire length:

$$A = 10\,\text{ft} \times 8\,\text{ft} + \frac{(8\,\text{ft} + 8\,\text{ft}) \times 8\,\text{ft}}{2} = 144\,\text{ft}^2$$

The length of the centerline at the ground level may be slightly longer than at the top of the dam. Assume that the length of the ground surface between the two ends of the dam is measured at 185 ft. The volume of the cutoff is

$$V = \frac{144\,\text{ft}^2 \times 185\,\text{ft}}{27\,\text{ft}^3/\text{yd}^3} = 987\,\text{yd}^3$$

Assuming a cut factor of 1.15, $987 \times 1.15 = 1{,}135\,\text{yd}^3$ of the fill material will be required for the cutoff.

The plan calls for the top of the dam to extend to 116-ft elevation. However, to allow for settlement, the dam will be built 10% higher at each point of measurement. The cutoff will be 8 ft deep, and the dam will be made an additional 0.8 ft higher over its entire length to account for the settlement of the cutoff. Obtain the presettled height and cross-sectional area of the embankment at each station as follows:

Presettled height:

$$H = \text{settled height} + 0.1(\text{settled height} + \text{cutoff depth})$$

Cross-sectional area (equation (13-9)):

$$A = 10H + 2 \times \left(\frac{3H \times H}{2}\right)$$

Volume of segment $(\text{yd}^3) =$

$$\text{segment length (ft)} \times \frac{\begin{array}{c}\text{sum of cross-sectional}\\ \text{areas of both ends of segment (ft}^2)\end{array}}{2 \times 27\,\text{ft}^3/\text{yd}^3}$$

At station 85, where the final height of dam is to be 16 ft high (ground surface elevation = 100 ft), the presettled height is $16 + 0.1(16 + 8) = 18.4$ ft. Calculation of the presettled volume of the embankment is given in Table 13-4. The total volume of embankment is $3{,}149.0\,\text{yd}^3$. Assuming a cut factor of 1.15, the amount of fill material for the dam is $3{,}149.0\,\text{yd}^3 \times 1.15 = 3{,}622.0\,\text{yd}^3$. Add to this the volume of earth fill for the cutoff, and the total amount of required fill material is $4{,}757.0\,\text{yd}^3$.

A vegetated emergency spillway will be installed just beyond one end

TABLE 13-4   Calculation of the Presettled Volume of an Embankment Pond Design Illustrated in Section 13.10.2

Station	Height (ft)		Cross-Sectional Area (ft^2)	Length of Segment (ft)	Volume of Segment (yd^3)
	Settled	Presettled			
0	0	0.8	9.9		
				20	24.8
20	2	3.0	57.0		
				10	35.2
30	4	5.2	133.1		
				10	68.8
40	6	7.4	238.3		
				10	113.1
50	8	9.6	372.5		
				10	168.2
60	10	11.8	535.7		
				10	234.0
70	12	14.0	728.0		
				10	310.6
80	14	16.2	949.3		
				10	398.0
85	16	18.4	1,199.7		
				10	444.3
95	16	18.4	1,199.7		
				10	398.0
100	14	16.2	949.3		
				10	310.6
110	12	14.0	728.0		
				10	234.0
120	10	11.8	535.7		
				10	168.2
130	8	9.6	372.5		
				10	113.1
140	6	7.4	238.3		
				10	68.8
150	4	5.2	133.1		
				10	35.2
160	2	3.0	57.0		
				20	24.8
180	0	0.8	9.9		
Total					3,149.7

of the embankment to carry a storm peak flow. In practice, the peak flow would be calculated from watershed features, rainfall data, and equations given in Chapter 4. Design dimensions of the emergency spillway may be determined by the method discussed in Chapter 8. The elevation of the bottom of the spillway will be 113 ft at the entrance of the control section. The bottom of the control section will slope toward the exit section at a rate of 0.5%. The bottom of the entrance section will slope toward the pond at a rate of 2%. The entrance section will be flared to favor smooth flow. The exit section will have a minimum slope of 4%. To conserve space, a drawing of the spillway will not be furnished, but it will conform to the specifications provided in Figure 13-11.

The edges of the pond should be deepened so that all water is deeper than 2 ft (Fig. 13-27). Of course, the edges should be sloped no less than 2:1 to prevent sloughing. The embankment above the waterline and the sides and bottom of the emergency spillway should be covered with grass to prevent erosion. To drain excess water a drop-inlet pipe spillway may be installed as a principal spillway. Determining the size of the principal spillway is found in Chapter 9. Even though the dam will be made of soil with a high clay content, antiseep collars are recommended.

13.10.3 Excavated Pond

An excavated pond is to be constructed in an area where the water table depth occurs 3 ft below the ground surface. The desired pond area at the water level is 120 ft by 60 ft; a water depth of 6 ft is required. Side slopes will be 2:1 for the clay–loam soil at the site.

Design dimensions of the pond at different depths are provided in Figure 13-28. Areas at the ground surface, and mid-depth and bottom of the excavation are 9,504 ft^2, 6,156 ft^2, 3,456 ft^2, respectively. The volume of the excavation is calculated by equation (13-13):

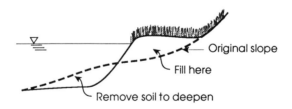

Deepening pond edges by cut and fill method to avoid weed problem

FIGURE 13-27.   A deepening method of fish pond edges.

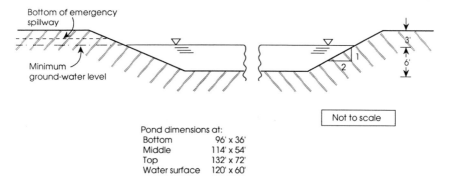

Pond dimensions at:
Bottom   96' x 36'
Middle   114' x 54'
Top   132' x 72'
Water surface   120' x 60'

FIGURE 13-28.   Final dimensions of an excavated pond in Section 13.10.3.

$$V = \frac{9{,}504\,\text{ft}^2 + 4 \times 6{,}156\,\text{ft}^2 + 3{,}456\,\text{ft}^2}{6} \times 9\,\text{ft} \times \frac{1}{27\,\text{ft}^3/\text{yd}^3}$$

$$= 2{,}088.0\,\text{yd}^3$$

The bottom of the emergency spillway entrance will be 1 ft above the normal water level. The spillway will be 2 ft deep and have 2:1 side slopes. The bottom of the spillway will be sloped at 0.5%, and it will be excavated to a nearby drainage ditch. The channel will be a vegetated waterway to prevent erosion. A method discussed in Chapter 8 may be used to determine design dimensions of the waterway. The waste material may be used as described in Figure 13-24.

14

Pond Effluents

14.1 INTRODUCTION

Effluents from aquacultural ponds contain living and dead particulate organic matter, dissolved organic matter, ammonia, nitrate, phosphate, suspended soil particles, and other substances that can be considered potential pollutants. Therefore, agencies responsible for water pollution abatement consider the aquaculture industry as a polluter. In the United States some states have developed standards and criteria for aquaculture effluents; they require permits for discharging pond effluents into natural water, and effluent characteristics must comply with specifications set forth in the permit. Other states are considering implementation of a regulatory process. Canada, Australia, and most European nations have also implemented regulations about aquaculture effluents. Even some countries in tropical regions are developing regulations about pond effluents. For example, in Thailand, a rule requiring shrimp farms to register for a permit was issued in 1991. The regulation states that farms larger than 50 rai (about 20 acres) must have a waste treatment or sediment pond not less than 10% of the total size of the farm, effluents discharged to natural waters cannot have a biochemical oxygen demand (BOD) greater than 10 ppm, mud or silt from ponds cannot be disposed of in natural waters, and saltwater cannot be drained into freshwater courses. It is clear that the aquaculture industry in most places will come under some type of regulation within the next few years.

Commercial aquaculturists generally disagree with government regulation of any type, but regulation of effluents from ponds and other aquacultural production facilities, either by the aquaculture industry itself or by government, is a necessary activity. Where many production units are located in a small area, effluents may seriously impact environmental conditions in receiving water bodies. In brackish-water aquaculture, the same body of water that is the recipient of aquaculture effluents is the water supply for the production facilities. Thus, brackish-water aquaculture is self-polluting.

We are including a brief discussion of effluents because this topic has become a major issue that affects water supply and water use in aquaculture.

14.2 WATER QUALITY IN PONDS

14.2.1 Effects of Feeding

When a pond is supplied with water of reasonably high quality, processes causing water-quality impairment are driven by feed input (Boyd 1990). The species under culture eat most of the feed applied to a pond; the remainder accumulates on the pond bottom. Of the feed consumed by the culture species, part is absorbed across the gut and part becomes feces. Of the absorbed portion, part is converted to animal flesh, and part is used in metabolism and excreted into the water as carbon dioxide and ammonia. The uneaten feed and feces that accumulate on the pond bottom decompose, and carbon dioxide, ammonia, phosphate, and other plant nutrients enter the pond water. Plant nutrients released by decomposition and aquatic animal excrement stimulate the production of organic matter in the form of phytoplankton cells. Phytoplankton has a life span of one to two weeks, and when individuals die they settle to the bottom and are decomposed by microorganisms. As the feeding rate increases, standing crops of the culture species, phytoplankton, and other organisms in the pond water and on the pond bottom increase. Organic matter concentrations on the bottom and dissolved organic matter concentrations in the water also increase. Although the amount of dissolved oxygen produced by phytoplankton photosynthesis is greater at higher feeding rates, the demand for oxygen by respiratory processes also increases. Natural supplies of dissolved oxygen often are inadequate at feeding rates above 25–30 lb/acre per day to maintain minimum, permissible dissolved oxygen concentrations in the water. Aeration may be applied to permit greater feeding rates. Ammonia concentrations increase with increasing feeding rates, and unless water can be flushed through ponds to maintain ammonia concentrations within acceptable ranges, ammonia will impose limits on production even when high rates of aeration are used (Cole and Boyd 1986).

Even with heavy aeration and water exchange, feed input and aquatic animal production have limits. A feeding rate will finally be reached, usually around 100–120 lb/acre per day, where microbial activity will deplete dissolved oxygen concentrations in bottom soils and sediment. Reduction reactions in the anaerobic zones of the bottom will release ferric iron, manganous manganese, nitrite, hydrogen sulfide, and other reduced substances. Many times the reduced substances cannot be detected in the water above the sediment or bottom soil because they are quickly oxidized in the oxygenated water. However, some species (e.g., shrimp) live on the pond bottom, and the presence of reduced substances, especially hydrogen sulfide, can cause chronic or acute toxicity. Stress resulting from exposure to poor environmental conditions weakens aquatic animals and makes them more susceptible to disease.

Water currents produced by large inputs of aeration may actually intensify water-quality problems in ponds. Soil and overlying organic sediment are eroded from areas of the bottom with high-velocity currents. This erosion causes turbidity in the water, and the suspended particles finally are deposited in areas with low-velocity currents to form large mounds of sediment with a high oxygen demand. Oxygen from the water cannot enter the sediment fast enough to maintain aerobic conditions. The erosion and sedimentation caused by excessive aeration is also undesirable because it reshapes the pond bottom. The pond will no longer drain properly, and it may not be possible to dry the pond bottom properly between crops.

A pond has a finite capacity to process plant nutrients and organic matter originating from feed. The first evidence of overfeeding is low concentrations of dissolved oxygen. Aeration augments dissolved oxygen supplies and permits greater feed input. The next evidence of overfeeding comes when ammonia concentrations become too high. Where water supply permits, water may be flushed through ponds to alleviate ammonia limitations and allow further increases in feed input. With aeration and water exchange at high rates, continued increase in the stocking and feeding rate results in deterioration of conditions at the pond bottom. Because of the nature of the physical factors regulating delivery of dissolved oxygen into the sediment and bottom soil, we doubt that a simple method for eliminating this third limitation is possible. Water quality and soil condition simply worsen as feeding rates are raised, and with the current knowledge of environmental factors and their manipulation there is no feasible way of further increasing animal production in ponds. In fact, many farmers are stocking too heavily and applying more feed than their ponds can process, and they are causing the water-quality problems. Of course, as water quality deteriorates in ponds, the quality of effluents from ponds also deteriorates. Once a pond becomes so eutrophic from feeding that aeration and water exchange cannot be depended on to maintain adequate water quality and soil condition for good animal growth, the only known feasible solution is to reduce the nutrient and organic matter input through lowering the stocking and feeding rate. This is not a very popular assessment, because most commercial producers want to obtain the highest production possible.

14.2.2 Material Budget

Information from channel catfish culture will be used to illustrate the amount of wastes that results from an aquacultural crop. It is not unusual to obtain an annual harvest weight of 5,000 lb/acre of channel catfish with a feed input of 8,000 lb/acre. The feed is about 5% nitrogen, 1% phosphorus, and 90% dry matter. Channel catfish are about 2% nitrogen, 0.4% phosphorus, and 25% dry matter. The dry matter of feed and fish is about 10% and 15% ash, respectively. The data were used to calculate differences between dry matter, carbon, nitrogen, and phosphorus applied in feed and harvested in fish (Table 14-1). Data from this table suggest that for each 1,000 lb of live channel catfish produced, around 540 lb carbon, 60 lb nitrogen, and 12 lb phosphorus will enter the water in metabolic wastes. These wastes consist primarily of organic matter in feces, ammonia, phosphate, and carbon dioxide. The nutrients in the waste stimulate phytoplankton productivity, and for each 1,000 lb of live channel catfish produced about 3,000 lb of organic matter (dry weight) are produced by phytoplankton photosynthesis (Boyd 1985c).

Nutrients and organic matter added to pond water as the result of feeding may be lost to the atmosphere (carbon dioxide and nitrogen), adsorbed on bottom soils (phosphorus), deposited on the pond bottom, or lost in outflow. These calculations are based on channel catfish produc-

TABLE 14-1   Material Budget for Channel Catfish Pond

Variable	Input in Feed (lb/acre)	Removal in fish (lb/acre)	Difference (lb/acre)	Waste (lb/1,000 lb)
Dry matter	6,422	1,115	5,307	1,190
Carbon[a]	2,890	474	2,416	542
Nitrogen	357	89	268	60
Phosphorus	71	18	53	12

[a] Assume 50% carbon in organic matter.

tion, but roughly the same proportions of wastes would be produced per ton of other species with feeding.

14.2.3 Quality of Effluents

During normal overflow in response to storms or water exchange, effluents will be of the same quality as water in the pond from which they originate. Such water usually contains adequate dissolved oxygen for aquatic life, but it has elevated concentrations of nitrogen, phosphorus, organic matter, and other potential pollutants (Table 14-2). The organic matter will contain most of the nitrogen and phosphorus, and living phytoplankton cells comprise the majority of the organic matter. Settleable solids will seldom be present in measurable quantities. Effluents discharged during harvest of culture species are more concentrated in dissolved and suspended substances than water discharged from ponds during normal overflow or water exchange (Boyd 1978). In deep, stratified ponds that are drained from the deepest point, the initial discharge may be oxygen-depleted and contain high concentrations of reduced substances such as ferrous iron and hydrogen sulfide. As the draining process continues, water quality will improve as the deeper, anoxic water is drained out and the surface, oxygenated water appears in the discharge. In all ponds, the last part of the draining procedure will yield water of relatively low quality that has especially high concentrations of settleable solids (Table 14-3). This deterioration in effluent quality results from the seining operation to remove the aquatic animals.

Pond effluents normally do not contain toxic substances such as are found in some industrial and agricultural wastes. They contain plant nutrients that can accelerate phytoplankton growth and eutrophication of

TABLE 14-2 Normal Ranges of Concentrations in Parts per Million (ppm) of Selected Water Quality Variables in Surface Waters of Aquaculture Ponds

Variable	Range
Dissolved oxygen, ppm	3–15
pH	6–9
Carbon dioxide, ppm	0–20
Total ammonia nitrogen, ppm	0.5–5.0
Chemical oxygen demand, ppm	20–200
Biochemical oxygen demand, ppm	2–5
Total suspended solids, ppm	10–100
Total phosphorus, ppm	0.1–1.0
Total nitrogen, ppm	1–15
Total settleable solids, mL/L	0–0.5

TABLE 14-3   Average Effluent Quality from Eight, Shallow (3–5 ft average depth), Channel Catfish Ponds during Harvest

Variable	Initial[a] Draining	Water Discharged during Fish Removal by Seining
Settleable matter, mL/L	0.08	28.5
Biochemical oxygen demand, ppm	4.31	28.9
Chemical oxygen demand, ppm	30.2	0.342
Dissolved orthophosphorus, ppb	16	59
Total phosphorus, ppm	0.11	0.49
Total ammonia nitrogen, ppm	0.98	2.34
Nitrate nitrogen, ppm	0.16	0.14

Source: Boyd (1978).

[a] About 95% of the total pond volumes were discharged during initial draining.

receiving waters, organic matter that creates an oxygen demand in receiving waters, settleable matter that can form sediment in receiving waters, and some effluents may be vectors of aquatic animal disease.

Sometimes sediment is removed from ponds and disposed of through effluent canals. This material usually has a high oxygen demand, and it contaminates the water supply.

14.3   VOLUME OF EFFLUENTS

The volume of effluents from ponds was discussed in Chapter 5. The chapter contains equations for preparing water budgets for ponds and methods for estimating pond discharge.

14.3.1   Watershed Ponds

Watershed ponds overflow in response to storm runoff on watersheds, and they will generally have continuous overflow during the wetter parts of the year. For example, watershed ponds in the southeastern United States typically have overflow from early winter until later spring. The volume of overflow depends on the volume of the pond in relation to area and runoff-producing features of the watershed and the amount of rainfall. Shelton and Boyd (1993) demonstrated that the volume of overflow from watershed ponds in Alabama was approximately equal to the volume of runoff entering the ponds. If ponds are drained for harvest, effluent equal to pond volume will occur. However, draining will not increase the amount of effluent, because an amount of runoff equal to the volume of water drained from the pond will be stored in refilling the pond.

The amount of effluent from watershed ponds is highly variable. Most ponds in the southeastern United States will annually discharge water equal to 1.5 to 3 times the pond volume. Some ponds with large watersheds or with inflow from springs or permanent streams may discharge a much greater volume of water.

The relationship between runoff entering ponds and volume of effluent may not occur in other climatic regions. In Alabama, the amount of precipitation falling into ponds roughly balances the amount of water lost through seepage and evaporation. This means that once ponds are full, runoff simply flows through ponds.

14.3.2 Levee Ponds

Levee ponds usually are supplied by water from an external source and do not receive significant runoff. Therefore, they have much less overflow from storms than do watershed ponds. In fact, if water levels are maintained a few inches below the intake of the overflow structure, rain falling into ponds can be conserved and there will be no overflow after most storms (Boyd 1982a). In most climates, it is possible to operate levee ponds with only a few inches of overflow annually. Of course, if ponds are drained for fish harvest, an amount of effluent equal to pond volume must be released.

14.3.3 Water Exchange

Obviously, water exchange increases the amounts of effluent from ponds. There are no standard practices on water exchange. However, in many freshwater ponds, water exchange is not attempted, because the necessary volume of water to permit it is unavailable. In brackish-water ponds, water exchange is frequently used, and exchange rates normally range from 2% to 5% per day in extensive aquaculture to as high as 30% to 40% per day in intensive aquaculture. A water-exchange rate of 1% of pond volume per day is equal to $436 \, ft^3$/acre of effluent per foot of average pond depth or $100 \, m^3$/ha of effluent per meter of average pond depth.

14.4 DILUTION AND ASSIMILATION BY RECEIVING WATERS

Effluents from freshwater ponds often are more concentrated in organic matter, settleable matter, and nutrient concentrations than receiving streams or lakes. When a pond effluent enters a natural body of water, it is diluted. Natural bodies of water also have the capacity to assimilate organic matter and nutrients. Organic matter is broken down by microbial processes, nitrogen is denitrified and lost as a gas, phosphorus is bound in sediment, and solids settle from the water. Nevertheless, it is possible to overload a natural body of water with plant nutrients, sediment, and organic matter. The results are low-dissolved-oxygen concentrations, high-plant-nutrient concentrations, phytoplankton blooms, and poor bottom soil conditions—the same symptoms produced by overfeeding in an aquaculture pond.

Industrial, domestic, and agricultural wastes cause eutrophication of water bodies by increasing concentrations of nitrogen, phosphorus, and organic matter. Many nations have devoted a large effort to mitigating eutrophication in natural water bodies. The only practical solution found has been to reduce the pollution load to a eutrophic water body. Therefore, where aquacultural effluents are polluting natural bodies of water, the only way to protect the natural waters is to reduce the amount and pollutant content of the aquaculture effluents.

Effluents from brackish-water ponds are discharged into brackish-water reaches of rivers and canals or released directly into the sea. The degree of dilution of the effluent in coastal water depends on many factors, the most important of which are volume ratios of effluents to receiving bodies, flow of freshwater into estuaries, mixing rates of waters within estuaries, tidal exchange with the sea, and transportation of effluent-polluted waters away from the shrimp-farming area by ocean currents. Of course, the natural processes of assimilation mentioned for freshwater also occur in brackish water.

14.5 EFFECTS OF EFFLUENTS ON WATER SUPPLY

Pollution from freshwater aquaculture ponds usually does not affect the water supply of other ponds. The pollution load is of concern because it affects the quality of the receiving water for other uses.

Brackish-water ponds, on the other hand, often are self-polluting, because they use the same body of water for water supply and effluent recipient. Effluent discharged by one farm may contaminate the water source of another farm. Of course, brackish-water ponds are not the only sources of pollution to coastal waters. Ponds often are built in regions where coastal waters were highly polluted with domestic, industrial, and agricultural effluents before the advent of aquaculture. However, a high concentration of semi-intensive or intensive ponds in an area contributes to further degradation of coastal water.

Water in ponds becomes highly enriched with nutrients and organic matter from feed input. However, the amount of feed wastes that can be assimilated by a pond before water quality deteriorates to an unacceptable level depends on the quality of the water used for filling the pond and for pond water exchange. Therefore, if water sources for ponds are highly polluted when they enter ponds, water-quality problems may occur at very low feeding rates or even in ponds without feeding. In addition, water sources contaminated by effluents from other ponds may spread fish and shrimp disease.

14.6   MITIGATION

14.6.1   Treatment

There is a tendency for those involved in the aquaculture pond effluent problem to propose various wastewater treatment techniques. One of the most popular ideas is the use of native or created wetlands for natural biological treatment of effluents. Another idea is the production of other aquacultural species, such as filter-feeding fish and molluscs, in the effluent to remove particulate matter and the cultivation of commercially valuable plants to remove nitrogen and phosphorus. It seems doubtful that such measures will be efficient and economically practical. Space requirements for wetlands generally would be prohibitive, and the discharges from freshwater ponds are sporadic and highly variable in quality and quantity. This variation in effluent characteristics would complicate the operation of wetland waste treatment systems. Ponds have large volumes of effluents and, as a result, release substantial quantities of organic matter, nitrogen, and phosphorus in the environment. However, the concentrations of dissolved and particulate substances are quite low in pond effluents, and past experience suggests that nutrient concentrations are not high enough to make effluents good media for producing marketable plant crops. Also, other aquacultural species can be produced in effluents, but work in this area has not yet resulted in practical systems.

The most feasible treatment appears to be the use of small sedimentation ponds. When ponds are drained for harvest, the water, which is laden with suspended particles of soil and organic matter, could be clarified by retention for a few hours in sedimentation ponds. Suspended particles settle more rapidly from brackish water than from freshwater, so a longer retention time would be necessary in freshwater. The major source of suspended particles in overflow from ponds is phytoplankton. Sedimentation ponds are less useful for phytoplankton removal.

14.6.2   Pond Management

The best approach to the pond effluent problem is to strive to improve water quality in ponds and to reduce the volume of effluents. This would enhance environmental conditions in ponds and favor more efficient aquacultural production, improve effluent quality, reduce effluent volume, and lower water consumption. There would be benefits to aquaculture beyond mitigation of effluent problems. Consideration should be given to the following:

1. Use high-quality feeds and conservative feeding practices. This will improve the feed conversion ratio and reduce quantities of metabolic waste and uneaten feed.
2. Aerate and circulate pond water. Maintenance of high-dissolved-oxygen concentrations encourages good feed conversion ratios. Circulation prevents stratification and enhances degradation of organic matter in pond bottoms.
3. Minimize water exchange. Studies to determine optimal rates of water exchange have not been conducted. Most water exchange is employed as a safety measure, and much water is exchanged needlessly (Wang 1990).
4. Harvest fish without draining ponds. Many farmers use this practice, and experience suggests that water quality is not greatly impaired. Of course, ponds have to be drained every few years for repairs and to remove large fish that have escaped capture. Shrimp harvest is not efficient unless ponds are drained.
5. Treat the pond bottom. When ponds are drained for harvest, pond bottoms should be dried and tilled to provide better contact of soil with air and accelerate organic matter decomposition. Acidic soils in pond bottoms should be limed to improve bacterial activity.
6. Reuse water. When ponds are drained for fish harvest, water could be pumped to a storage pond and then reused in the same or other ponds.
7. Remove sediment. Where sediment is removed from settling ponds or grow-out ponds, do not dispose of the sediment in natural bodies of water.
8. Control saline water. Water from brackish-water ponds should not be discharged into bodies of freshwater.

14.6.3   Regulation

There are many possible schemes for regulating aquaculture pond effluents, but most regulations involve some combination of the following:

1. Limits on volume and quality of effluents.
2. Limits on production area of each producer.
3. Limits on feed input.
4. Specification of best management practices; e.g., effluents would have to be handled and treated according to specified procedures.
5. Zoning of areas for ponds.
6. Limits on expansion in a locality.

For a regulation to be successful, it would have to be based on sound principles so that its implementation would bring about an improvement in water quality or prevent deterioration of water quality in the recipient water body. Also, the regulatory agency would have to develop a permitting procedure to include all producers, and ponds would have to be monitored to ensure compliance with regulations. Such endeavors tend to become very political and are often not fairly applied to all producers. A much better approach to solving the problem would be to educate producers on the benefits of improving the quality and reducing the volume of effluents through better pond management procedures.

14.7   ENVIRONMENTAL IMPACT OF AQUACULTURE

14.7.1   Shrimp Farming

High profits have been possible from the cultivation of marine shrimp in ponds in several countries in tropical regions. In some instances, producers have been able to pay back the entire cost of farms in one or two years. In Southeast Asia, annual profits from some intensive farms have exceeded $15,000 per acre. The profit motive has caused a "shrimp rush" equal to "land rushes" and "gold rushes" that have occurred at other times.

In many locations suitable for shrimp culture, local landowners and investors have, within periods of two or three years, converted all of the available land into shrimp ponds. Tremendous expanses of mangrove and other coastal wetlands have been destroyed or badly damaged. Many freshwater habitats near the coast have been damaged by high-salinity effluents from ponds. Producers raised stocking and feeding rates to maximum possible levels, and sometimes beyond, in hopes of reaping tremendous profits. In some places, producers made large profits for two or four years, after which the shrimp ponds were plagued with water-quality and disease problems and production and profits fell. These failures are related to water quality caused by contamination of the coastal water, which also serves as water supply for the shrimp farms, with shrimp pond effluents. There are places where shrimp production is no longer possible because the water supply is no longer of adequate quality to support shrimp culture.

An investor who obtains a lot of profit in a short time may not be upset when shrimp farming fails in an area. He has made his money, and he can invest it in shrimp ponds at another location or in some other enterprise. However, the greed of such investors is very destructive to the environment and detrimental in the long run to the shrimp-farming industry.

Long-term sustainability of shrimp farming requires changes in current practices. In order to prevent water-quality deterioration in ponds and water supplies, restrictions must be placed on stocking density, feeding rate, and the number of ponds in an area so that the pollution load created by the feed does not exceed the assimilative capacity of the ponds and water supply. There are not quick solutions to the problem that will not place any restrictions on current practices. The shrimp-farming industry must move rapidly to develop "conservation shrimp farming," or shrimp farming will collapse in many more places worldwide.

14.7.2 Cage Culture in Large, Deep Lakes

In Indonesia there are several large, deep lakes on Java where there has been a proliferation of cage culture of carp. The input of feed to the cages has led to eutrophication of the lakes. Dense phytoplankton blooms in surface waters cause shallow thermal and chemical stratification. The surface stratum (epilimnion) is usually 6–12 ft thick, contains plenty of dissolved oxygen, and is a suitable environment for fish. Below the surface layer, water temperature drops quickly and there is sufficient density difference between the warm epilimnion and the cooler, deeper water (hypolimnion) that the two strata do not normally mix. The bottom water becomes depleted of dissolved oxygen as microorganisms decompose wastes that settle downward from the cages. Under anaerobic conditions, partially decomposed organic matter and reduced inorganic compounds (nitrite, ammonia, hydrogen sulfide, ferrous iron, etc.) accumulate. Heavy winds and rains frequently cause upwelling of the deep, oxygen-depleted water. When this happens, the upwelling hypolimnetic water dilutes the dissolved oxygen concentration. Further, organic matter and other reduced material in the upwelling water exerts an immediate and large oxygen demand. Fish kills are common in cages during periods of upwelling (Schmittou 1991). The only feasible solution to this problem is to reduce the amount of feed applied to the lake.

Appendixes

Appendix A

Conversion Tables

TABLE A-1   International System (SI metric) and U.S. Customary Unit Conversion Table[a]

To Convert From	To	Multiply By
	Length	
in.	mm	25.4
	m	0.0254
	mils	1,000
	ft	0.083
ft	mm	305.0
	m	0.305
	in.	12
	yd	0.33
yd	m	0.91
	in.	36
	ft	3
miles	m	1610.0
	km	1.61
	ft	5280.0
	yd	1760.0
	ft	0.00328
mm	μm	1,000
	cm	0.1
	m	0.001
	mils	39.4
	in.	0.0394
cm	mm	10
	m	0.01
	mils	394.0

To Convert From	To	Multiply By
	Length	
cm	in.	0.394
	ft	0.0328
m	mm	1,000
	km	0.001
	in.	39.4
	ft	3.28
	yd	1.1
	mile	0.00062
km	m	1,000
	ft	3280.0
	mile	0.62
	Area	
$in.^2$	mm^2	645.0
	cm^2	6.45
ft^2	m^2	0.093
	$in.^2$	144
yd^2	m^2	0.836
	ft^2	99
acres	m^2	4050.0
	ha	0.405
	ft^2	43,560
$mile^2$	ha	259.0
	km^2	2.59
	acres	640.0
cm^2	$in.^2$	0.155
	mm^2	100
	m^2	0.0001
m^2	ha	0.0001
	ft^2	10.8
	yd^2	1.2
ha	m^2	10,000
	acre	2.47
km^2	ha	100
	acre	247.0
	$mile^2$	0.386
	Volume	
$in.^3$	cm^3 or mL	16.4
	ft^3	0.007
	L	0.0164
gal	L	3.80
	ft^3	0.134
	$in.^3$	231.0
ft^3	L	28.3

To Convert From	To	Multiply By
	Length	
ft^3	$in.^3$	1,728
	gal	7.48
yd^3	m^3	0.764
	ft^3	27.0
cm^3	L	0.001
	mL	1.0
ml	L	1,000
	cm^3	1.0
L	mL	1,000
	m^3	0.001
	ft^3	0.035
	gal	0.264.0
m^3	L	1,000
	gal	264
	ft^3	35.3
	yd^3	1.31
acre-ft	m^3	1,233
	ft^3	43,560
	gal	325,850
	Velocity	
fps or ft/sec	m/sec	0.305
	km/hr	1.1
	miles/hr	0.68
mph or miles/hr	km/hr	1.61
	m/sec	0.45
	ft/sec	1.47
m/sec	km/hr	3.60
	ft/sec	3.3
	miles/hr	2.22
km/hr	m/sec	0.28
	ft/sec	0.91
	miles/hr	0.62
	Volume Per Unit Time (Flow Rate)	
cfs or ft^3/sec	L/sec	28.3
	m^3/sec	0.028
	gpm	449.0
	acre-ft/day	1.98
	ft^3/min	60.0
gpm	L/sec	0.063
	cfs	0.0022
	acre-ft/day	0.0042
acre-ft/day	m^3/sec	0.0143
	cfs	0.504

To Convert From	To	Multiply By
	Weight	
oz	g	28.3
	kg	0.0283
	lb	0.063
lb	kg	0.45
	oz	16.0
slugs	kg	14.6
short tons	kg	907.0
	metric tons	0.907
	lb	2,000
long tons	kg	1,016.0
	metric tons	1.016
	short tons	1.12
	lb	2,240.0
g	kg	0.001
	oz	0.035
kg	g	0.001
	short tons	0.0011
	metric tons	1,000
	lb	2.24
t or Mg	kg	1,000
	lb	2,240.0
	short tons	1.12
	Power	
ft · 1b/sec	W	1.36
	hp	550
hp	W	746.0
	kW	0.746
W	kW	1,000
	ft · 1b/sec	0.74
	Energy	
Btu	J	1,055
	ft · 1b	778
	Acceleration[a]	
ft/sec^2	m/sec^2	0.305
m/sec^2	ft/sec^2	3.28

To Convert From	To	Multiply By
	Force Per Unit Area (Pressure–Stress)[b]	
1b/in.^{2} or psi	kPa	6.9
	m of water	0.7[c]
	mm Hg	51.7[d]
	in. Hg	2.04[d]
	ft of water	2.31[c]
	1b/ft^2	144.0
	std. atm	0.068
1b/ft^2	kPa	0.048
	m of water	0.0049[c]
	mm Hg	0.36[d]
	ft of water	0.016[c]
	psi	0.0069
	std. atm	0.00047
ft of water	kPa	3.0
	m of water	0.305[c]
	mm Hg	22.4[d]
	in. Hg	0.88[d]
	psi	0.43
	1b/ft^2	62.4[c]
m of water	kPa	9.8
	mm Hg	73.6[d]
	ft of water	3.28[c]
	psi	1.42
	1b/ft^2	205.0
kPa	N/m^2	0.001
	mm Hg	7.5[d]
	m of water	0.102[c]
	in. Hg	0.29[c]
	1b/ft^2	20.9[c]
	psi	0.145
	std. atm	0.00987
	Mass Per Unit Volume (Density)[e]	
1b/ft^3	kg/m^3	16.0
	slug/ft^3	0.031
	1b/gal	0.133
1b/gal	kg/m^3	119.8
	slug/ft^3	0.23
1b/yd^3	kg/m^3	0.59
	1b/ft^3	0.037
g/cm^3	kg/m^3	1,000
	1b/yd^3	1,686

To Convert From	To	Multiply By
	Mass Per Unit Volume (Density)[e]	
kg/m^3	g/cm^3	0.001
	metric tons/m^3	0.001
	$1b/yd^3$	1.69

Note: To convert in reverse order, divide by the same factor.
[a] Standard gravitational acceleration (g): $9.8 \, m/sec^2$, $32.2 \, ft/sec^2$.
[b] Standard atmosphere at mean sea level: 101.3 kPa, 760.0 mm Hg, 29.9 in. Hg, 14.70 psi, 33.9 ft of water, 10.34 m of water.
[c] at 50°F.
[d] at 0°C.
[e] Weight of water at 50°F: $62.4 \, 1b/ft^3$, $8.3 \, 1b/gal$, $9.81 \, kN/m^3$, $1,000 \, kg/m^3$.

TABLE A-2   Temperature Conversions

From °F to °C		From °C to °F	
°F	°C	°C	°F
30	−1.11	−5	23.0
32	0	0	32.0
35	1.67	2	35.6
40	4.44	4	39.2
45	7.22	6	42.8
50	10.0	8	46.4
55	12.8	10	50.0
60	15.6	12	53.6
65	18.3	14	57.2
70	21.1	16	60.8
75	23.9	18	64.4
80	26.7	20	68.0
85	29.4	22	71.6
90	32.2	24	75.2
95	35.0	26	78.8
100	37.8	28	82.4
105	40.5	30	86.0
110	43.3	34	93.2
115	46.1	38	100.4
120	48.9	42	107.6
125	51.7	46	114.8
130	54.4	50	122.0
135	57.2	55	131.0
140	60.0	60	140.0

$$°F = \frac{9}{5} \times °C + 32° \qquad °C = \frac{5}{9}(°F - 32°)$$

Appendix B

Determining Friction Head Losses due to Pipe and Pipe Fittings

TABLE B-1 Friction Head Loss in Pipes from Hazen–Williams Equation for Coefficient of 100 Corresponding to 10-Year-Old Steel or 18-Year-Old Cast-Iron Pipe (flow in gpm, velocity in ft/sec, and friction head loss in ft per 100 ft of pipe length)

Flow	Velocity	Velocity Head	Head Loss	Flow	Velocity	Velocity Head	Head Loss	Flow	Velocity	Velocity Head	Head Loss
1/2" Pipe (0.622" ID)				1" Pipe (1.049" ID)				2" Pipe (2.067" ID)			
0.5	0.52	0.00	0.6	2	0.74	0.01	0.6	10	0.96	0.01	0.4
1.0	1.06	0.02	2.1	3	1.11	0.02	1.3	12	1.15	0.02	0.6
1.5	1.58	0.04	4.4	4	1.49	0.03	2.1	14	1.34	0.03	0.8
2.0	2.11	0.07	7.6	5	1.86	0.05	3.2	16	1.53	0.04	1.0
2.5	2.64	0.11	11.4	6	2.23	0.08	4.5	18	1.72	0.05	1.3
3.0	3.17	0.16	16.0	8	2.97	0.14	7.7	20	1.91	0.06	1.6
3.5	3.70	0.21	21.3	10	3.71	0.21	11.7	22	2.10	0.07	1.9
4.0	4.23	0.28	27.3	12	4.46	0.31	16.4	24	2.29	0.08	2.2
4.5	4.75	0.35	33.9	14	5.20	0.42	21.8	26	2.49	0.10	2.5
5.0	5.28	0.43	41.2	16	5.94	0.55	27.9	28	2.68	0.11	2.9
5.5	5.81	0.52	49.2	18	6.68	0.69	34.7	30	2.87	0.13	3.3
6.0	6.34	0.62	57.8	20	7.43	0.86	42.1	35	3.35	0.17	4.4
6.5	6.87	0.73	67.0	22	8.17	1.04	50.2	40	3.82	0.23	5.6
7.0	7.39	0.85	76.8	24	8.91	1.23	59.0	45	4.30	0.29	7.0
7.5	7.92	0.97	87.3	26	9.66	1.45	68.4	50	4.78	0.36	8.5
8.0	8.45	1.11	98.3	28	10.4	1.7	78.5	55	5.26	0.43	10.1
8.5	8.45	1.25	110.	30	11.1	1.9	89.2	60	5.74	0.51	11.9
9.0	9.51	1.4	122.	35	12.0	2.6	119.	65	6.21	0.60	13.7
9.5	10.0	1.6	135.	40	14.9	3.5	152.	70	6.69	0.70	15.8
10	10.6	1.7	149.	45	16.7	4.3	189.	75	7.17	0.80	17.9

	3" Pipe (3.068" ID)				4" Pipe (4.026" ID)				5" Pipe (5.047" ID)		
30	1.30	0.03	0.48	60	1.51	0.04	0.5	100	1.60	0.04	0.4
35	1.52	0.04	0.64	80	2.02	0.06	0.8	120	1.92	0.06	0.6
40	1.74	0.05	0.82	100	2.52	0.10	1.2	160	2.56	0.10	1.0
45	1.95	0.06	1.0	120	3.02	0.14	1.7	200	3.20	0.16	1.4
50	2.17	0.07	1.2	140	3.53	0.19	2.2	250	4.02	0.25	2.2
60	2.60	0.11	1.7	160	4.03	0.25	2.8	300	4.81	0.36	3.0
70	3.04	0.14	2.3	180	4.54	0.32	3.5	350	5.61	0.49	4.0
80	3.47	0.19	3.0	200	5.05	0.40	4.3	400	6.41	0.64	5.2
90	3.99	0.24	3.7	220	5.55	0.48	5.1	450	7.22	0.81	6.4
100	4.34	0.29	4.5	240	6.05	0.57	6.0	500	8.02	1.00	7.8
120	5.21	0.42	6.3	260	6.55	0.67	7.0	550	8.82	1.21	9.3
140	6.08	0.57	8.3	280	7.06	0.77	8.0	600	9.62	1.49	10.9
160	6.94	0.75	10.7	300	7.57	0.89	9.1	650	10.4	1.7	12.6
180	7.81	0.95	13.2	320	8.07	1.01	10.2	700	11.2	1.9	14.5
200	8.68	1.17	16.1	340	8.58	1.14	11.5	750	12.0	2.2	16.5
220	9.55	1.42	19.2	360	9.08	1.28	12.7	800	12.8	2.5	18.6
240	10.4	1.7	22.6	380	9.59	1.43	14.1	850	13.6	2.9	20.8
260	11.3	2.0	26.2	400	10.1	1.6	15.5	900	14.4	3.2	23.1
280	12.2	2.3	30.0	420	10.6	1.7	16.9	950	15.2	3.6	25.5
300	13.0	2.6	34.1	460	11.6	2.1	20.0	1,000	16.0	4.0	28.1
320	13.9	3.0	38.4	500	12.6	2.5	23.4	1,100	17.6	4.8	33.5
340	14.8	3.4	43.0	550	13.9	3.0	27.9	1,200	19.2	5.7	39.3
360	15.6	3.8	47.8	600	15.1	3.5	32.8	1,300	20.8	6.7	45.6
380	16.5	4.2	52.8	650	16.4	4.2	38.0	1,400	22.4	7.8	52.3
400	17.4	4.7	58.0	700	17.6	4.8	43.6	1,500	24.0	9.0	59.4
420	18.2	5.1	63.5	750	18.9	5.6	49.5	1,600	25.6	10.2	66.9

Source: Reprinted with permission of Sta-Rite Industries, Inc.; © 1987 Sta-Rite Industries, Inc.

461

TABLE B-2   Friction Head Loss in Pipes from Hazen–Williams Equation for Coefficient of 150 Corresponding to Plastic Pipes (Schedule 40) (flow in gpm, velocity in ft/sec, and friction head loss in ft per 100 ft of pipe length)

Flow	Velocity	Head Loss	Velocity	Head Loss	Velocity	Head Loss	Velocity	Head Loss	Velocity	Head Loss
	½″ Pipe		¾″ Pipe							
2	2.10	3.47	1.20	0.89	1″ Pipe		1¼″ Pipe		1½″ Pipe	
4	4.23	12.7	2.41	3.29	1.49	1.01	0.86	0.27	0.63	0.12
6	6.34	26.8	3.61	6.91	2.23	2.14	1.29	0.57	0.94	0.26
8	8.45	46.1	4.82	11.8	2.98	3.68	1.72	0.95	1.26	0.45
10	10.6	69.1	6.02	17.9	3.72	5.50	2.14	1.44	1.57	0.67
12			7.22	24.9	4.46	7.71	2.57	2.02	1.89	0.94
15			9.02	37.6	5.50	11.8	3.21	3.05	2.36	1.41
20	5″ Pipe		12.0	63.9	7.74	19.7	4.29	5.21	3.15	2.44
30	0.49	0.02			11.15	41.8	6.43	10.8	4.72	5.17
40	0.65	0.03	6″ Pipe		14.88	71.4	8.58	18.8	6.30	8.83
50	0.82	0.05	0.57	0.02			10.72	28.2	7.87	13.3
60	0.98	0.07	0.68	0.03			12.87	40.0	9.44	18.6
70	1.14	0.10	0.79	0.04			15.01	53.1	11.02	24.9
80	1.31	0.13	0.91	0.05			17.16	68.2	12.69	32.0
90	1.47	0.16	1.02	0.07			19.30	84.6	14.71	39.5
100	1.63	0.19	1.13	0.08					15.74	47.9
120	1.96	0.27	1.36	0.11	8″ Pipe				18.89	67.2
140	2.29	0.36	1.59	0.15	0.90	0.04			22.04	89.3
160	2.61	0.46	1.80	0.19	0.96	0.04				
180	2.94	0.57	2.04	0.24	1.15	0.06				
200	3.27	0.70	2.27	0.29	1.28	0.07	10″ Pipe			
240	3.92	0.98	2.67	0.41	1.53	0.10	0.98	0.03		
280	4.50	1.30	3.11	0.54	1.79	0.13	1.15	0.04		
320	5.13	1.66	3.56	0.69	2.05	0.17	1.31	0.06	12″ Pipe	
400	6.44	2.5	4.43	1.03	2.60	0.25	1.63	0.09	1.14	0.03
500	8.02	3.8	5.56	1.36	3.19	0.39	2.04	0.13	1.42	0.05
600	9.62	5.3	6.65	2.19	3.85	0.54	2.45	0.18	1.70	0.07
700	11.2	7.1	7.78	2.92	4.46	0.72	2.86	0.24	1.99	0.10
800	12.8	9.1	8.90	3.74	5.10	0.89	3.26	0.31	2.27	0.13
900	14.4	11.3	10.0	4.75	5.75	1.16	3.67	0.39	2.56	0.16
1,000	16.0	13.7	11.1	5.56	6.38	1.40	4.08	0.48	2.84	0.19
1,500	24.0		16.67	10.6	8.95	2.59	5.71	0.88	3.98	0.37
2,000			22.2		12.78	5.03	8.16	1.69	5.68	0.70
2,600							10.61	2.73	7.38	1.14
3,200							13.05	3.7	9.10	1.65
3,800							15.51	6.3	10.80	2.30
4,500									12.78	3.24
5,000									14.20	3.95

Source: Reprinted with permission of Sta-Rite Industries, Inc.; © 1987 Sta-Rite Industries, Inc.

Flow	Velocity	Head Loss	Velocity	Head Loss	Velocity	Head Loss	Velocity	Head Loss	Velocity	Head Loss
2										
4	2″ Pipe									
6	0.57	0.09	2½″ Pipe							
8	0.77	0.16	0.52	0.05	3″ Pipe					
10	0.96	0.24	0.65	0.08	0.43	0.03				
12	1.15	0.37	0.78	0.11	0.52	0.05	3½″ Pipe			
15	1.50	0.51	0.98	0.17	0.65	0.07	0.49	0.03	4″ Pipe	
20	1.01	0.86	1.31	0.29	0.87	0.12	0.65	0.05	0.51	0.03
30	2.89	1.80	1.96	0.61	1.30	0.25	0.97	0.11	0.77	0.06
40	3.82	3.10	2.68	1.03	1.74	0.43	1.30	1.19	1.02	0.10
50	4.78	4.65	3.35	1.65	2.17	0.65	1.62	0.29	1.28	0.16
60	5.74	6.53	4.02	2.19	2.60	0.90	1.95	0.40	1.53	0.22
70	6.69	8.64	4.69	2.91	3.04	1.21	2.27	0.54	1.79	0.30
80	7.65	11.1	5.36	3.71	3.49	1.54	2.60	0.69	2.04	0.38
90	8.61	13.8	6.03	4.61	3.91	1.92	2.92	0.85	2.30	0.47
100	9.56	16.8	6.70	5.64	4.34	2.33	3.25	1.03	2.55	0.57
120	11.5	23.5	8.04	7.89	5.21	3.29	3.99	1.45	3.06	0.81
140	13.4	31.5	9.38	10.5	6.08	4.32	4.54	1.93	3.57	1.07
160	15.3	40.4	10.7	13.6	6.94	5.54	5.19	2.47	4.08	1.37
180	17.2	50.3	12.1	16.8	7.81	6.58	5.85	3.07	4.60	1.70
200	19.1	60.6	13.4	20.3	8.68	8.36	6.50	3.73	5.11	2.06
240	22.9	85.5	16.1	28.7	10.4	11.8	7.79	5.22	6.13	2.91
280			18.8	38.1	12.2	15.7	9.09	6.95	7.15	3.85
320			21.6	48.4	13.9	20.1	10.04	8.88	8.17	4.93
400			26.8	73.3	17.4	30.6	13.0	13.4	10.10	7.52
500					21.7	46.1	16.2	20.3	12.6	11.3
600					26.0	64.4	19.5	28.5	15.10	15.8
700							22.7	37.9	17.60	21.1
800							26.0	48.4	20.20	26.8
900									22.7	33.4
1,000										
1,500										
2,000										
2,600										
3,200										
3,800										
4,500										
5,000										

TABLE B-3 Friction Head Loss in Pipes from Hazen–Williams Equation for Coefficient of 150 Corresponding to Plastic Pipes (Schedule 80) (flow in gpm, velocity in ft/sec, and friction head loss in ft per 100 ft of pipe length)

Flow	½″ Pipe		¾″ Pipe		1″ Pipe		1¼″ Pipe		1½″ Pipe		2″ Pipe		2½″ Pipe		3″ Pipe		3½″ Pipe		4″ Pipe		5″ Pipe		6″ Pipe		8″ Pipe	
	Velocity	Head Loss	Velocity	Head Loss	Velocity	Head Loss	Velocity	Head Loss	Velocity	Head Loss	Velocity	Head Loss	Velocity	Head Loss	Velocity	Head Loss	Velocity	Head Loss	Velocity	Head Loss	Velocity	Head Loss	Velocity	Head Loss	Velocity	Head Loss
2	2.74	6.72	1.48	1.51																						
4	5.48	24.2	2.97	5.45	1.79	1.54	1.00	0.39	0.73	0.177																
6	8.23	51.2	4.45	11.5	2.68	3.34	1.50	0.82	1.09	0.375	0.65	0.107														
8	11.0	86.9	5.94	19.6	3.57	5.69	2.00	1.39	1.45	0.64	0.87	0.183	0.61	0.077												
10	13.7	132.0	7.42	29.6	4.46	8.60	2.50	2.10	1.82	0.96	1.09	0.276	0.76	0.115	0.485	0.039										
12			8.91	41.5	5.36	12.0	3.00	2.94	2.18	1.35	1.30	0.387	0.91	0.161	0.572	0.055										
15			11.1	62.7	6.7	22.9	3.76	4.45	2.72	2.04	1.63	0.585	1.14	0.243	0.727	0.083	0.54	0.035								
20			14.8	107	8.92	30.9	5.00	7.57	3.63	3.47	2.17	0.996	1.51	0.414	0.97	0.140	0.72	0.068	0.56	0.037						
30					13.4	65.3	7.50	16.0	5.45	7.38	3.26	2.11	2.27	0.874	1.44	0.297	1.08	0.145	0.84	0.077	0.53	0.025				
40					17.9	111	10.0	27.3	7.26	12.5	4.35	3.59	3.03	1.49	1.94	0.507	1.44	0.246	1.12	0.132	0.71	0.043				
50							12.5	41.3	9.08	18.9	5.43	5.41	3.79	2.25	2.42	0.766	1.80	0.372	1.40	0.199	0.88	0.065	0.62	0.027		
60							15.0	57.8	10.9	26.5	6.52	7.61	4.54	3.16	2.92	1.07	2.17	0.522	1.67	0.279	1.06	0.091	0.74	0.039		
70							17.5	77.1	12.7	35.3	7.61	10.1	5.30	4.20	3.38	1.43	2.53	0.691	1.95	0.371	1.23	0.121	0.86	0.051		
80							20.0	98.2	14.5	45.2	8.69	12.9	6.05	5.36	3.88	1.83	2.89	0.888	2.23	0.475	1.41	0.155	0.98	0.065		
90							22.5	122	16.3	55.9	9.78	16.1	6.81	6.53	4.33	2.27	3.25	1.10	2.51	0.592	1.59	0.193	1.11	0.080		
100									18.2	68.2	10.9	19.6	7.57	8.13	4.85	2.76	3.67	1.34	2.79	0.719	1.76	0.234	1.23	0.098		
120									21.8	95.4	13.0	27.4	9.08	11.4	5.80	3.87	4.33	1.88	3.35	1.00	2.11	0.329	1.48	0.137		
140									25.4	127	15.2	36.5	10.6	15.1	6.80	5.12	5.05	2.50	3.91	1.33	2.47	0.437	1.72	0.182	0.98	0.047
160											17.4	46.7	12.1	19.4	7.75	6.58	5.78	3.20	4.47	1.71	2.82	0.559	1.97	0.234	1.12	0.059
180											19.6	58.3	13.6	24.1	8.60	8.18	6.50	3.97	5.02	2.12	3.16	0.696	2.22	0.290	1.26	0.074
200											21.7	70.5	15.1	29.3	9.70	9.96	7.22	4.84	5.38	2.58	3.52	0.846	2.46	0.353	1.41	0.090
240											26.1	98.7	18.2	41.0	11.6	13.9	8.66	6.77	6.70	3.62	4.23	1.18	2.96	0.484	1.69	0.126
280													21.2	54.5	13.5	18.6	10.1	9.02	7.82	4.79	4.94	1.57	3.45	0.658	1.97	0.168
320													24.2	69.9	15.5	23.7	11.5	11.5	8.94	6.16	5.64	2.01	3.94	0.747	2.24	0.215
400													30.3	106.0	19.4	35.9	14.4	17.4	11.2	9.31	7.05	3.05	4.93	1.27	2.81	0.325
500															23.2	54.1	18.1	26.3	14.0	14.1	8.82	4.61	6.16	1.92	3.51	0.493
600															29.1	76.1	21.7	36.9	16.7	19.7	10.6	6.44	7.39	2.69	4.22	0.686
700																	25.3	48.9	19.5	26.2	12.3	8.60	8.63	3.58	4.92	0.916
800																	28.9	61.6	22.3	33.6	14.1	11.0	9.85	4.58	5.62	1.17
900																			25.1	41.8	15.9	13.7	11.1	5.69	6.32	1.46
1,000																					17.6	16.6	12.3	6.91	7.03	1.77
1,500																									10.05	3.75
2,000																									14.1	6.39

Source: Reprinted with permission of Sta-Rite Industries, Inc.; © 1987 Sta-Rite Industries, Inc.

TABLE B-4   Values of Minor Loss Coefficient (k_m) for Various Fittings, Valves, and Other Fixtures of Standard Pipes

Fitting or Valve	Nominal Pipe Diameter (in.)						
	3.0	4.0	5.0	6.0	7.0	8.0	10.0
Elbows							
Regular, flanged 90°	0.34	0.31	0.30	0.28	0.27	0.26	0.25
Long radius, flanged 90°	0.25	0.22	0.20	0.18	0.17	0.15	0.14
Long radius, flanged 45°	0.19	0.18	0.18	0.17	0.17	0.17	0.16
Regular, screwed 90°	0.80	0.70					
Long radius, screwed 90°	0.30	0.23					
Regular, screwed 45°	0.30	0.28					
Bends							
Return, flanged	0.33	0.30	0.29	0.28	0.27	0.25	0.24
Return, screwed	0.80	0.70					
Tees							
Flanged, line flow	0.16	0.14	0.13	0.12	0.11	0.10	0.09
Flanged, branch flow	0.73	0.68	0.65	0.60	0.58	0.56	0.52
Screwed, line flow	0.90	0.90					
Screwed, branch flow	1.20	1.10					
Valves							
Globe, flanged	7.0	6.3	6.0	5.8	5.7	5.6	5.5
Globe, screwed	6.0	5.7					
Gate, flanged	0.21	0.16	0.16	0.13	0.11	0.09	0.075	0.06
Gate, screwed	0.14	0.12					
Swing check, flanged	2.0 for all diameters						
Swing check, screwed	2.1	2.0					
Angle, flanged	2.2	2.1	2.0	2.0	2.0	2.0	2.0
Angle, screwed	1.3	1.0					
Foot valves	0.80 for all diameters						
Strainers, basket type	1.25	1.05	0.95	0.85	0.80	0.75	0.67
Inlets or entrances							
Inward projecting	0.80 for all diameters						
Sharp cornered	0.50 for all diameters						
Slightly rounded	0.23 for all diameters						
Bell-mouth shape	0.04 for all diameters						
Exits							
Free flow	0.0 for all diameters						
Submerged	1.0 for all diameters						
Sudden enlargements[a]	$k_m = (1 - d_1/d_2)^2$ for all diameters						
Sudden contractions[a]	$k_m = 0.7(1 - d_1^2/d_2^2)^2$ for all diameters						

Source: U.S. Soil Conservation Service 1968.

[a] d_1 = smaller pipe diameter and d_2 = larger pipe diameter.

TABLE B-5   Conduit Head Loss Coefficients, k_c^a for Circular Culverts Flowing Full

Pipe Dia. (in.)	Flow Area (ft^2)	Manning's Coefficient (n)							
		0.010	0.012	0.014	0.016	0.018	0.020	0.022	0.025
6	0.196	0.0467	0.0672	0.0914	0.1194	0.151	0.187	0.266	0.292
8	0.349	0.0318	0.0458	0.0623	0.0814	0.1030	0.1272	0.154	0.199
10	0.545	0.0236	0.0340	0.0463	0.0604	0.0765	0.0944	0.1143	0.148
12	0.785	0.0185	0.0267	0.0363	0.0474	0.0600	0.0741	0.0896	0.1157
15	1.23	0.0138	0.0198	0.0270	0.0309	0.0446	0.0550	0.0666	0.0859
18	1.77	0.01078	0.0155	0.0211	0.0276	0.0349	0.0431	0.0522	0.0674
24	3.14	0.00889	0.01058	0.0144	0.0188	0.0238	0.0294	0.0356	0.0459
30	4.91	0.00546	0.00786	0.01070	0.0140	0.0177	0.0218	0.0264	0.0341
36	7.07	0.00428	0.00616	0.00839	0.01096	0.0139	0.0171	0.0207	0.0267
48	12.57	0.00292	0.00420	0.00572	0.00747	0.00945	0.01166	0.0141	0.0182
60	19.63	0.00217	0.00312	0.00424	0.00554	0.00702	0.00866	0.01048	0.0135

Source: U.S. Soil Conservation Service 1979a.

[a] Coefficients calculated by equation (9-25):

$$k_c = \frac{5,087n^2}{d^{4/3}}, \quad \text{where } d = \text{diameter (in.)}$$

TABLE B-6   Conduit Head Loss Coefficients, k_c^a for Square Culverts Flowing Full

Pipe Size (ft)	Flow Area (ft^2)	Hydraulic Radius (ft)	Manning's Coefficient (n)			
			0.012	0.014	0.016	0.020
2 × 2	4.00	0.5	0.01058	0.01440	0.01880	0.0294
3 × 3	9.00	0.75	0.00616	0.00839	0.01096	0.0171
4 × 4	16.00	1.0	0.00420	0.00493	0.00746	0.0117
5 × 5	25.00	1.25	0.00312	0.00425	0.00554	0.00866
6 × 6	36.00	1.5	0.00245	0.00333	0.00435	0.00680
7 × 7	49.00	1.75	0.00199	0.00271	0.00354	0.00553
8 × 8	64.00	2.0	0.00167	0.00227	0.00296	0.00463
9 × 9	81.00	2.25	0.00142	0.00194	0.00253	0.00395
10 × 10	100.00	2.5	0.00124	0.00168	0.00220	0.00344

Source: U.S. Soil Conservation Service 1979a.

[a] Coefficients calculated by equation (9-26).

$$k_c = \frac{29.16n^2}{R^{4/3}}, \quad \text{where } R = \text{hydraulic radius (ft)}$$

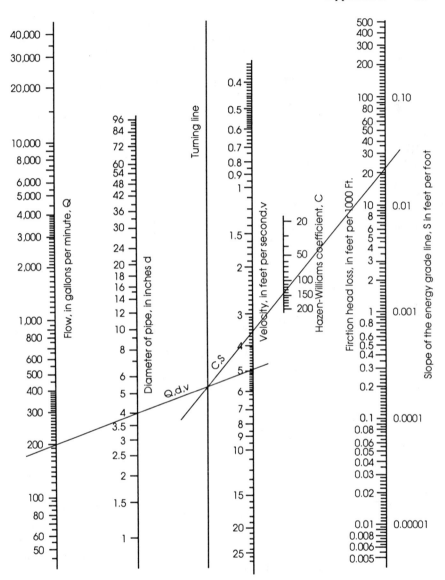

FIGURE B-1.   Graphical solution of the Hazen–Williams equation.

A simple way to account for the resistance offered to flow by valves and fittings is to add to the length of pipe in the line a length which will give a pressure drop equal to that which occurs in the valves and fittings in the line.

Example: The dotted line shows that the resistance of a 6-inch Standard Elbow is equivalent to approximately 16 feet of 6-inch Standard Steel Pipe.

Note: For sudden enlargements or sudden contractions, use the smaller diameter on the nominal pipe size scale.

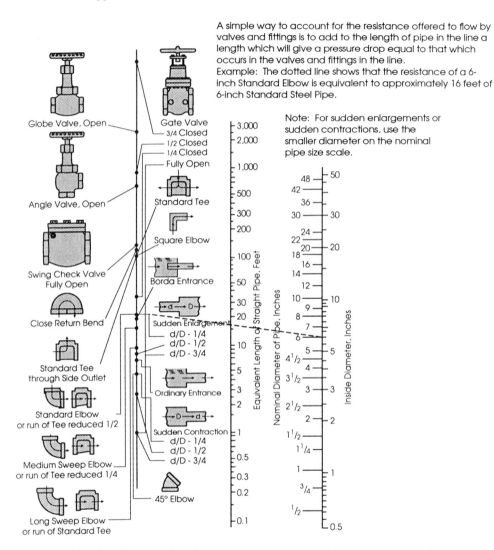

FIGURE B-2.   Nomograph to determine minor losses of various fittings and other fixtures in pipelines in terms of equivalent lengths of the same size straight pipes. (*Courtesy of Crane Co.*)

References

Aboukhaled, A., A. Alfard, and M. Smith. 1982. *Lysimeters*. FAO Irrigation and Drainage Paper 39, Food and Agriculture Organization of the United Nations, Rome, Italy.

Ackers, P., W. R. White, S. A. Perkins, and A. J. M. Harrison. 1978. *Weirs and Flumes for Flow Measurement*. New York: Wiley.

Aisenbrey, A. J., Jr., R. B. Hayes, H. J. Warren, D. L. Winsett, and R. B. Young. 1978. *Design of Small Canal Structures*. The Bureau of Reclamation. U.S. Department of the Interior, Denver, CO.

Allred, E. R., P. W. Manson, G. M. Schwartz, P. Golany, and J. W. Reinke. 1971. *Continuation of Studies on the Hydrology of Ponds and Small Lakes*. Minnesota Agricultural Experiment Station Technical Bulletin 274.

American Society of Agricultural Engineers. 1982. Trapezoidal flumes for irrigation flow measurement. Standard ASAE S359.1. In *1982–1983 Agricultural Engineers Yearbook*. St. Joseph, MI: ASAE.

Anderson, K. E. 1984. *Water Well Handbook*. Belle, MO: Missouri Water Well & Pump Contractors Association.

Baier, W., and G. W. Robertson. 1965. Estimation of latent evaporation from simple weather observations. *Canadian J. Plant Sci.* **45**:276–284.

Baumgartner, A., and E. Reichel. 1975. *The World Water Balance*. Amsterdam, The Netherlands: Elsevier.

Beasley, R. P. 1960. Canopy inlet for closed conduits. *Agr. Engr.* **41**:226–228.

Beasley, R. P., J. M. Gregory, and T. R. McCarty. 1984. *Erosion and Sediment Pollution Control*, 2d ed. Ames, IA: Iowa State University Press.

Berkeley Pump Co. 1987. *Pump Catalogue*. Berkeley, CA: Berkeley Pump Co.

Blaney, H. F., and W. D. Criddle. 1952. *Determining Water Requirements in Irrigated Areas from Climatological and Irrigation Data*. USDA, SCS, Technical Publication 96.

Bos, M. G. (ed). 1989. *Discharge Measurement Structures*. 3d ed. ILRI Publication 20. Wageningen, The Netherlands: International Institute for Land Reclamation and Improvement/ILRI.

Bos, M. G., J. A. Replogle, and A. J. Clemens. 1984. *Flow Measuring Flumes for Open Channel Systems*. New York: Wiley.

Bosch, J. M., and J. D. Hewlett. 1982. A review of catchment experiments to determine the effect of vegetation changes on water yield and evapotranspiration. *Hydrology* **55**:3–23.

Bouthillier, P. H. 1981. *Hydraulic Tables for Water Supply and Drainage*. Ann Arbor, MI: Ann Arbor Science Pub.

Boyd, C. E. 1975. *Competition for Light by Aquatic Plants in Fish Ponds*. Circular 215. Auburn University, AL: Alabama Agricultural Experiment Station.

Boyd, C. E. 1976. Water chemistry and plankton in unfertilized ponds in pastures and in woods. *Trans. Amer. Fish. Soc.* **105**:634–636.

Boyd, C. E. 1978. Effluents from catfish ponds during fish harvest. *J. Environmental Quality* **7**:59–62.

Boyd, C. E. 1982a. Hydrology of small experimental fish ponds at Auburn, Alabama. *Trans. Amer. Fish. Soc.* **111**:638–644.

Boyd, C. E. 1982b. *Water Quality Management for Pond Fish Culture*. Amsterdam, The Netherlands: Elsevier.

Boyd, C. E. 1983. Seepage control in a pond with "Soil-Crete Geo-Lining System." *Farm Pond Harvest* **83**(fall):5–7.

Boyd, C. E. 1985a. Pond evaporation. *Trans. Amer. Fish. Soc.* **114**:299–303.

Boyd, C. E. 1985b. Hydrology and pond construction. In *Channel Catfish Culture*, C. S. Tucker, ed., pp. 109–133. Amsterdam, The Netherlands: Elsevier.

Boyd, C. E. 1985c. Chemical budgets for channel catfish ponds. *Trans. Amer. Fish. Soc.* **114**:291–298.

Boyd, C. E. 1986. Influence of evaporation excess on water requirements for fish farming. In *Proceedings of the Conference on Climate and Water Management—A Critical Era*, pp. 62–64. Boston: American Meteorological Society.

Boyd, C. E. 1987a. Evapotranspiration/evaporation (E/E_o) ratios for aquatic plants. *J. Aquatic Plant Management* **25**:1–3.

Boyd, C. E. 1987b. Water conservation measures in fish farming. In *Proceedings of the Fifth Conference on Applied Climatology*, pp. 88–91. Boston: American Meteorological Society.

Boyd, C. E. 1990. *Water Quality in Ponds for Aquaculture*. Auburn University, AL: Alabama Agricultural Experiment Station.

Boyd, C. E. 1992. Shrimp pond bottom soils and sediment management. In *Proceedings of Shrimp Farming Workshop*, J. A. Wyban, ed., pp. 166–181. Baton Rouge, LA: World Aquaculture Society.

Boyd, C. E., and J. L. Shelton, Jr. 1984. *Observations on the Hydrology and Morphometry of Ponds on the Auburn University Fisheries Research Unit*. Bulletin 558. Auburn University, AL: Alabama Agricultural Experiment Station.

Boyd, C. E., and C. S. Tucker. 1992. *Water Quality and Pond Soil Analyses for Aquaculture*. Auburn University, AL: Alabama Agricultural Experiment Station.

Brenzy, O., I. Mehta, and R. K. Sharma. 1973. Studies on evapotranspiration of some aquatic weeds. *Weed Sci*. **21**:197–204.

Brinkman, R., and V. P. Singh. 1982. Rapid reclamation of brackish water fishponds in acid sulfate soils. In *Proceedings of the Bangkok Symposium on Acid Sulfate Soils*, H. Dost and N. van Breeman, eds., pp. 318–330. Publication 18. Wageningen, The Netherlands: International Institute Land Reclamation and Improvement.

Chow, V. T. 1959. *Open-Channel Hydraulics*. New York: McGraw-Hill.

Cole, B. A., and C. E. Boyd. 1986. Feeding rate, water quality, and channel catfish production in ponds. *Progressive Fish-Culturist* **81**:25–29.

Critchfield, H. J. 1974. *General Climatology*. Englewood Cliffs, NJ: Prentice-Hall.

Daugherty, R. L., J. B. Franzini, and E. J. Finnemore. 1985. *Fluid Mechanic with Engineering Application*. 8th ed. New York: McGraw-Hill.

Davenport, D. C., P. E. Martin, and R. M. Hagan. 1982. Evapotranspiration from riparain vegetation: Conserving water by reducing saltcedar transpiration. *J. Soil & Water Conservation* **37**:237–239.

Debusk, T., J. H. Ryther, and L. D. Williams. 1983. Evapotranspiration of *Eichhornia crassipes* (Mart.) Solms and *Lemna minor* L. in central Florida: Relation to canopy structure and season. *Aquatic Bot*. **16**:31–39.

Douglass, J. E., and W. T. Swank. 1975. Effects of management practices on water quality and quantity: Cowetta Hydrologic Laboratory, North Carolina. In *Municipal Watershed Management*, pp. 1–13. USDA Forest Service General Technical Report NE-13.

Driscoll, F. G. (ed). 1986. *Groundwater and Wells*. 2d ed. St. Paul, MN: Johnson Filtration Systems.

Eisenlohr, W. S., Jr. 1966. Water loss from a natural pond through transpiration by hydrophytes. *Water Resources Res*. **2**:443–453.

Evans, J. O., and J. H. Patric. 1983. Harvest trees, reap water. *J. Soil & Water Conservation* **38**:390–392.

Fairbridge, R. W. (ed). 1967. *Encyclopedia of Atmospheric Sciences and Astrogeology*. New York: Reinhold.

Farnsworth, R. K., and E. S. Thompson. 1982. *Mean Monthly, Seasonal, and Annual Pan Evaporation for the United States*. U.S. Dept. Commerce, NOAA Technical Report NWS 34.

Fleming, J. F., and L. T. Alexander. 1961. Sulfur acidity in South Carolina tidal marsh soils. *Soil Sci. Soc. America Proc*. **25**:94–95.

Frasier, G. W., and L. E. Myers. 1983. *Handbook of Water Harvesting*. USDA, ARS, Agricultural Handbook 600.

Gaviria, J. I., H. R. Schmittou, and J. H. Grover. 1986. Acid sulfate soils: Identification, formation, and implications for aquaculture. *J. Aquaculture Tropics* **1**:99–109.

Gilbert, M. J., and C. H. M. van Bavel. 1954. A simple field installation for measuring maximum evapotranspiration. *Trans. Amer. Geophysical Union* **35**:937–942.

Gray, D. M. (ed). 1970. *Handbook on the Principles of Hydrology.* Port Washington, NY: Water Information Center.

Greeve, F. W. 1928. *Measurement of Pipe Flow by the Coordinate Method.* Purdue Engineering Expt. Sta. Bulletin 32, Purdue Univ., Purdue, IN.

Gustafson, R. J. 1988. *Fundamentals of Electricity for Agriculture.* 2d ed. St. Joseph, MI: American Society of Agricultural Engineers.

Gwinn, W. R., and W. O. Ree. 1979. *Maintenance Effects on the Hydraulic Properties of a Vegetation-Lined Channel.* ASAE Paper No. 79-2063, St. Joseph, MI.

Hajek, B. F., and C. E. Boyd. 1990. *Rating Soil and Water Information for Aquaculture.* Fisheries Research and Development Project, Central Research Institute for Fisheries, USAID/Indonesia, Jakarta, Indonesia.

Hansen, V. E., O. W. Israelson, and G. E. Stringham. 1980. *Irrigation Principles and Practices.* New York: Wiley.

Harrold, L. L., and F. R. Dreibelbis. 1958. *Evaluation of Agricultural Hydrology by Monolith Lysimeters 1944–1955.* USDA Technical Bulletin 1179.

Heath, R. C. 1984. *Basic Ground Water Hydrology.* U.S. Geological Survey Water Supply Paper 2220. Washington, D.C.

Hewlett, J. D. 1982. *Principles of Forest Hydrology.* Athens, GA: University of Georgia Press.

Hewlett, J. D., G. B. Cunningham, and C. A. Troendle. 1977. Predicting stormflow and peakflow from small basins in humid areas by the R-index method. *Water Resources Bull.* **13**:231–253.

Hicks, T. G., and T. W. Edwards. 1971. *Pump Application Engineering.* New York: McGraw-Hill.

Hjelmfelt, A. T., Jr., and J. J. Cassidy. 1975. *Hydrology for Engineers and Planners.* Ames, IA: Iowa State University Press.

Hollerman, W. D., and C. E. Boyd. 1985. Effects of annual draining on water quality and production of channel catfish in ponds. *Aquaculture* **46**:45–54.

Holtan, H. N. 1950a. Sealing farm ponds. *Agr. Engr.* **31**:125–134.

Holtan, H. N. 1950b. *Holding Water in Farm Ponds.* USDA, SCS, Technical Publication 93.

Holtan, H. N., N. E. Minshall, and L. L. Harrold. 1962. *Field Manual for Research in Agricultural Hydrology.* USDA, ARS, Agriculture Handbook No. 224.

Hounam, C. E. 1973. *Comparisons between Pan and Lake Evaporation.* World Meteorological Organization Technical Note No. 126, Geneva, Switzerland.

Hudson, N. 1981. *Soil Conservation.* 2d ed. Ithaca, NY: Cornell University Press.

Idso, S. B., and J. M. Foster. 1974. Light and temperature relations in a small desert pond as influenced by phytoplanktonic density variations. *Water Resources Res.* **10**:129–132.

James, L. G. 1988. *Principles of Farm Irrigation System Design.* New York: John Wiley.

Karassik, I. J., W. C. Krutzcsh, W. H. Fraser, and J. P. Messina. 1986. *Pump Handbook*. New York: McGraw-Hill.

Kemmer, F. N. 1979. *Water: The Universal Solvent*. Oak Brook, IL: Nalco Chemical Co.

Kohler, M. A., T. J. Nordensen, and W. E. Fow. 1955. *Evaporation from Pans and Lakes*. U.S. Weather Bureau Res. Paper No. 38.

Kohler, M. A., T. J. Nordensen, and D. R. Baker. 1959. *Evaporation Maps of the United States*. U.S. Weather Bureau Technical Paper No. 37.

Kraatz, D. B., and I. K. Mahajan. 1975. *Small Hydraulic Structures*. FAO Irrigation and Drainage Paper No. 26. Food and Agriculture Organization of the United Nations, Rome.

Kramer, P. J. 1969. *Plant and Soil Water Relationships: A Modern Synthesis*. New York: McGraw-Hill.

Lawson, T. B., and F. W. Wheaton. 1983. Crawfish culture systems and their management. *J. World Mariculture Soc.* **14**:325–335.

Leopold, L. B. 1974. *Water: A Primer*. San Francisco: W. H. Freeman.

Linsley, R. K., and J. B. Franzini. 1979. *Water-Resources Engineering*. 3d ed. New York: McGraw-Hill.

Lloyd, T. C. 1969. *Electric Motors and Their Applications*. New York: Wiley-Interscience.

Lowrance, R. R., R. A. Leonard, and L. E. Asmussen. 1985. Nutrient budgets for agricultural watersheds in the southeastern Coastal Plain. *Ecology* **66**:287–296.

M&W Iron Works, Inc. *M&W Pump Catalogue. Manufacturers and Fabricators*. Deerfield Beach, FL: M&W Pump Corporation.

McCarthy, D. F. 1981. *Essentials of Soil Mechanics and Foundations*. Reston, VA: Reston.

McCarthy, D. F. 1988. *Essentials of Soil Mechanics and Foundations: Basic Geotechnics*. Englewood Cliffs, NJ: Prentice-Hall.

McIlroy, I. C., and D. E. Angus. 1964. Grass, water and soil evaporation at Aspendale. *Agric. Meteorol.* **1**:201–224.

Manson, P. W., G. M. Schwartz, and E. R. Allred. 1968. *Some Aspects of the Hydrology of Ponds and Small Lakes*. Technical Bulletin 257, University of Minnesota, Minnesota Agricultural Experiment Station.

Miller, D. H. 1977. *Water at the Surface of the Earth*. New York: Academic Press.

Moran, R. J., and P. J. O'Shaughnessy. 1984. Determination of the evapotranspiration of *E. regnans* forested catchments using hydrological measurements. *Agric. Water Management* **8**:57–76.

Morris, D. A., and A. I. Johnson. 1967. *Summary of Hydrologic and Physical Properties of Rocks and Soil Materials as Analyzed by the Hydrologic Laboratory of the U.S. Geological Survey, 1948–60*. Water Supply Paper 1839-D, U.S. Geological Survey, U.S. Department of the Interior, Washington, D.C.

Mustonen, S. E., and J. L. McGuinness. 1968. *Estimating Evapotranspiration in a Humid Region*. USDA, ARS, Technical Bulletin 1389.

The F. E. Myers & Bro. Co. 1966. *Electric Motor Technical Manual*. Ashland, OH: Subsidiary of McNeil Corp.

National Fire Protection Association. 1987. *National Electrical Code*. Quincy, MA.

Nelson, D. W., and L. E. Sommers. 1982. Total carbon, organic carbon, and organic matter. In *Methods of Soil Analysis, Part II. Chemical and Microbiological Properties*, A. L. Page, ed., pp. 539–579. Madison, WI: American Society of Agronomy, Soil Science Society of America.

Pair, C. H., W. H. Hinz, K. R. Frost, R. E. Sneed, and T. J. Schiltz. 1983. *Irrigation*, 5th ed. Silver Spring, MD: Irrigation Association.

Palmer, W. C., and A. V. Havens. 1958. A graphical technique for determining evapotranspiration by the Thornthwaite method. *Monthly Weather Rev.* **86**:123–128.

Pansini, A. J. 1989. *Basics of Electric Motors*. Englewood Cliffs, NJ: Prentice Hall.

Parmakian, J. 1955. *Waterhammer Analysis*. Englewood Cliffs, NJ: Prentice-Hall.

Parsons, D. A. 1949. *The Hydrology of a Small Area near Auburn, Alabama*. USDA, SCS, Technical Publication No. 85.

Penman, H. L. 1948. Natural evaporation from open water, bare soil, and grass. *Proc. Royal Soc.* **193**:120–145.

Penman, H. L. 1956. *Vegetation and Hydrology*. Commonwealth Bureau Soils. Harpenden, Technical Communication No. 53.

Plasti-Fab. Inc. 1980. *General Product Bulletin*. Tualatin, OR: Plasti-Fab, Inc.

Renfro, G. 1952. *Sealing Leaking Ponds and Reservoirs*. USDA, SCS, Technical Publication 150.

Rickard, D. T. 1973. Sedimentary iron formation. In *Proc. Internat. Symp. Acid Sulfate Soils*, Vol. I, H. Dost, ed., pp. 28–65. Wageningen, The Netherlands: International Institute Land Reclamation and Improvement.

Robinson, A. R., and A. R. Chamberlain. 1960. Trapezoidal flumes for open channel flow measurement. *Trans. ASAE* 3(2):120–124, 128.

Rohwer, C., and O. V. P. Stout. 1948. *Seepage Losses from Irrigation Channels*. Colorado Agricultural Experimental Station Technical Bulletin 38.

Rosenberry, R. 1990. *World Shrimp Farming*. San Diego: Aquaculture Digest.

Satterlund, D. 1972. *Wildland Watershed Management*. New York: Ronald Press.

Schmittou, H. R. 1991. *Cage Culture: A Method of Fish Production in Indonesia*. Jakarta, Indonesia: Fisheries Research and Development Project, Central Research Institute for Fisheries.

Schwab, G. O., R. K. Frevert, T. W. Edminster, and K. K. Barnes. 1981. *Soil and Water Conservation Engineering*. New York: Wiley.

Sharma, M. L. 1984. Evapotranspiration from a Eucalyptus community. *Agric. Water Management* **8**:41–56.

Shelton, J. L., Jr., and C. E. Boyd. 1993. Water budgets for aquaculture ponds supplied by runoff with reference to effluent volume. *J. Applied Aquaculture* **2** (in press).

Shen, J. 1981. *Discharge Characteristics of Triangular-Notch Thin-Plate Weirs*. U.S. Geological Survey Water-Supply Paper 1617-B, Washington, D.C.

Simon, A. L. 1976. *Practical Hydraulics*. New York: Wiley.

Simon, A. L. 1981. *Basic Hydraulics*. New York: Wiley.

Singh, V. P. 1980. Management of fishponds with acid-sulfate soils. *Asian Aquaculture* 5:4–6.

Skogerboe, G. V. 1967. *Design and Calibration of Submerged Open Channel Flow Measurement Structures: Part 1, Submerged flow*. Ref. WG31-2. Logan, UT: Utah Water Research Laboratory. Utah State University.

Skogerboe, G. V., R. S. Bennett, and W. R. Walker. 1973. *Selection and Installation of Cutthroat Flumes for Measuring Irrigation and Drainage Water*. Technical Bulletin 120, Colorado State Univ., Fort Collins, CO.

Smeaton, R. W. 1987. *Motor Application and Maintenance Handbook*. 2d ed. New York: McGraw-Hill.

Snyder, R. L., and C. E. Boyd. 1987. Evapotranspiration by *Eichhornia crassipes* (Mart.) Solms and *Typha* latifolia L. *Aquatic Bot*. 27:217–227.

Soderberg, R. W. 1986. *Flowing Water Fish Culture*. Mansfield University Biology Department, Mansfield, PA.

Soil Survey Staff. 1990. *Keys to Soil Taxonomy*. Virginia Polytechnic Institute and State University, SMSS Technical Monograph No. 19, Blacksburg, VA.

Sorensen, D. L., W. A. Knieb, D. B. Porcella, and B. Z. Richardson. 1980. Determining the lime requirement for the blackbird mine spoil. *J. Environmental Quality* 9:162–166.

Stepanoff, A. J. 1957. *Centrifugal and Axial Flow Pumps: Theory, Design, and Application*. New York: Wiley.

Stewart, J. B. 1984. Measurement and prediction of evaporation from forested and agricultural catchments. *Agric. Water Management*. 8:1–28.

Stone, N. M. 1988. Factors affecting hydrologic parameters used in modeling fishpond water requirements. Unpublished Ph.D. dissertation, Auburn University, Auburn, AL.

Stone, N. M., and C. E. Boyd. 1989. *Seepage from Fish Ponds*. Bulletin 599, Auburn University, AL: Alabama Agricultural Experiment Station.

Swingle, H. S. 1955. Storing water for use in irrigation. In *Proceedings of the Water Resources and Supplemental Irrigation Workshop*, pp. 1–6. Auburn University, AL: Alabama Agricultural Experiment Station.

Thailand Ministry of Communications. 1981. *Climatological Data of Thailand 30-Year Period (1951–1980)*. Bangkok, Thailand: Royal Meteorol. Dept.

Thiessen, A. N. 1911. Precipitation averages for large areas. *Monthly Weather Res*. 39:1082–1084.

Thornthwaite, C. W. 1948. An approach toward a rational classification of climate. *Geogr. Rev*. 38:55–94.

Thornthwaite, C. W., and B. Holzman. 1939. The determination of evaporation from land and water surfaces. *Monthly Weather Rev*. 67:4–11.

Todd, D. K. 1970. *The Water Encyclopedia*. Port Washington, NY: Water Information Center.

Todd, D. K. 1980. *Groundwater Hydrology*. 2d ed. New York: Wiley.

U.S. Bureau of Reclamation. 1952. *Linings for Irrigation Canals*. Washington, D.C.: The Bureau of Reclamation, U.S. Department of the Interior.

U.S. Bureau of Reclamation. 1975. *Water Measurement Manual.* Washington, D.C.: The Bureau of Reclamation, U.S. Department of the Interior.

U.S. Bureau of Reclamation. 1981. *Ground Water Manual—A Water Resources Technical Publication.* Denver, CO: Engineering and Research Center, Water, and Power Resources Service, U.S. Department of the Interior.

U.S. Department of Agriculture. 1979. *Field Manual for Research in Agricultural Hydrology.* Agriculture Handbook No. 224, Washington, D.C.

U.S. Department of Agriculture, Agriculture Research Service. 1978. *Predicting Rainfall Erosion Losses.* Agricultural Handbook No. 537, Washington, D.C.

U.S. Department of Agriculture, Agriculture Research Service and Environmental Protection Agency. 1975. *Control of Water Pollution from Cropland,* Vol. I. Washington, D.C.

U.S. Environmental Protection Agency. 1976a. *Quality Criteria for Water.* Washington, D.C.

U.S. Environmental Protection Agency, 1976b. *Manual of Water Well Construction Practices.* Rept. EPA 570/9-75-001. Washington D.C.

U.S. National Academy of Sciences. 1974. *More Water for Acid Lands.* Washington, D.C.

U.S. National Weather Service. 1972. *National Weather Service Observing Handbook No. 2.* Substation observation. Washington, D.C.

U.S. Soil Conservation Service. 1964. Soil-plant-water relationships. In *SCS National Engineering Handbook*, Chap. 1, Sec. 15. Washington, D.C.

U.S. Soil Conservation Service. 1966. *Handbook of Channel Design for Soil and Water Conservation.* SCS-TP61, Washington, D.C.

U.S. Soil Conservation Service. 1968. Sprinkler irrigation. In *SCS National Engineering Handbook*, Chap. 11, Sec. 15, Irrigation. Washington, D.C.

U.S. Soil Conservation Service. 1970. Irrigation pumping plants. In *SCS National Engineering Handbook*, Chap. 8, Sec. 15, Irrigation. Washington, D.C.

U.S. Soil Conservation Service. 1972. Hydrology. In *SCS National Engineering Handbook*, Sec. 4. Washington, D.C.

U.S. Soil Conservation Service. 1973a. Measurement of irrigation water. In *SCS National Engineering Handbook*, Chap. 9, Sec. 15, Irrigation. Washington, D.C.

U.S. Soil Conservation Service. 1973b. *Drainage of Agricultural Land.* Port Washington, NY: Water Information Center.

U.S. Soil Conservation Service, 1975. Structures. In *SCS Engineering Field Manual*, Chap. 6. Washington, D.C.

U.S. Soil Conservation Service. 1979a. Hydraulics. In *SCS National Engineering Handbook*, Sec. 5. Washington, D.C.

U.S. Soil Conservation Service. 1979b. Ponds and Reservoirs. In *SCS Engineering Field Manual*, Chap. 11. Washington, D.C.

U.S. Soil Conservation Service. 1982. *Ponds—Planning, Design and Construction.* U.S. Dept. of Agriculture, Agriculture Handbook No. 590, Washington, D.C.

U.S. Soil Conservation Service. 1983. Furrow irrigation. In *SCS National Engineering Handbook*, Chap. 5, Sec. 15. Washington, D.C.

U.S. Soil Conservation Service. 1986. Grassed Waterways. In *SCS Engineering Field Manual*, Chap. 7. Washington, D.C.

U.S. Soil Conservation Service. 1990. Soil Engineering. In *SCS Engineering Field Manual*, Chap. 4. Washington, D.C.

U.S. Weather Bureau. 1955. *Rainfall Intensity-Duration-Frequency Curves*. Technical Paper No. 25, Washington, D.C.

U.S. Weather Bureau. 1961. *Rainfall Frequency Atlas of the United States*. Technical Paper No. 40, Washington, D.C.

Utah Power and Light Co. 1980. *Energy Efficient Pumping Standards*. Salt Lake City, UT.

van der Leeden, F., F. L. Troise, and D. K. Todd. 1990. *The Water Encyclopedia*. Chelsea, MI: Lewis.

van Keuren, R. W., J. L. McGuinness, and F. W. Chichester. 1979. Hydrology and chemical quality of flow from small pastured watersheds: I. Hydrology. *J. Environ. Quality* 8:163–179.

van Olphen, H. 1977. *An Introduction to Clay Colloid Chemistry*. New York: Wiley.

Viessman, W., J. W. Knapp, G. L. Lewis, and T. E. Harbaugh. 1972. *Introduction to Hydrology*. New York: Harper and Row.

Walker, R. 1972. *Pump Selection: A Consulting Engineer's Manual*. Stoneham, MA: Butterworth Pub.

Walker, W. R., and G. V. Skogerboe. 1987. *Surface Irrigation: Theory and Practice*. Englewood Cliffs, NJ: Prentice-Hall.

Wallis, A. L., Jr. 1977. *Comparative Climatic Data Through 1976*. NOAA, Environmental Data Service, National Climatic Center, Asheville, NC.

Wang, J. K. 1990. Managing shrimp pond water to reduce discharge problems. *Aquacultural Engineering* 9:61–73.

Weber, J. B. 1977. Soil properties, herbicide sorption and model soil systems. In *Research Methods in Weed Science*, 2d ed., B. Truelove, ed., pp 59–72. Auburn, AL: Southern Weed Sci. Soc.

Welch, P. S. 1948. *Limnological Methods*. New York: McGraw-Hill.

Wenzel, L. K. 1942. *Methods for Determining Permeability of Water-Bearing Materials with Special Reference to Discharging-Well Methods*. U.S. Geological Survey Water-Supply Paper 887, Washington, D.C.

Wetzel, R. G. 1975. *Limnology*. Philadelphia: W. B. Sanders Company.

Wheaton, F. W. 1977. *Aquacultural Engineering*. New York: Wiley-Interscience.

Wheaton, F. W. 1981. *Aquacultural Engineering*. Malabar, FL: Krieger Pub.

Whitwell, T., and D. R. Bayne. Undated. *Weed Control in Lakes and Farm Ponds*. Circular ANR-48, Auburn University, AL: Alabama Cooperative Extension Service.

Wyban, J. A., and J. N. Sweeney. 1989. Intensive shrimp growout trials in a round pond. *Aquaculture* 76:215–225.

Wyban, J. A., J. N. Sweeney, and R. A. Kanna. 1988. Shrimp yields and economic potential of intensive round pond systems. *J. World Aquaculture Soc.* 19:210–217.

Index